OT 12:
Operator Theory: Advances and Applications
Vol. 12

Birkhäuser Verlag
Basel · Boston · Stuttgart

Topics in Operator Theory Systems and Networks

Workshop on Applications of Linear Operator Theory to Systems and Networks, Rehovot (Israel), June 13–16, 1983

Edited by

H. Dym
I. Gohberg

1984

Birkhäuser Verlag
Basel · Boston · Stuttgart

Volume Editorial Office

Department of Theoretical Mathematics
The Weizmann Institute of Science
Rehovot 76100, Israel

Library of Congress Cataloging in Publication Data

Workshop on Applications of Linear Operator Theory to
Systems and Networks (1983 : Rehovot, Israel)
Topics in operator theory systems and networks.
1. System analysis — Congresses. 2. Linear operators —
Congresses. 3. Electric network analysis — Congresses.
I. Dym, H. (Harry), 1938- . II. Gohberg, I. (Israel),
1928- . III. Title.
QA402.W65 1983 003 84-3076
ISBN 3-7643-1550-4

CIP-Kurztitelaufnahme der Deutschen Bibliothek

Topics in operator theory systems and networks /
Workshop on Applications of Linear Operator
Theory to Systems and Networks, Rehovot
(Israel), June 13–16, 1983. Vol. ed.
H. Dym & I. Gohberg. – Basel ; Boston ;
Stuttgart : Birkhäuser, 1984.
(Operator theory ; Vol. 12)
ISBN 3-7643-1550-4

NE: Dym, Harry [Hrsg.]; Workshop on Applications of Linear Operator Theory to Systems and Networks <1983, Rehobot>; GT

Printed in Germany
ISBN 3-7643-1550-4

This volume is dedicated to

M. S. LIVŠIČ

in recognition of his pioneering role in the development of the theory of operators and systems and his fundamental contributions to both.

CONTENTS

PREFACE

This volume contains the proceedings of the Workshop on applications of linear operator theory to systems and networks, which was held at the Weizmann Institute of Science in the third week of June, 1983, just before the MTNS Conference in Beersheva.

For a long time these subjects were studied independently by mathematical analysts and electrical engineers. Nevertheless, in spite of the lack of communication, these two groups often developed parallel theories, though in different languages, at different levels of generality and typically quite different motivations. In the last several years each side has become aware of the work of the other and there is a seemingly ever-increasing involvement of the abstract theories of factorization, extension and interpolation of operators (and operator/matrix valued functions) to the design and analysis of systems and networks. Moreover, the problems encountered in electrical engineering have generated new mathematical problems, new approaches, and useful new formulations.

The papers contained in this volume constitute a more than representative selection of the presented talks and discussion at the workshop, and hopefully will also serve to give a reasonably accurate picture of the problems which are under active study today and the techniques which are used to deal with them.

It is a pleasure to thank the Weizmann Institute for extending its facilities and hospitality to the participants in this Workshop. Particular thanks are due to the Maurice and Gabriela Goldschleger Conference Foundation at the Weizmann Institute of Science, for providing financial support, and to Mr. Yitzhak Berman who took over much of the administrative burden and planned the social activities which contributed so much to the good spirits which pervaded the conference. We would also like to express our appreciation to Mrs. Ruby Musrie who handled all the secretarial work associated with the conference, and the preparation of this volume.

Thanks are also due to Tel-Aviv University for hosting one of the afternoon sessions, and for its hospitality thereafter, through funds which were available because of the generous support of Nathan and Lily Silver.

Finally, it is a special pleasure to dedicate this volume to Professor M. S. Livšič of the Ben-Gurion University of the Negev, who was perhaps the first to see the connections between linear systems and operator theory, and contributed so much to the development of both.

Harry Dym , Israel Gohberg

Rehovot, Israel
December 25, 1983
Tevet 19, 5744

Operator Theory:
Advances and Applications, Vol. 12
© 1984 Birkhäuser Verlag Basel

INVARIANT SUBSPACE REPRESENTATIONS, UNITARY INTERPOLANTS AND FACTORIZATION INDICES

Joseph A. Ball

The goal of this paper is to describe and analyze the set of all unitary $n \times n$ matrix valued functions $F(\zeta) = \sum_{j=-\infty}^{\infty} F_j \zeta^j$ on the unit circle $\mathbb{T}$ with prescribed matrix Fourier coefficients $F_j = K_j$ for $j < 0$. It is known that $\|H_K\| \leq 1$ is a necessary condition for such an F to exist, where H_K is the infinite Hankel matrix $[K_{-(j+k-1)}]_{j,k=1,2,\ldots}$ acting on ℓ_n^2. With the added assumption that $I - H_K^* H_K$ is Fredholm, there is a linear fractional map $T_\Xi : BL^\infty_{s\times s} \to BL^\infty_{n\times n}$ (B = unit ball) such that the set of unitary interpolants is precisely all F of the form $F = T_\Xi(G)$ for a $s \times s$ matrix inner function G. The linear fractional map T_Ξ arises from a $2n \times 2n$ matrix function Ξ which satisfies an identity $\Xi^* J \Xi = \hat{D}$ a.e. on $\mathbb{T}$. Here J is the signature matrix $\begin{bmatrix} I_n & 0 \\ 0 & -I_n \end{bmatrix}$ and $\hat{D}$ has the form $\begin{bmatrix} 0 & & d^* \\ & \begin{bmatrix} I_s & 0 \\ 0 & -I_s \end{bmatrix} & \\ d & & 0 \end{bmatrix}$ where d is a $(n-s) \times (n-s)$ matrix of the form $[d]_{ij} = \zeta^{\kappa_j} \delta_{n+1-i,j}$ $(i,j = 1,\ldots,n-s)$ for negative integers $\kappa_1 \leq \kappa_2 \leq \ldots \leq \kappa_{n-s} < 0$. The function Ξ together with the indices $\{\kappa_1,\ldots,\kappa_{n-s}\}$ can be computed directly from the original data $\{K_j\}_{j<0}$. In particular there is a unique unitary interpolant if and only if $s = 0$, in which case all factorization indices are negative. Moreover, a unitary interpolant F has a generalized Wiener-Hopf factorization if and only if $F = T_\Xi(G)$ where G is an $s \times s$ matrix Blaschke product. In this case the negative factorization indices of F are necessarily equal to the integers $\{\kappa_1,\kappa_2,\ldots,\kappa_{n-s}\}$ mentioned above, while the remaining s nonnegative factorization indices are identical to the set of factorization indices for G. If

$\sum_{j=1}^{\infty} \|K_j\| < \infty$, then the set of all $T_{\Xi}(G)$ with G a $s \times s$ matrix Blaschke product also coincides with the set of all unitary interpolants in the Wiener algebra. The results arise by combining the invariant subspace representation ideas of Ball and Helton developed separately before for interpolation and factorization problems. All the results here are suggested by the recent work of Dym and Gohberg on factorization indices for unitary interpolants.

1. INTRODUCTION

Let $L^p_{m\times n}$ $(1 \leq p \leq \infty)$ be the space of $m \times n$ matrix functions $F(\zeta)$ on the unit circle $\mathbb{T}$ with all matrix entries $[F(\zeta)]_{i,j}$ in L^p of the unit circle. We take $H^p_{m\times n}$ to be the Hardy subspace of all such F with negative Fourier coefficients equal to zero $(F(\zeta) \sim \sum_{j=0}^{\infty} F_j\zeta^j)$, while $K^p_{m\times n}$ will be the complementary space of all such F whose nonnegative Fourier coefficients vanish $(F \sim \sum_{j=-1}^{-\infty} F_j\zeta^j)$. We abbreviate $L^p_{n\times 1}$ to L^p_n and similarly for H^p_n and K^p_n. We let $\mathcal{U}L^{\infty}_{n\times n}$ denote the set of *unitary valued* elements in $L^{\infty}_{n\times n}$, and similarly $\mathcal{U}H^{\infty}_{n\times n}$ is the set of unitary valued functions of $H^{\infty}_{n\times n}$; elements of $\mathcal{U}H^{\infty}_{n\times n}$ are also known as $n \times n$ matrix inner functions. Let $\mathcal{B}L^{\infty}_{n\times n}$ denote the set of $L^{\infty}_{n\times n}$ functions F with $\|F\|_{\infty} \leq 1$. For $K \in L^{\infty}_{n\times n}$ a given function, we let $\mathcal{H}_K$ denote the Hankel operator which maps H^2_n into K^2_n defined by

$$\mathcal{H}_K(f) = P_{K^2_n}(Kf), \quad f \in H^2_n .$$

Here $P_{K^2_n}$ is the orthogonal projection of L^2_n onto K^2_n. For computational purposes it is useful to note that the matrix representation for $\mathcal{H}_K$ with respect to the block orthonormal basis $\{\zeta^{j-1}\mathbb{C}^n\}_{j=1,2,\ldots}$ for H^2_n and the block orthonormal basis $\{\zeta^{-i}\mathbb{C}^n\}_{i=1,2,\ldots}$ for K^2_n is given by the Hankel matrix $[K_{-(i+j-1)}]_{i,j=1,2,\ldots}$, where $\{K_j\}_{j=0,\pm 1,\ldots}$ are the matrix Fourier coefficients for K. Note in particular that $\mathcal{H}_K$ Is determined by the negative Fourier coefficients of K.

We recall at this point some elements of the theory of generalized Wiener-Hopf factorization for a function $F \in L^\infty_{n\times n}$ in the sense of Clancey and Gohberg [CG]. The function $F \in L^\infty_{n\times n}$ is siad to have a <u>generalized right Wiener-Hopf factorization</u> if $F = X_- D X_+$ where $X_+^{\pm 1} \in H^2_{n\times n}$, $X_-^{\pm 1} \in \zeta K^2_{n\times n}$ and D is a diagonal matrix function of the form

$$D(\zeta) = \mathrm{diag}\{\zeta^{\kappa_1}, \zeta^{\kappa_2}, \ldots, \zeta^{\kappa_n}\}$$

where $\kappa_1 \leq \kappa_2 \leq \ldots \leq \kappa_n$ are integers. It is also demanded that the operator $X_- P_{H^2_n} X_-^{-1}$, which a priori is defined only as an operator from L^∞_n into L^1_n, extend by continuity to define a bounded operator from L^2_n into L^2_n.

Let us use the notation $GL^2_{n\times n}$ for those functions X with $X^{\pm 1} \in L^2_{n\times n}$ and such that $XP_{H^2_n}X^{-1}$ extends to define a bounded operator on L^2_n. The integers $\{\kappa_1, \kappa_2, \ldots, \kappa_n\}$ are called the <u>right factorization indices</u> of F, and are uniquely determined by F. The factorization is said to be <u>right canonical</u> if all the indices are zero. A <u>generalized left Wiener-Hopf factorization</u> is one of the form $F = X_+ D X_-$; this gives rise to <u>left factorization indices</u>. Since we shall restrict ourselves to right factorizations, we shall refer to <u>generalized Wiener-Hopf factorization</u>, <u>canonical factorization</u>, and <u>factorization indices</u> with the understanding that all of these are of the right type. It is known that a given function $F \in L^\infty_{n\times n}$ has a generalized Wiener-Hopf factorization if and only if the Toeplitz operator $T_F\colon H^2_n \to H^2_n$ defined by $T_F(f) = P_{H^2_n}(Ff)$ is Fredholm, and that the factorization is canonical if and only if T_F is invertible. Let us say more simply that a given F is <u>factorable</u> if it has a generalized Wiener-Hopf factorization.

The problem of interest in this paper is the following:

(Ia) For a given $K \in L^\infty_{n\times n}$, describe the set $[K + H^\infty_{n\times n}] \cap \mathcal{U}L^\infty_{n\times n}$.

(Ib) Describe all the factorable functions in $[K + H^\infty_{n\times n}] \cap \mathcal{U}L^\infty_{n\times n}$.

(Ic) Describe all the factorable functions in $[K + H^\infty_{n\times n}] \cap \mathcal{U}L^\infty_{n\times n}$ having preassigned factorization indices $\{\kappa_1, \ldots, \kappa_n\}$.

It is known by the matrix Nehari theorem (see [AAK]) that $\|H_K\| \leq 1$ is necessary for $[K + H^\infty_{n\times n}] \cap \mathcal{U}L^\infty_{n\times n}$ to be non-empty. We can complete results with the added restriction

(*) $I - H_K^* H_K$ is Fredholm on H^2_n .

With this assumption in force, it follows from the main result of [BH1] that there is a $2n \times 2n$ matrix function $\Xi \in GL^2_{2n\times 2n}$, a nonnegative integer r between 0 and n inclusive, and r negative integers $\kappa_1 \leq \kappa_2 \leq \ldots \leq \kappa_r < 0$ such that

$$\begin{bmatrix} K & I \\ I & 0 \end{bmatrix} H^2_{2n} = \{\Xi\, H^\infty_{2n}\}^- \tag{1.1}$$

and

$$\Xi^* J\, \Xi = \hat{D} \tag{1.2}$$

identically on $\mathbb{T}$, where

$$J = \begin{bmatrix} I_n & 0 \\ 0 & -I_n \end{bmatrix}, \qquad \hat{D} = \begin{bmatrix} & & d^* \\ & \begin{bmatrix} I_s & 0 \\ 0 & -I_s \end{bmatrix} & \\ d & & \end{bmatrix}, \tag{1.3}$$

$$s = n-r\,, \quad d(\zeta) = \begin{bmatrix} & & & \zeta^{\kappa_r} \\ & & \cdot & \\ & \cdot & & \\ & \zeta^{\kappa_2} & & \\ \zeta^{\kappa_1} & & & \end{bmatrix} \in L^\infty_{r\times r}$$

This matrix function Ξ , together with r and the r negative indices $\{\kappa_1, \kappa_2, \ldots, \kappa_r\}$, in principle, is computable from the data $\{K_j\}_{j<0}$ of the problem; we indicate how this is done below. Once we have the matrix function Ξ , a complete solution to problems (Ia-c) is given in terms of a linear fractional map

$T_\Xi \colon \mathcal{B}L^\infty_{s\times s} \to \mathcal{B}L^\infty_{n\times n}$ defined as follows. Write Ξ in block form $\begin{bmatrix} \psi & \alpha & \beta & \psi' \\ \omega & \kappa & \gamma & \omega' \end{bmatrix}$ where the blocks are compatible with the operator of multiplication by $\Xi(\zeta)$ thought of as acting from $\mathbb{C}^r \oplus \mathbb{C}^s \oplus \mathbb{C}^s \oplus \mathbb{C}^r$ to $\mathbb{C}^n \oplus \mathbb{C}^n$. Let i^* be the $s \times n$ matrix given by

$$i^* = [I_r \quad 0_{r,s}] \quad (0_{r,s} = \text{the } r\times s \text{ zero matrix}$$
$$I = \text{the } r\times r \text{ identity matrix})$$

and let j^* be the $r \times n$ matrix given by

$$j^* = [0_{s,r} \quad I_s] .$$

The map $T_\Xi \colon \mathcal{B}L^\infty_{s\times s} \to \mathcal{B}L^\infty_{n\times n}$ is defined by

$$(1.4) \qquad T_\Xi(G) = (\alpha G j^* + \psi i^* + \beta j^*)(\kappa G j^* + \omega i^* + \gamma j^*)^{-1} .$$

Note that only the first three columns of Ξ are needed to define the map T_Ξ.

By a matrix Blaschke product we mean a matrix function $\Psi \in \mathcal{U}H^\infty_{n\times n}$ which is the product $\Psi = \overset{\rightarrow}{\prod}_{j=1}^{\ell} b_j$ of finitely many elementary Blaschke factors b_j of the form

$$b_j(\zeta) = U_j[(I - P_j) + \frac{\zeta - w_j}{1-\zeta\bar{w}_j} P_j\}$$

where U_j is a unitary matrix, P_j is a rank 1 orthogonal projection, and $|w_j| < 1$. It can be shown that any such Ψ is factorable and all its factorization indices are nonnegative.

We are now ready to state the main result of this paper.

THEOREM 1. *Let* $K \in L^\infty_{n\times n}$ *be given and suppose that* $I - H_K^* H_K$ *is nonnegative definite and Fredholm. Let* $\Xi \in GL^2_{2n\times 2n}$ *be a matrix function satisfying* (1.1) *and* (1.2), *with negative indices* $\kappa_1 \leq \kappa_2 \leq \dots \leq \kappa_r < 0$ *for some* r, $0 \leq r \leq n$, *and set* $s = n-r$. *Then*

$$[K + H^\infty_{n\times n}] \cap \mathcal{U}L^\infty_{n\times n} = \{T_\Xi(G) \colon G \in \mathcal{U}H^\infty_{s\times s}\} .$$

Moreover a given $F \in [K + H^\infty_{n\times n}] \cap \mathcal{U}L^\infty_{n\times n}$ *is factorable if and only if* $F = T_{\Xi}(G)$ *for* G *a matrix Blaschke product. The negative factorization indices of any such* F *coincide with* $\{\kappa_1, \kappa_2, \ldots, \kappa_r\}$ *, while the* s *remaining nonnegative factorization indices of* F *coincide with the factorization indices of* G *, and thus can be arbitrarily prescribed.*

In §4 we give analogues of these results for the Wiener-like algebras studied by Dym and Gohberg [DG1].

We now mention connections with other work. Adamyan, Arov and Krein [AAK] obtain a linear fractional map parametrization of $[K + H^\infty_{n\times n}] \cap \mathcal{B}L^\infty_{n\times n}$ (i.e. of contractive rather than unitary interpolants) with the assumption that $\|H_K\| < 1$. Arsene, Ceausescu and Foias [ACF] obtained an analogous result with the weaker assumption $\|H_K\| \leq 1$; they worked in the more general setting of the Sz. Nagy-Foias lifting theorem. The first to analyze systematically the nature of the factorization indices for factorable unitary interpolants were Dym and Gohberg, first in a Banach algebra setting [DG1,2] and then in the present L^2 setting [DG3]. Using a refinement of the one-step extension method of Adamyan, Arov and Krein, they identified which unitary interpolants are factorable, they showed how the negative (in terms of our conventions here) factorization indices of a unitary interpolant are determined by the data of the problem, they were able to show that the nonnegative indices can be prescribed arbitrarily, and in the case of uniqueness obtained explicit formulas for the unitary interpolant and the factors in its Wiener-Hopf factorization. Our result here, which gives a parametrization of all factorable unitary interpolants with any prescribed set of indices, is a refinement which their work strongly suggested.

The proof of Theorem 1 rests on applying techniques of invariant subspace representations which have recently been developed by the author and Helton [BH1-4]. In [BH2] it was shown how invariant subspace techniques and ideas from the theory of spaces with an indefinite metric can be used to study interpolation problems; among other things there followed a linear fractional map parametrization of $[K + H^\infty_{n\times n}] \cap \mathcal{B}L^\infty_{n\times n}$ for the case that $\|H_K\| \leq 1$ and $I - H_K^* H_K$ has closed range. Here we refine the ideas there for the special case where $I - H_K^* H_K$ is Fredholm; we use instead a later invariant subspace representation theorem from [BH1] so as not to obliterate information about indices. With regard to factorization, on the other hand, in [BH3] it was shown how generalized Wiener-Hopf factorization fits into a framework of invariant subspace representations. The contribution of this paper is to combine these two frameworks to obtain the above results on factorization of unitary interpolants. For the Wiener-like algebra setting of §4, the needed invariant subspace representations are obtained by reversing the procedure of [BH3]; the known results on Wiener-Hopf factorization over a Wiener-like algebra (see [BG]) are used to prove the invariant subspace theorems required. This gives refinements of some of the results on factorization of unitary interpolants in a Wiener-like algebra from [DG1].

The organization of the paper is as follows. In §2, we review the results and point of view on invariant subspaces needed for our application here from [BH1-3], first with regard to factorization, secondly with regard to interpolation. In §3 we show how these two frameworks can be combined to prove Theorem 1. In §4 we indicate how the same ideas can be used to get the analogous results for a Wiener-like algebra.

2. INVARIANT SUBSPACE REPRESENTATIONS

2a. Wiener-Hopf factorization

We recall here the main results from [BH3]. A subspace M of L_n^2 is said to be <u>full range simply invariant</u> (FRSI) if

i) $\zeta M \subset M$

ii) $\bigcap_{j \geq 0} \zeta^j M = \{0\}$

and

iii) $\bigcup_{j \geq 0} \zeta^{-j} M$ is dense in L_n^2.

Recall that an $n \times n$ matrix function Ξ is said to be in the class $GL^2_{n\times n}$ if $\Xi^{\pm 1} \in L^2_{n\times n}$ and the operator $\Xi P_{H_n^2} \Xi^{-1}$ extends to define a bounded operator on L_n^2. By a <u>winding matrix</u> D we mean a diagonal matrix function $D(\zeta)$ of the form $\operatorname{diag.}\{\zeta^{\kappa_1},\ldots,\zeta^{\kappa_n}\}$ where $\kappa_1 \leq \ldots \leq \kappa_n$ are integers. A subspace $M^x \subset L_n^2$ is said to be <u>full range simply *-invariant</u> (FRSI)$_*$ if $\overline{M^x}$ (the complex conjugates of the elements of M^x) is FRSI. A pair of subspaces $\{M^x, M\}$ is said to be <u>Fredholm</u> if

i) $\operatorname{codim}(M^x + M) < \infty$

ii) $\dim(M^x \cap M) < \infty$.

We emphasize that in i) we do not take the closure of $M^x + M$, so part of the definition is that $M^x + M$ be closed. For N a subset of L_n^2, let N^- denote its closure in L_n^2.

THEOREM 2.1. (see [BH3]). *Let* M^x *and* M *be a given pair of subspaces of* L_n^2. *Then there is a matrix function* $\Xi \in GL^2_{n\times n}$ *and a winding matrix* D *such that* $M^x = \{\Xi K_n^\infty\}^-$ *and* $M = \{\Xi DH_n^\infty\}^-$ *if and only if*

i) M *is FRSI*

ii) M^x *is (FRSI)*$_*$

iii) $\{M^x, M\}$ *is a Fredholm pair of subspaces of* L_n^2.

Moreover, the indices $\kappa_1 \leq \kappa_2 \leq \ldots \leq \kappa_n$ *of* D *are uniquely determined by the pair* $\{M^x, M\}$ *by backsolving either the system of equations*

$$\dim(\zeta^k M^x \cap M) = \sum_{\kappa_j < k} (k - \kappa_j) , \quad k = 0,\pm 1,\ldots$$

or

$$\operatorname{codim}(\zeta^k M^x + M) = \sum_{\kappa_j > k} (\kappa_j - k) , \quad k = 0,\pm 1,\ldots .$$

(If there are no κ_j 's satisfying the indicated condition, it is understood that the sum on the right is zero.)

We sketch here how the indices can be constructed by analyzing the fine structure of the subspace $M^x \cap M$ and a complement for $M^x + M$. We first consider the negative indices. If $M^x \cap M = \{0\}$, all indices are nonnegative; otherwise consider a basis for $M^x \cap M$ of the form

$$(2.1)\qquad \begin{aligned}&\{\zeta^{-1}x_1,\zeta^{-2}x_1,\ldots,\zeta^{\kappa_1}x_1\ ;\\ &\zeta^{-1}x_2,\zeta^{-2}x_2,\ldots,\zeta^{\kappa_2}x_2\ ;\\ &\ldots\ldots\ldots\\ &\zeta^{-1}x_r,\zeta^{-2}x_r,\ldots,\zeta^{\kappa_r}x_r\}\end{aligned}$$

where $\kappa_1 \leq \kappa_2 \leq \ldots \leq \kappa_r < 0$ are negative integers, and $\zeta^{\kappa_j - 1} x_j \notin M^x \cap M$ for $1 \leq j \leq r$. Then it can be shown that while the basis vectors $\{\zeta^{\ell}x_j\colon \kappa_j \leq \ell \leq -1,\ 1 \leq \ell \leq r\}$ are not uniquely determined by the subspace $M^x \cap M$, the number r of chains and the lengths $-\kappa_j$ $(1 \leq j \leq r)$ of these chains is uniquely determined by $M^x \cap M$. Then the winding matrix D in the above representation has precisely r negative indices and these are given by (-1) times the lengths $-\kappa_j$ of these r chains. The determination of the positive indices is a dual analysis. If $M^x + M = L_n^2$, then all indices are nonpositive. Otherwise construct a basis for a subspace complementary to $M^x + M$ of the form

$$(2.2)\qquad \begin{aligned}&\{y_1,\zeta y_1,\ldots,\zeta^{\mu_1-1}y_1\ ;\\ &y_2,\zeta y_2,\ldots,\zeta^{\mu_2-1}y_2\ ;\\ &\ldots\ldots\ldots\ldots\\ &y_t,\zeta y_t,\ldots,\zeta^{\mu_t-1}y_t\}\end{aligned}$$

where $\mu_1 \geq \mu_2 \geq \ldots \geq \mu_t > 0$ and $\zeta^{\mu_j}y_j \in M^x + M$ for $1 \leq j \leq t$. Then t of the indices for the winding matrix D are positive and coincide with the set $\{\mu_1,\mu_2,\ldots,\mu_t\}$. To conclude, it can be shown that necessarily $0 \leq r+t \leq n$, and the remaining $n-r-t$ indices are zero. In particular, the so-called canonical case where all indices are zero matches up with the case where $\dim(M^x \cap M) = \operatorname{codim}(M^x + M) = 0$, that is, where $L_n^2 = M^x \dotplus M$ (direct sum decomposition).

The construction of the matrix function Ξ proceeds as follows. By considering the pair $\{M^x, \zeta^{\kappa}M\}$ instead of $\{M^x, M\}$, we may suppose that $M^x + \zeta M = L_n^2$, so all indices are negative. Then $r = n$ in any basis of the form (2.1) for $M^x \cap M$, and a choice for a representing function Ξ is to let the j-th column of Ξ equal the vector function x_j determined by the basis (2.1)

$$\Xi = [x_1\ x_2\ \dots\ x_n]\ .$$

Conversely, given a representing function $\Xi = [x_1\ x_2\ \dots\ x_n]$ for a Fredholm pair $\{M^x, M\}$ such that all the indices $\kappa_1 \leq \kappa_2 \leq \dots \leq \kappa_n < 0$ are negative, then the columns $\{x_j : j = 1,\dots,n\}$, of Ξ determine a basis as in (2.1) for the subspace $M^x \cap M$. The construction of a basis of the form (2.1) amounts to solving a succession of classical Riemann-Hilbert barrier problems (see [BH3] for details).

Now suppose F is a given function in $L^\infty_{n\times n}$. We apply the above analysis to $M^x = K^2_n$ and $M = FH^2_n$. Clearly M^x is $(\mathrm{FRSI})_*$. If we assume also that $F^{-1} \in L^\infty_{n\times n}$, then certainly M is FRSI. One can easily check that $\{M^x, M\}$ is Fredholm if and only if the Toeplitz operator $T_F\colon H^2_n \to H^2_n$ defined by $T_F(f) = P_{H^2_n}(Ff)$ is Fredholm. Thus, by Theorem 2.1 we see that there is a function $\Xi \in GL^2_{n\times n}$ and a winding matrix D such that

$$\{\Xi\, K^\infty_n\}^- = K^2_n$$

and

$$\{\Xi\, DH^\infty_n\}^- = FH^2_n$$

if and only if $F^{\pm 1} \in L^\infty_{n\times n}$ and T_F is Fredholm. From the above representations, we see that

$$X_-: = \Xi \quad \text{has} \quad X_-^{\pm 1} \in K^2_{n\times n} \qquad \text{and}$$

$$X_+: = D^{-1}\,\Xi^{-1}F \quad \text{has} \quad X_+^{\pm 1} \in H^2_{n\times n}\ , \qquad \text{and}$$

trivially, $F = X_-DX_+$. Thus we see that $F(F^{\pm 1} \in L^\infty_{n\times n})$ has a generalized (right) Wiener-Hopf factorization if and only if T_F is Fredholm, as a direct consequence of Theorem 2.1. This factorization result is due to Simonenko (See [CG]).

We shall be particularly interested in the case where F is unitary valued. In this case note that $x \in K^2_n \cap FH^2_n$ if and only if

$$x = Fy = P_{K^2_n}(Fy)$$

for some $y \in H^2_n$, and thus

$$\|H_F y\| = \|Fy\| = \|y\|$$

so $y \in \mathrm{Ker}\, I - H_F^* H_F$. We see that the negative factorization indices $\kappa_1 \leq \kappa_2 \leq \ldots \leq \kappa_r < 0$ of a unitary valued F (which are well defined under the assumption that T_F is Fredholm) are had by determining a basis for $\mathrm{Ker}\, I - H_F^* H_F$ of the form (2.1), since this becomes a basis for $K_n^2 \cap FH_n^2$ of the form (2.1) upon multiplication by F . How the number of chains r and the indices $\kappa_1 \leq \kappa_2 \leq \ldots \leq \kappa_r < 0$ can be computed explicitly in terms of of $\dim[\mathrm{Ker}(I - H_F^* H_F) \cap \zeta^j H_n^2] = \dim \mathrm{Ker}(I - H^*_{\zeta^j F} H_{\zeta^j F})$, $(j \geq 0)$ is given by Dym and Gohberg [DG1-3]. We also note at this point that the positive factorization indices $\mu_1 \geq \mu_2 \geq \ldots \geq \mu_s > 0$ for a $n \times n$ matrix valued function F are had by determining a basis of the form (2.2) for a subspace complementary to $K_n^2 + FH_n^2$.

2b. Interpolation

We now review that approach to interpolation in [BH2]. We are given $K + L^\infty_{n\times n}$ and wish to describe the set of unitary interpolants $[K + H^\infty_{n\times n}] \cap \mathcal{U}L^\infty_{n\times n}$. (In [BH2] the emphasis was on contractive interpolants $[K + H^\infty_{n\times n}] \cap \mathcal{B}L^\infty_{n\times n}$; we adapt the discussion there to the unitary case.) Introduce the following subspaces of L^2_{2n} :

$$(2.3) \qquad \mathcal{K} = \begin{bmatrix} L_n^2 \\ H_n^2 \end{bmatrix} , \qquad \mathcal{M} = \begin{bmatrix} K & I_n \\ I_n & 0 \end{bmatrix} H^2_{2n} .$$

Instead of a matrix function F in $[K + H^\infty_{n\times n}] \cap \mathcal{U}L^\infty_{n\times n}$, we look instead for their graphs $\mathcal{G}_F := \begin{bmatrix} F \\ I_n \end{bmatrix} H_n^2$ as a subspace of $\mathcal{K}$. Characterization of subspaces of the form $\mathcal{G}_F$ necessitates the introduction of an indefinite metric on L^2_{2n} :

$$[f,g] = \langle Jf,g\rangle_{L^2_{2n}}$$

where $J = \begin{bmatrix} I_n & 0 \\ 0 & -I_n \end{bmatrix}$. This makes L^2_{2n} a Krein space (see [B] for definitions and background) and it is easy to see that $\mathcal{K}$ also is a Krein space in the inner product [,] .

We now review some facts and definitions we need concerning general Krein spaces. For any Krein space K , a subspace $M \subset K$ is said to be nonpositive if $[x,x] \leq 0$ for all $x \in M$, and negative if equality holds only if $x = 0$. (This terminology differs from that of [BH2] where "negative" and "strictly negative" were used in lieu of "nonpositive" and "negative" respectively; the present terminology is more standard.) We define a subspace to be nonnegative [positive] if it is nonpositive [negative] with respect to $-[\ ,]$. If $G \subset M$ are two subspaces, G is said to be M-maximal nonpositive if G itself is nonpositive and any other nonpositive subspace N with $G \subset N \subset M$ must be equal to G . A subspace G is said to be isotropic if $[x,y] = 0$ for all $x,y \in G$; equivalently G is both nonpositive and nonnegative. Any Krein space has a $[\ ,\ ]$-orthogonal direct sum decomposition $K = K_1 \dotplus K_2$ where K_1 and K_2 are Hilbert spaces, and

$$[k_1 \dotplus k_2 , k_1 \dotplus k_2]_K = \langle k_1,k_1\rangle_{K_1} - \langle k_2,k_2\rangle_{K_2}$$

for $k_j \in K_j$ $(j = 1,2)$. If we fix such a decomposition, then a subspace N is K-maximal nonpositive if and only if N has the form

$$N = \{Ck \dotplus k \mid k \in K_2\}$$

for some contraction operator C mapping K_2 into K_1 . A subspace N has codimension ℓ in some K-maximal negative subspace if and only if N has the form

$$N = \{Ck \dotplus k \mid k \in K_2'\} \tag{2.4}$$

where K_2' is a subspace of K_2 of codimension ℓ and $C\colon K_2' \to K_1$ is a contraction.

We now return to our concrete situation where L^2_{2n} is a Krein space with indefinite inner product induced by $J = \begin{bmatrix} I_n & 0 \\ 0 & -I_n \end{bmatrix}$ and K and M are the subspaces given by (2.3). For $N \subset L^2_{2n}$, denote by $N^{\perp J}$ the orthogonal complement of N with respect to the J-inner product. The main reduction from

[BH2] is the following.

LEMMA 2.2. *A subspace* $G \subset L^2_{2n}$ *has the form* $\begin{bmatrix} F \\ I_n \end{bmatrix} H^2_n$ *for some* $F \in [K + H^\infty_{n\times n}] \cap \mathcal{U}L^\infty_{n\times n}$ if and only if

(2.5i) $G \subset M$

(2.5ii) G *is* K*-maximal nonpositive.*

(2.5iii) G *is isotropic.*

(2.5iv) $\zeta G \subset G$ *(i.e.* G *is invariant under multiplication by* ζ *).*

LEMMA 2.3. *A* M*-maximal nonpositive subspace* G *is also* K*-maximal nonpositive if and only if* $M^{\perp J} \cap K$ *is non-negative, or equivalently* $\|H_K\| \leq 1$.

By Lemma 2.3, $\|H_K\| \leq 1$ is necessary for $[K + H^\infty_{n\times n}] \cap \mathcal{U}L^\infty_{n\times n}$ to be nonempty. With this assumption in force, condition ii) in Lemma 2.2 can be switched to

(2.5ii') G is M-maximal nonpositive, and the whole problem is localized to M . In [BH2] it was assumed at this point that $I - H^*_K H_K$ was semi-Fredholm, or equivalently, that $M^{\perp J} + M$ is closed. We get finer results by assuming that $M^{\perp J} + M$ has finite codimension, or equivalently, that $I - H^*_K H_K$ is Fredholm. Then we may apply the main invariant subspace representation theorem of [BH1]; for a proof based on Theorem 2.1, see [BH4]. By a self-adjoint winding matrix $\hat{D}$ we mean a $2n \times 2n$ matrix function as in (1.3).

THEOREM 2.4. *Suppose* $I - H^*_K H_K$ *is Fredholm. Then there is a matrix function* $\Xi \in GL^2_{2n\times 2n}$ *and a uniquely determined self-adjoint winding matrix* $\hat{D}$ *such that*

i) $M = \{\Xi H^\infty_{2n}\}^-$ *and*

ii) $\Xi(\zeta)^* J \Xi(\zeta) = \hat{D}(\zeta)$ *for a.e.* $\zeta \in \mathbb{T}$.

We remark that from the construction in the proof of Theorem 2.4 from [BH1], it is clear that the negative indices $\kappa_1 \leq \kappa_2 \leq \dots \leq \kappa_r < 0$ for the self-adjoint winding matrix $\hat{D}$ in the above representation for M are had by determining a basis for $M \cap M^{\perp J}$ of the form (2.1). For M of the form $\begin{bmatrix} K & I_n \\ I_n & 0 \end{bmatrix} H^2_{2n}$, one computes that $x \in M \cap M^{\perp J}$ if and only if $x = \begin{bmatrix} -H_K \\ I \end{bmatrix} y$ for $y \in \operatorname{Ker}(I - H_K^* H_K)$. Thus we see that the negative indices $\kappa_1 \leq \kappa_2 \leq \ldots \leq \kappa_r < 0$ for $\hat{D}$ alternatively arise from a basis for $\operatorname{Ker}(I - H_K^* H_K)$ of the form (2.1), since one obtains a basis for $M \cap M^{\perp J}$ from this simply by multiplying by $\begin{bmatrix} -H_K \\ I \end{bmatrix}$. In particular, if $F \in [K + H^\infty_{n\times n}] \cap UL^\infty_{n\times n}$ is a unitary interpolant of K, then $H_F = H_K$; from the above combined with the discussion in §2a, we see that the negative factorization indices of any factorable unitary interpolant coincide with the negative indices of the self-adjoint winding matrix $\hat{D}$ in the representation for the subspace $M = \begin{bmatrix} K & I \\ I & 0 \end{bmatrix} H^2_{2n}$ given by Theorem 2.4. A choice of representing function Ξ is computable from the negative Fourier coefficients of K, but we do not give the details here. We refer the reader to [BH1] for a constructive proof of Theorem 2.4, from which explicit formulas can be obtained with more work.

We now want to consider L^2_{2n} with the alternative indefinite inner product induced by $\hat{D}$:

$$[f,g]_{\hat{D}} = \langle \hat{D} f, g\rangle_{L^2_{2n}} \quad \text{for } f,g \in L^2_{2n}.$$

Then L^2_{2n} is also a Krein space in this inner product. Let us say that a subspace is $\hat{D}$-nonpositive if it is nonpositive in this $\hat{D}$-inner product. The following lemma can be proved in the same way as an analogous statement in [BH2]. Note that the result is trivial if $\Xi^{\pm 1} \in L^\infty_{2n\times 2n}$.

LEMMA 2.5. *Suppose* $M \subset L^2_{2n}$ *has the representation* $M = \{\Xi H^\infty_{2n}\}^-$ *where* $\Xi \in GL^2_{2n}$ *and* $\Xi^* J \Xi = \hat{D}$. *Then a subspace* $G \subset M$ *is invariant* M*-maximal nonpositive if and only if* $G = \{\Xi[G_1 \cap H^\infty_{2n}]\}^-$ *where* G_1 *is an invariant* H^2_{2n}*-maximal* $\hat{D}$*-nonpositive subspace of* H^2_{2n}.

When we combine all these reductions, we arrive at the following: A subspace $G \subset L^2_{2n}$ has the form $G = \begin{bmatrix} F \\ I_n \end{bmatrix} H^2_n$ with $F \in [K + H^\infty_{n\times n}] \cap \mathcal{U}L^\infty_{n\times n}$ if and only if $G = \{\Xi[G_1 \cap H^\infty_{2n}]\}^-$ where $G_1 \subset H^2_{2n}$ satisfies

(2.6.i) G_1 is H^2_{2n}-maximal $\hat{D}$-nonpositive

(2.6.ii) G_1 is $\hat{D}$-isotropic

(2.6.iii) $\zeta G_1 \subset G_1$.

Such subspaces G_1 are easy to characterize. Indeed let us decompose H^2_{2n} in accordance with the block decomposition of $\hat{D}$:

$$\hat{D} = \begin{bmatrix} 0 & & & d^* \\ & \begin{bmatrix} I_s & 0 \\ 0 & -I_s \end{bmatrix} & & \\ d & & & 0 \end{bmatrix}, \quad H^2_{2n} = \begin{bmatrix} H^2_r \\ H^2_s \\ H^2_s \\ H^2_r \end{bmatrix}, \quad d = \begin{bmatrix} & & \zeta^{\kappa_r} \\ & \cdot^{\cdot^{\cdot}} & \\ \zeta^{\kappa_1} & & \end{bmatrix}$$

where $\kappa_1 \leq \kappa_2 \leq \ldots \leq \kappa_r < 0$ are integers.

One computes that the isotropic subspace $H^2_{2n} \cap H^{2\,\perp\hat{D}}_{2n}$ is spanned by $\{\zeta^k e_j : 0 \leq k \leq -\kappa_j - 1,\ 1 \leq j \leq r\}$, where e_j is the j-th standard basis vector. Any H^2_{2n}-maximal $\hat{D}$-nonpositive subspace G_1 must contain this subspace. If G_1 is also invariant then necessarily G_1 contains $\mathrm{col}[H^2_r \ 0 \ 0 \ 0]$. Next check that any $\hat{D}$-nonpositive subspace G_1 containing $\mathrm{col}[H^2_r \ 0 \ 0 \ 0]$ must itself be contained in $\mathrm{col}[H^2_r \ H^2_s \ H^2_s \ 0]$, and is therefore of the form $\mathrm{col}[H^2_r \ \hat{G}_1 \ 0]$, where $\hat{G}_1$ is $\begin{bmatrix} I_s & 0 \\ 0 & -I_s \end{bmatrix}$-nonpositive subspace of $\begin{bmatrix} H^2_s \\ H^2_s \end{bmatrix}$. The end result is that a subspace $G_1 \subset H^2_{2n}$ satisfies i), ii) and iii) above if and only if

$$G_1 = \begin{bmatrix} I_r & 0 \\ 0 & G \\ 0 & I_s \\ 0 & 0 \end{bmatrix} \begin{bmatrix} H^2_r \\ H^2_s \end{bmatrix}$$

where $G \in \mathcal{U}H^\infty_{s\times s}$ is an $s \times s$ inner function.

Let $\Xi = \begin{bmatrix} \psi & \alpha & \beta & \psi' \\ \omega & \kappa & \gamma & \omega' \end{bmatrix}$ be the block decomposition of Ξ consistent with the decomposition $H^2_{2n} = \mathrm{col}[H^2_r \; H^2_s \; H^2_s \; H^2_r]$ of its domain and the decomposition $L^2_{2n} = \mathrm{col}[L^2_n \; L^2_n]$ of its range. We conclude that a matrix function F is a unitary interpolant of K (i.e. $F \quad [K + H^\infty_{n\times n}] \cap \mathcal{U}L^\infty_{n\times n}$) if and only if

$$\begin{bmatrix} F \\ I \end{bmatrix} H^2_n = \left\{ \begin{bmatrix} \psi & \alpha & \beta & \psi' \\ \omega & \kappa & \gamma & \omega' \end{bmatrix} \begin{bmatrix} I_r & 0 \\ 0 & G \\ 0 & I_s \\ 0 & 0 \end{bmatrix} \begin{bmatrix} H^\infty_r \\ H^\infty_s \end{bmatrix} \right\}^-$$

$$= \left\{ \begin{bmatrix} \psi i^* + \alpha G j^* + \beta j^* \\ \omega i^* + \kappa G j^* + \gamma j^* \end{bmatrix} H^\infty_n \right\}^-$$

(where $i^* = [I_r \quad 0_{r,s}]$ and $j^* = [0_{s,r} \quad I_s]$) for some $s \times s$ inner function G .

From the second components of this equation, we see that $(\omega i^* + \kappa G j^* + \gamma j^*) H^\infty_n$ is dense in H^2_n , and hence in particular must be invertible a.e. on $\mathbb{T}$. If we now solve for F , we see that $F = T_\Xi(G)$ where T_Ξ is the linear fractional map defined by (1.4). This establishes the first part of Theorem 1.

3. FACTORIZATION OF UNITARY INTERPOLANTS

It remains to establish the assertions concerning factorization in Theorem 1. By the discussion immediately after Theorem 2.4, we know that the r negative factorization indices of any factorable unitary interpolant are completely determined by the interpolation data. We also have a formula $F = T_\Xi(G)$ $(G \in \mathcal{U}H_{s\times s})$ for the most general unitary interpolant. It remains to characterize which of these are factorable, and which of these factorable interpolants have a given set $\{\mu_1 \geqslant \mu_2 \geqslant \dots \geqslant \mu_s\}$ of nonnegative factorization indices.

We begin with the correspondence between unitary

interpolants $F \in [K + H^{\infty}_{n\times n}] \cap UL^{\infty}_{n\times n}$ and certain subspaces $G = \begin{bmatrix} F \\ I \end{bmatrix} H^2_n$ of L^2_{2n} established by Lemma 2.2. The following lemma gives a useful invariant geometric characterization of which subspaces G as in Lemma 2.2 come from factorable F.

LEMMA 3.1. *Assume* $K \in L^{\infty}_{n\times n}$ *is given and* $I - H^*_K H_K$ *is nonnegative definite and Fredholm. Let* $G \subset L^2_{2n}$ *satisfy conditions* (2.5) *of Lemma 2.2. Then* $G = \begin{bmatrix} F \\ I \end{bmatrix} H^2_n$ *where* F *is a factorable unitary interpolant if and only if in addition*

(2.5v) *G has finite codimension in an M-maximal J-nonnegative subspace P of* L^2_{2n}.

Moreover the unitary interpolant F corresponding to G has precisely t positive factorization indices $\mu_1 \geq \mu_2 \geq \dots \geq \mu_t$ *if and only if there is a basis of the form* 2.2 *for a subspace F complementary to G inside a M-maximal J-nonnegative subspace $P \supset G$:*

$P = G \dot{+} F$ where F has basis of the form (2.2) *and P is M-maximal J-nonnegative.*

PROOF. We are given that $I - H^*_k H_k$ is nonnegative definite and Fredholm and that $F \in [K + H^{\infty}_{n\times n}] \cap UL^{\infty}_{n\times n}$ is a unitary interpolant. Thus $H_F = H_K$, so $I - H^*_F H_F$ is also Fredholm. For $f \in H^2_n$,

$$\begin{aligned}(I - H^*_F H_F) f &= f - P_{H^2_n} F^* P_{K^2_n} F f \\ &= f - P_{H^2_n} F^* (I - P_{H^2_n}) F f \\ &= T^*_F T_F f\end{aligned}$$

where we used that F is unitary valued in the last step. We conclude that in general $I - H^*_F H_F = T^*_F T_F$ for unitary valued F, and thus $I - H^*_F H_F$ is Fredholm if and only if T_F is semi-Fredholm (i.e. $\dim \operatorname{Ker} T_F < \infty$ and T_F has a closed range). Equivalently, $\dim(K^2_n \cap FH^2_n) < \infty$ and the subspace $K^2_n + FH^2_n$ is closed in L^2_n. It remains to distinguish those unitary interpolants F for which $K_n^2 + FH_n^2$ has finite codimension in L_n^2, but in terms of its graph space $\begin{bmatrix} F \\ I_n \end{bmatrix} H_n^2$.

For this we need the following basic result.

LEMMA 3.2. (Lemma 1.1 from [BH2]) *Suppose K is a Krein space, $M \subset K$ is a subspace, and $M^{\perp}$ is the orthogonal complement of M with respect to K's indefinite inner product and that $M + M^{\perp}$ is closed. Then if $P_1 \subset M$ is M-maximal nonnegative and $P_2 \subset M^{\perp}$ is $M^{\perp}$-maximal nonnegative, then $P_1 + P_2$ is dense in a K-maximal nonnegative subspace.*

Now let us suppose that $K_n^2 + FH_n^2$ has codimension ℓ in L_n^2 where $F \in [K + H_{n\times n}^{\infty}] \cap UL_{n\times n}^{\infty}$. Set $M = \begin{bmatrix} K & I \\ I & 0 \end{bmatrix} H_{2n}^2$. Then one computes that $M^{\perp J} = \begin{bmatrix} I_n & 0 \\ K^* & I_n \end{bmatrix} K_{2n}^2$. From this representation it is easy to see that $\begin{bmatrix} I_n \\ K^* \end{bmatrix} K_n^2$ is a $M^{\perp J}$-maximal nonnegative subspace. Note that $\begin{bmatrix} F \\ I_n \end{bmatrix} H_n^2$ is a nonnegative subspace of M, so $Q = \begin{bmatrix} I_n \\ K^* \end{bmatrix} K_n^2 + \begin{bmatrix} F \\ I_n \end{bmatrix} H_n^2$ $\left(= \begin{bmatrix} I_n \\ K^* \end{bmatrix} K_n^2 + \begin{bmatrix} I_n \\ F^* \end{bmatrix} FH_n^2\right)$ is a J-nonnegative subspace of L_{2n}^2. Since $\begin{bmatrix} L_n^2 \\ 0 \end{bmatrix} \dotplus \begin{bmatrix} 0 \\ L_n^2 \end{bmatrix}$ is a Krein space decomposition of L_{2n}^2 into a maximal positive and maximal negative subspace, we see from the characterization (2.4) that the codimension of Q in a L_{2n}^2- maximal nonnegative subspace P is equal to the codimension ℓ of $K_n^2 + FH_n^2$ as a subspace of L_n^2. On the other hand, since $\begin{bmatrix} I_n \\ K^* \end{bmatrix} K_n^2$ is $M^{\perp}$-maximal nonnegative and $\begin{bmatrix} F \\ I_n \end{bmatrix} H_n^2 \subset M$, by Lemma 3.2 it is clear that the codimension of $\begin{bmatrix} F \\ I \end{bmatrix} H_n^2$ in an M-maximal J-nonnegative subspace is also the codimension of Q in an L_{2n}^2-maximal J-nonnegative subspace. The first part of

Lemma 3.1 follows.

By earlier remarks we know that the positive factorization indices $\mu_1 \geq \mu_2 \geq \ldots \geq \mu_t > 0$ of F are obtained by determining a basis of the form (2.2) for a subspace complementary to $K_n^2 + FH_n^2$ in L_n^2. One can arrange moreover that all the basis vectors are in $FH_n^2 + H_n^2$. Upon multiplying these basis vectors by $\begin{bmatrix} I_n \\ 0_n \end{bmatrix}$, one obtains a basis of the form (2.2) for a subspace complementary to $[\begin{smallmatrix} F \\ I \end{smallmatrix}] H_n^2$ inside a M-maximal J-nonnegative subspace P. Conversely, given a basis of the form (2.2) for a finite dimensional subspace F such that $P = [\begin{smallmatrix} F \\ I \end{smallmatrix}] H_n^2 \dot{+} F$ is M-maximal J-nonnegative, multiplication of each basis element by $[I_n \;\; 0_n]$ gives a basis of the form (2.2) for a subspace of L_n^2 complementary to FH_n^2 inside $FH_n^2 + H_n^2$, and hence also complementary to $K_n^2 + FH_n^2$ inside L_n^2. This completes the proof of Lemma 3.1.

The achievement of Lemma 3.1 is again to localize the problem to M. The next step is to use the representing function Ξ to pull the problem from M back to H_{2n}^2 where it is simpler. This is the point of the next lemma.

LEMMA 3.3. *Assume* $K \in L_{n\times n}^\infty$ *is given such that* $I - H_K^* H_K$ *is nonnegative definite and Fredholm. Let* $\Xi \in GL_{2n\times 2n}^2$ *and* $\hat{D} \in L_{2n\times 2n}^\infty$ *be as in Theorem* 2.4. *Then a subspace* $G \subset L_{2n}^2$ *satisfies conditions i) - iii) of Lemma* 2.2 *and condition iv) of Lemma* 3.1 *if and only if*

$$G = \{\Xi[G_1 \cap H_{2n}^\infty]\}^-$$

where $G_1 \subset H_{2n}^2$ *satisfies conditions* (2.6) *of Lemma* 2.5 *together with*

(2.6.iv) *G_1 has finite codimension in an H_{2n}^2-maximal $\hat{D}$-nonnegative subspace P_1.*

Moreover the unitary interpolant F *corresponding to such a* G *has precisely* t *positive factorization indices* $\mu_1 \geq \mu_2 \geq \ldots \geq \mu_t$ *if and only if there is a basis of the form* (2.2) *for a subspace* F_1 *complementary to* G_1 *inside a* H_{2n}^2-*maximal* $\hat{D}$-*nonnegative subspace* P_1 :

$P_1 = G_1 \dot{+} F_1$ where F_1 has a basis of the form (2.2) and P is H^2_{2n}-maximal $\hat{D}$-nonnegative.

PROOF. By Lemma 2.5, applied to the (-J)-inner product rather than the J-inner product, the correspondence

$$G = \{\Xi[G_1 \cap H^\infty_{2n}]\}^-$$

establishes a one - to - one correspondence between invariant M-maximal J-nonnegative subspaces G of M and invariant H^2_{2n}-maximal $\hat{D}$-nonnegative subspaces G_1 of H^2_{2n}. By the Beurling-Lax theorem, if L_1 and L_2 are simply invariant subspaces of L^2_{2n} such that L_1 has finite codimension ℓ in L_2 then a complementary ℓ dimensional subspace $F \subset L^\infty_n$ exists such that $L_2 = L_1 \dot{+} F$. This fact moreover can be refined so that we can arrange that F has a basis of the form (2.2). From this it is clear that the above correspondence $G \longleftrightarrow G_1$ extends to give a correspondence between invariant J-nonnegative subspaces G having finite codimension ℓ in an M-maximal J-nonnegative subspace P and invariant $\hat{D}$-nonnegative subspaces G_1 having finite codimension ℓ in an H^2_{2n}-maximal $\hat{D}$-nonnegative subspace P_1, and that the positive indices $\mu_1 \geq \mu_2 \geq \ldots \geq \mu_t > 0$ for G are the same as those for G_1 in this correspondence. This completes the proof of Lemma 3.1.

Thus it suffices to analyze subspaces G_1 which satisfy (2.6i-iv). We already know that subspaces satisfying (2.6i-iii) are of the form (2.7) where $G \in UH^\infty_{s\times s}$. By arguments similar to some of those above, one sees that a subspace G_1 as in (2.7) also satisfies (2.6iv) if and only if $K^2_s + GH^2_s$ has finite codimension in L^2_s, that is, if and only if G is factorable. Moreover, certain bases of the form (2.2) for a subspace complementary to G_1 inside a H^2_{2n}-maximal $\hat{D}$-nonnegative subspace P_1 match up with certain bases of the form (2.2) for a subspace complementary to $K^2_s + GH^2_s$ inside L^2_s. We conclude that the positive indices arising from the subspace G_1 in this way are the same as the positive factorization indices of G. Also, since G is inner, all factorization indices of G are nonnegative. Finally, if G is inner, then GH^2_s has finite codimension in H^2_s if and only if G is a finite matrix Blaschke product. Using the analysis of §2b, we conclude that a unitary interpolant $F \in [K + H^\infty_{n\times n}] \cap \mathcal{U}L^\infty_{n\times n}$ is factorable if and only if $F = T_\Xi(G)$ where $G \in \mathcal{U}H^\infty_{s\times s}$ is a finite matrix Blaschke product, and that the nonnegative factorization indices of F agree with those of G. This completes the proof of Theorem 1.

4. WIENER-LIKE ALGEBRA

In this section we show how our methods can be modified to prove an analogue of Theorem 1 for matrix functions over a certain type of Banach algebra of continuous functions. An easy example is the Wiener algebra $\mathcal{W}$ of continuous functions f on the unit circle $\mathbb{T}$ with absolutely summable Fourier coefficients

$$\mathcal{W} = \{f(\zeta) = \sum_{-\infty}^{\infty} f_j\zeta^j \in C(\mathbb{T}) : \sum_{-\infty}^{\infty} |f_j| < \infty\} .$$

More generally, we work with a "Wiener-like" algebra Λ of continuous functions on $\mathbb{T}$ as defined by Dym and Gohberg [DG1]. We refer the reader there for the precise definition and other examples; to keep the paper self-contained the reader is welcome to substitute $\mathcal{W}$ for Λ in the following.

We shall use the following properties and notations concerning our Wiener-like algebra Λ . Let $\Lambda_{m\times n}$ be the matrix function Banach algebra of $m \times n$ matrix functions with entries in Λ . Abbreviate $\Lambda_{n\times 1}$ to Λ_n . We set

$$(\Lambda_{m\times n})^+ := \Lambda_{m\times n} \cap H^\infty_{m\times n}$$

$$(\Lambda_{m\times n})^+_0 = \Lambda_{m\times n} \cap \zeta H^\infty_{m\times n}$$

$$(\Lambda_{m\times n})^-_0 = \Lambda_{m\times n} \cap K^\infty_{m\times n}$$

$$(\Lambda_{m\times n})^- = \Lambda_{m\times n} \cap \zeta K^\infty_{m\times n} .$$

It is known that if $F \in \Lambda_{n\times n}$ and $\det F(\zeta) \neq 0$ for all $\zeta \in \mathbb{T}$ then $F^{-1} \in \Lambda_{n\times n}$. We shall need the following basic factorization results.

THEOREM 4.1. (see [BG]). *Let* $F \in \Lambda_{n\times n}$. *Then* F *has a Wiener-Hopf factorization* $F = X_- D X_+$, *where* $X_-^{\pm 1} \in (\Lambda_{n\times n})^-$, $X_+^{\pm 1} \in (\Lambda_{n\times n})^+$ *and* $D(\zeta) = \text{diag.}\{\zeta^{\kappa_1}, \zeta^{\kappa_2}, \ldots, \zeta^{\kappa_n}\}$ ($\kappa_1 \leq \kappa_2 \leq \ldots \leq \kappa_n$ *integers)* , *if and only if* $\det F(\zeta) \neq 0$ *for all* $\zeta \in \mathbb{T}$.

THEOREM 4.2. *Let* $F \in \Lambda_{n\times n}$. *Then* F *can be factored in the form* $F = A^* D A$, *where*

i) $A^{\pm 1} \in (\Lambda_{n\times n})^+$

and

$$ii)\quad \hat{D}(\zeta) = \begin{bmatrix} & & d(\zeta)^* \\ & \begin{bmatrix} I_s & 0 \\ 0 & -I_{s'} \end{bmatrix} & \\ d(\zeta) & & \end{bmatrix}$$

$$\text{where}\quad d(\zeta) = \begin{bmatrix} & & & \zeta^{\kappa_r} \\ & & \cdot^{\cdot^{\cdot}} & \\ & \zeta^{\kappa_2} & & \\ \zeta^{\kappa_1} & & & \end{bmatrix} \in \Lambda_{r\times r}$$

for negative integers $\kappa_1 \leq \kappa_2 \leq \ldots \leq \kappa_r$ *and* $r = n-s-s'$, *if and only if* $F(\zeta) = F(\zeta)^*$ *and* $\det F(\zeta) \neq 0$ *for all* $\zeta \in \mathbb{T}$.

PROOF. The necessity is obvious, so we consider only the proof of sufficiency. By the theorem of Nikolaičuk and Spitkowski (see [CG] or [BH1]), such a factorization exists for F with $A^{\pm 1} \in H^2_{n\times n}$. But such a factorization can be modified in a trivial way to produce a factorization of the form as in Theorem 4.1. Since the factors X_- and X_+ in this factorization are determined up to a certain type of invertible polynomial factor, we conclude that in fact $A^{\pm 1} \in (\Lambda_{n\times n})^+$ by Theorem 4.1.

We now state the analogue of Theorem 1 for a Wiener-like algebra Λ .

THEOREM 4.3. *Let* $K \in \Lambda_{n\times n}$ *be given. Then a necessary and sufficient condition that there exist a unitary valued* F *in* $K + (\Lambda_{n\times n})^+$ *is that* $\|H_K\| \leq 1$. *Moreover, in this case,* $I - H_K^* H_K$ *is automatically Fredholm and the* $2n \times 2n$ *matrix function* Ξ *given by Theorem* 2.4 *satisfies* $\Xi^{\pm 1} \in \Lambda_{2n\times 2n}$. *A unitary interpolant* F *is in* $\Lambda_{n\times n}$ *(i.e.* $F \in [K+(\Lambda_{n\times n})^+] \cap U\Lambda_{n\times n}$*) if and only if* F *has the form*

$$F = T_\Xi(G)$$

where G *is a finite matrix Blaschke product in* $U(\Lambda_{s\times s})^+$. *The negative factorization indices of* F *are determined by the interpolation data* $\{K_j\}_{j=-1,-2,\dots}$ *and the nonnegative factorization indices are the same as those of* G .

PROOF. Suppose $K \in \Lambda_{n\times n}$. Then

$$M = \begin{bmatrix} K & I_n \\ I_n & 0 \end{bmatrix} \in \Lambda_{2n\times 2n}$$

and clearly $\det M(\zeta) \neq 0$ for all $\zeta \in \mathbb{T}$. If $H(\zeta) = M(\zeta)^* J M(\zeta)$ $\left(J = \begin{bmatrix} I_n & 0 \\ 0 & -I_n \end{bmatrix}\right)$ then $H(\zeta) = H(\zeta)^*$ and $\det H(\zeta) \neq 0$ for all $\zeta \in \mathbb{T}$. By Theorem 4.2, we can factor $H(\zeta)$ as $H(\zeta) = A(\zeta)^* \hat{D}(\zeta) A(\zeta)$ where $A^{\pm 1} \in (\Lambda_{2n\times 2n})^+$ and $\hat{D}(\zeta)$ is a self-adjoint winding matrix as in Theorem 2.4 with negative indices $\kappa_1 \leq \kappa_2 \leq \dots \leq \kappa_r < 0$. If we set $\Xi = MA^{-1}$, then

$$\text{(4.1i)} \quad \Xi^{\pm 1} \in \Lambda_{2n\times 2n}$$

$$\text{(4.1ii)} \quad \begin{bmatrix} K & I \\ I & 0 \end{bmatrix} H^2_{2n} = \Xi H^2_{2n}$$

and

$$\text{(4.1iii)} \quad \Xi(\zeta)^* J \Xi(\zeta) = \hat{D}(\zeta) \quad \text{for all } \zeta \in \mathbb{T} .$$

That is, Ξ satisfies all the conclusions of Theorem 2.4 with the additional property (4.1i).

It is not difficult to show that the only inner functions $G \in UH^\infty_{s\times s}$ which are continuous are the finite matrix Blaschke products, and therefore the only inner functions G in our matrix Wiener algebra $\Lambda_{s\times s}$ are the finite Blaschke products. By the work of the previous sections, Theorem 4.3 will be completely proved once we prove the following.

LEMMA 4.4. *Suppose* Ξ *is a* $2n \times 2n$ *matrix function which satisfies conditions* (4.1) *above, and the linear fractional map* $T_\Xi \colon \mathcal{B}H^\infty_{s\times s} \to \mathcal{B}L^\infty_{n\times n}$ *is defined as in* (1.4). *Suppose* $F \in \mathcal{B}L^\infty_{n\times n}$ *and* $G \in \mathcal{B}H^\infty_{s\times s}$ *are related by*

$$F = T_\Xi(G) \ .$$

Then $F \in \Lambda_{n\times n}$ *if and only if* $G \in \Lambda_{s\times s}$.

PROOF. The relation $F = T_\Xi(G)$ is equivalent to the following relation between subspaces.

$$\begin{bmatrix} F \\ I_n \end{bmatrix} H^2_n = \Xi \begin{bmatrix} I_r & 0 \\ 0 & G \\ 0 & I_s \\ 0 & 0 \end{bmatrix} \begin{bmatrix} H^2_r \\ H^2_s \end{bmatrix}$$

$$= \begin{bmatrix} \psi & \alpha & \beta & \psi \\ \omega & \kappa & \gamma & \omega \end{bmatrix} \begin{bmatrix} I_r & 0 \\ 0 & G \\ 0 & I_s \\ 0 & 0 \end{bmatrix} \begin{bmatrix} H^2_r \\ H^2_s \end{bmatrix}$$

Here we used that $\Xi^{\pm 1}$ is bounded, so we can write H^2 without closures on the right. Suppose $G \in (\Lambda_{s\times s})^+ \cap \mathcal{B}L^\infty_{s\times s}$. Then the second component in this identity implies

$$H^2_n = (\omega i^* + \kappa G j^* + \gamma j^*) H^2_n \ .$$

Since all matrix entries of Ξ and of G are in Λ , we see that $B := \omega i^* + \kappa G j^* + \gamma j^*$ is in $\Lambda_{n\times n}$. The above identity forces B to be invertible with inverse in $H^\infty_{n\times n}$. Therefore $\det B(\zeta) \neq 0$ for all $\zeta \in \mathbb{T}$ and by a property of $\Lambda_{n\times n}$ mentioned above

$$B^{-1} \in \Lambda_{n\times n} \cap H^\infty_{n\times n} = (\Lambda_{n\times n})^+ \ .$$

Similarly, since Λ is an algebra, it is clear that $A := \psi i^* + \alpha G j^* + \beta j^*$ is in $\Lambda_{n\times n}$. Conclude that $F = T_\Xi(G) = AB^{-1}$ is in $\Lambda_{n\times n}$ as claimed.

Conversely, suppose $F \in \Lambda_{n\times n}$. Write (4.2) in the form

$$(4.3) \qquad \Xi^{-1}\begin{bmatrix} F \\ I_n \end{bmatrix} H_n^2 = \begin{bmatrix} I_r & 0 \\ 0 & G \\ 0 & I_s \\ 0 & 0 \end{bmatrix}\begin{bmatrix} H_r^2 \\ H_s^2 \end{bmatrix} .$$

Block decompose Ξ^{-1} as $\begin{bmatrix} a & b \\ c & d \\ e & f \\ g & h \end{bmatrix}$.

Conclude from (4.3) that

$$(4.4) \qquad \begin{bmatrix} cF + d \\ eF + f \end{bmatrix} H_n^2 = \begin{bmatrix} G \\ I_s \end{bmatrix} H_s^2 .$$

Since $\Xi^{-1} \in \Lambda_{2n\times 2n}$ and $F \in \Lambda_{n\times n}$ by assumption, we have $cF+d$ and $eF+f$ are in $\Lambda_{s\times n}$. From the second components in the identity (4.4), we see that $(eF+f)(\zeta)$ has full rank s for all $\zeta \in \mathbb{T}$. Identity (4.4) also implies that there is a function $\Psi \in H^\infty_{s,n}$ such that

$$(4.5) \qquad \begin{bmatrix} cF+d \\ eF+f \end{bmatrix} = \begin{bmatrix} G \\ I_s \end{bmatrix} \Psi .$$

From the identity of the second components, conclude that $\Psi = eF+f$, so $\Psi \in \Lambda_{s,n}$ and $\Psi(\zeta)$ has full rank s for all $\zeta \in \mathbb{T}$. Therefore $\Psi\Psi^* \in \Lambda_{s\times s}$ and $\det(\Psi\Psi^*)(\zeta) \neq 0$ for all $\zeta \in \mathbb{T}$, so $\Psi\Psi^*$ has an inverse $(\Psi\Psi^*)^{-1}$ in $\Lambda_{s\times s}$. From the first components of (4.5) we see that $G\Psi \in \Lambda_{s\times n}$, and hence also $G\Psi\Psi^* = (G\Psi)\Psi^* \in \Lambda_{s\times s}$. Finally conclude that $G = (G\Psi\Psi^*)(\Psi\Psi^*)^{-1} \in \Lambda_{s\times s}$ as needed. This completes the proof of Lemma 4.4, and hence also of Theorem 4.3.

In the same way we can parametrize the contractive interpolants $F \in [K + (\Lambda_{n\times n})^+] \cap \mathcal{B}L^\infty_{n\times n}$ if $K \in \Lambda_{n,n}$.

THEOREM 4.5. *Suppose* $K \in \Lambda_{n,n}$, $\|H_K\| \leq 1$ *and* Ξ *is as given by Theorem* 2.4. *Then the function* F *is a contractive* $\Lambda_{n,n}$*-interpolant of* F *(i.e.* $F \in K+(\Lambda_{n\times n})^+$ *and* $\|F\|_\infty \leq 1$) *if and only if* $F = T_\Xi(G)$ *for a matrix function* $G \in (\Lambda_{n,n})^+ \cap BL^\infty_{n\times n}$.

Certain corollaries are immediate from Theorem 4.3.

COROLLARY 4.6. *Suppose* $K \in \Lambda_{n\times n}$ *is given, and suppose* $F \in [K + H^\infty_{n\times n}] \cap UL^\infty_{n\times n}$ *is a unitary interpolant. Then the following are equivalent.*

i) F *is factorable*

ii) F *is continuous*

iii) F *is in* $\Lambda_{n\times n}$.

PROOF. i) $\Longleftrightarrow$ iii) is clear from combining Theorems 1 and 4.3 and iii) $\Rightarrow$ ii) is trivial. Suppose $F = T_\Xi(G) \in [K + H^\infty_{n\times n}] \cap UL^\infty_{n\times n}$ is continuous. By Theorem 1, $G \in UH^\infty_{s\times s}$. Since Ξ^{-1} is known to be continuous, an argument as in the proof of Lemma 4.4 shows that G is continuous on $\mathbb{T}$. Thus G is a continuous inner function and is therefore a finite matrix Blaschke product, and so $F \in \Lambda_{n\times n}$ as claimed. Corollary 4.4 follows.

The case where K is rational (i.e. all matrix entries of K are rational functions of ζ) can be analyzed similarly. In this case it follows from a result in [BH1] that the representing function Ξ for $\begin{bmatrix} K & I_n \\ I_n & 0 \end{bmatrix} H^2_{2n}$ given by Theorem 2.4 is rational. By an argument completely analogous to that given above to prove Corollary 4.6, we obtain

COROLLARY 4.7. *Suppose* $K \in L^\infty_{n\times n}$ *is rational and* $F \in [K + H^\infty_{n\times n}] \cap UL^\infty_{n\times n}$ *is a unitary valued interpolant. Then the following are equivalent.*

i) F *is factorable*

ii) F *is continuous*

iii) F *is rational*.

We should mention that Theorem 4.3 for the case of uniqueness (s=0), the parametrization of all unitary and all contractive interpolants in the Wiener-like matrix algebra $\Lambda_{n\times n}$ for the case $s = n$, as well as the equivalence of ii) and iii) in Corollaries 4.6 and 4.7, are due to Dym and Gohberg [DG1].

REFERENCES

[AAK] Adamjan, V.M.; Arov, D.Z.; Krein, M.G., Infinite Hankel block matrices and related extension problems, Amer. Math. Soc. Transl. (2) 111 (1978), 133-156.

[ACF] Arsene, G.; Ceausescu, Z.; Foias, C., On intertwining dilations VIII, J. Operator Theory 4 (1980), 55-91.

[BH1] Ball, J.A.; Helton, J.W., Factorization results related to shifts in an indefinite metric, Integral Equations and Operator Theory 5 (1982), 632-658.

[BH2] Ball, J.A.; Helton,J.W., A Beurling-Lax theorem for the Lie group U(m,n) which contains most classical interpolation theory, J. Operator Theory 9 (1983), 107-142.

[BH3] Ball, J.A.; Helton, J.W., Beurling-Lax representations using classical Lie groups with many applications II: $GL(n,\mathbb{C})$, preprint.

[BH4] Ball, J.A.; Helton, J.W., Beurling-Lax representations using classical Lie groups with many applications III: groups preserving forms, preprint.

[B] Bognar, J., Indefinite Inner Product Spaces, Springer-Verlag (1974).

[BG] Budjanu, M.A.; Gohberg, I.C., General theorems on the factorization of matrix valued functions, I. The fundamental theorem, Amer. Math. Soc. Transl. (2) 102 (1973) 1-14.

[CG] Clancey, K.; Gohberg, I., Factorization of Matrix Functions and Singular Integral Operators, Birkhauser Verlag (1981).

[DG1] Dym, H.; Gohberg, I., Unitary interpolants, factorization indices and infinite Hankel block matrices, J. Functional Analysis, to appear.

[DG2] Dym, H.; Gohberg, I., Hankel integral operators and isometric interpolants on the line, J. Functional Analysis, to appear.

[DG3] Dym, H.; Gohberg, I., On unitary interpolants and Fredholm infinite block Toeplitz matrices, Integral Equations and Operator Theory, to appear.

This work was partially supported by a grant from the U.S. National Science Foundation.

J. A. Ball
Department of Theoretical Mathematics
The Weizmann Institute of Science
Rehovot 76100, Israel

(Permanent Address)

Department of Mathematics
Virginia Polytechnic Institute & State University
Blacksburg, Virginia 24061
U.S.A.

Operator Theory:
Advances and Applications, Vol. 12

THE COUPLING METHOD FOR SOLVING INTEGRAL EQUATIONS

H. Bart, I. Gohberg and M.A. Kaashoek

This paper presents a new method to reduce integral operators of various classes to simpler operators, which often are just finite matrices. By this method the problem to find the inverse, generalized inverses, kernel and image of an integral operator is reduced for several cases to the corresponding problem for a finite matrix. The classes of integral operators dealt with include integral operators of the second kind with a finite rank or semi-separable kernel and also, which is more surprising, systems of Wiener-Hopf integral operators and singular integral operators with rational matrix symbols.

TABLE OF CONTENTS

0. INTRODUCTION

The method of reducing operators to simpler ones presented here is based on a notion of matricial coupling of operators, which is defined as follows. The operators $T : X_1 \to Z_1$ and $S : Z_2 \to X_2$, acting between Banach spaces, are said to be *matricially coupled* if they can be dilated to 2×2 operator matrices that are each other inverses in the following way:

$$(0.1) \qquad \begin{bmatrix} T & * \\ * & * \end{bmatrix}^{-1} = \begin{bmatrix} * & * \\ * & S \end{bmatrix}.$$

If (0.1) holds, we call S the *indicator* of T. The notion is of particular interest when S is more simple than T. As soon as one has all entries appearing in the coupling relation (0.1) explicit formulas can be given for the inverse, generalized inverses, kernel and image, etc. of the operator T in terms of the corresponding objects for its indicator S.

For integral operators of the second kind with a finite rank or semi-separable kernel and for systems of Wiener-Hopf integral operators and singular integral operators with rational matrix symbols we construct indicators which are finite matrices. This allows us to solve explicitly the corresponding integral equations.

The type of reduction which is given here, we used before for convolution equations on a finite interval ([5], Sections I.6,I.7) and on a half line ([8], Section 2). In this paper the reduction is presented (probably for the first time) as part of a general principle. In a more primitive and preliminary form the principle was used in [13], Section 4 of [14], and Section 8 of [15].

I. A GENERAL PRINCIPLE OF EQUIVALENCE FOR OPERATORS

In this chapter a general principle of reducing operators to simpler ones is introduced. All considerations are on an abstract level. Applications and further concretizations will appear in the next chapters.

I.1 Matricial coupling and indicator

Let $T : X_1 \to Z_1$ and $S : Z_2 \to X_2$ be bounded linear operators acting between Banach spaces. We call T and S *matricially coupled* if T and S are related in the following way:

$$(1.1)\qquad \begin{bmatrix} T & A_{12} \\ A_{21} & A_{22} \end{bmatrix}^{-1} = \begin{bmatrix} B_{11} & B_{12} \\ B_{21} & S \end{bmatrix}.$$

More precisely, this means that one can construct an invertible 2×2 operator matrix

$$(1.2)\qquad \begin{bmatrix} A_{11} & A_{12} \\ A_{21} & A_{22} \end{bmatrix} : X_1 \oplus X_2 \to Z_1 \oplus Z_2$$

with $A_{11} = T$, whose inverse is given by

$$(1.3)\qquad \begin{bmatrix} B_{11} & B_{12} \\ B_{21} & B_{22} \end{bmatrix} : Z_1 \oplus Z_2 \to X_1 \oplus X_2$$

where $B_{22} = S$. The 2×2 operator matrices appearing in (1.2) and (1.3) are called the *coupling matrices* and to (1.1) we shall refer as the *coupling relation*. If T and S are matricially coupled operators, then we say that S is an *indicator* of T (and reversely T is an indicator of S). This notion is of particular interest if S is more simple than T. Throughout this paper all spaces are assumed to be complex Banach spaces and all operators are bounded and linear. The identity operator on a Banach space X is denoted by I_X or simply by I.

EXAMPLE. Let $A : X \to Y$ and $B : Y \to X$ be given operators, and let D and K be invertible operators acting on the spaces X and Y, respectively. Then the operators $D - BK^{-1}A$ and $K - AD^{-1}B$ are matricially coupled operators. Indeed

$$(1.4)\qquad \begin{bmatrix} D-BK^{-1}A & -BK^{-1} \\ K^{-1}A & K^{-1} \end{bmatrix}^{-1} = \begin{bmatrix} D^{-1} & D^{-1}B \\ -AD^{-1} & K-AD^{-1}B \end{bmatrix}.$$

THEOREM 1.1 *Assume* $T : X_1 \to Z_1$ *and* $S : Z_2 \to X_2$ *are matricially coupled operators, and let the coupling relation be given by*

$$\begin{bmatrix} T & A_{12} \\ A_{21} & A_{22} \end{bmatrix}^{-1} = \begin{bmatrix} B_{11} & B_{12} \\ B_{21} & S \end{bmatrix}.$$

Then

$$(1.5)\qquad \begin{bmatrix} T & 0 \\ 0 & I_{X_2} \end{bmatrix} = E \begin{bmatrix} S & 0 \\ 0 & I_{Z_1} \end{bmatrix} F,$$

where E *and* F *are invertible* 2×2 *operator matrices*

$$E = \begin{bmatrix} -A_{12} & TB_{11} \\ I_{X_2} & B_{21} \end{bmatrix}, \quad F = \begin{bmatrix} A_{21} & A_{22} \\ T & A_{12} \end{bmatrix}$$

with inverses

$$E^{-1} = \begin{bmatrix} -B_{21} & SA_{22} \\ I_{Z_1} & A_{12} \end{bmatrix}, \quad F^{-1} = \begin{bmatrix} B_{12} & B_{11} \\ S & B_{21} \end{bmatrix}.$$

PROOF. By direct computation, using (1.1). □

Note that formula (1.5) says that after a simple extension the operators T and S are equivalent.

The conditions of Theorem 1.1 are symmetric with respect to A_{ij} and B_{ij}, but the 2×2 operator matrices E and F appearing in the equivalence relation (1.5) are not. This phenomenon is also reflected by the fact that under the conditions of Theorem 1.1 formula (1.5) may be replaced by

$$(1.6) \qquad \begin{bmatrix} T & 0 \\ 0 & I_{Z_2} \end{bmatrix} = G \begin{bmatrix} S & 0 \\ 0 & I_{X_1} \end{bmatrix} H,$$

where

$$G = \begin{bmatrix} A_{12} & T \\ A_{22} & A_{21} \end{bmatrix}, \quad G^{-1} = \begin{bmatrix} B_{21} & S \\ B_{11} & B_{12} \end{bmatrix},$$

$$H = \begin{bmatrix} -A_{21} & I_{X_2} \\ B_{11}T & B_{12} \end{bmatrix}, \quad H^{-1} = \begin{bmatrix} -B_{12} & I_{X_1} \\ A_{22}S & A_{21} \end{bmatrix}.$$

Theorem 1.1 is of particular interest when the operators T and S depend on a parameter. For example, if the entries of the coupling matrix (1.2) depend analytically on a parameter λ, for λ in some open subset of $\mathbb{C}$, then the same is true for the entries in its inverse (assuming it exists) and in that case the operators E and F appearing in Theorem 1.1 also depend analytically on λ.

The coupling relation is also stable under small perturbations. Thus, if S is an indicator of T and $\tilde{T}$ is a (sufficiently small) perturbation of T, then $\tilde{T}$ has an indicator $\tilde{S}$ close to S.

I.2 Invertibility of matricially coupled operators

In this section we compare the invertibility properties of matricially coupled operators. Generalized invertibility is used in a weak sense, i.e., an operator T is said to have a *generalized inverse* T^+ whenever $T = TT^+T$.

THEOREM 2.1 *Let* T *and* S *be matricially coupled operators, and let the coupling relation be given by*

$$(2.1)\qquad \begin{bmatrix} T & A_{12} \\ A_{21} & A_{22} \end{bmatrix}^{-1} = \begin{bmatrix} B_{11} & B_{12} \\ B_{21} & S \end{bmatrix}.$$

Then

$$(2.2)\qquad \operatorname{Ker} T = B_{12}\operatorname{Ker} S , \quad \operatorname{Ker} S = A_{21}\operatorname{Ker} T ,$$

$$(2.3)\qquad \operatorname{Im} T = B_{21}^{-1}\operatorname{Im} S , \quad \operatorname{Im} S = A_{12}^{-1}\operatorname{Im} T .$$

Further, T *has a generalized inverse (resp. left, right, two-sided inverse) if and only if* S *has a generalized inverse (resp. left, right, two-sided inverse). If* S^+ *is a generalized inverse of* S, *then*

$$(2.4)\qquad T^+ = B_{11} - B_{12}S^+B_{21}$$

is a generalized inverse of T, *and, conversely, if* T^+ *is a generalized inverse of* T, *then*

$$(2.5)\qquad S^+ = A_{22} - A_{21}T^+A_{12}$$

is a generalized inverse of S. *Also* T *is a (semi-) Fredholm operator if and only if* S *is a (semi-) Fredholm operator, and in that case* $\operatorname{ind} T = \operatorname{ind} S$.

PROOF. Since the first matrix in (2.1) is the inverse of the second matrix in (2.1), we know that $B_{21}T + SA_{21} = 0$. This shows that $\operatorname{Im} T \subset B_{21}^{-1}\operatorname{Im} S$. Now, assume that $B_{21}y = Sz$. Then

$$\begin{aligned} y &= TB_{11}y + A_{12}B_{21}y = TB_{11}y + A_{12}Sz \\ &= TB_{11}y - TB_{12}z \in \operatorname{Im} T . \end{aligned}$$

We have proved the first identity in (2.3). The second identity in (2.3) follows by interchanging the roles of T and S. All other statements in the theorem are straightforward consequences of the equivalence relation laid down in formula (1.5). □

From the relation (1.5) and the definition of the operator F in Theorem 1.1 it is clear that under the hypotheses of the previous theorem the operator B_{12} maps the space $\operatorname{Ker} S$ in a one-one manner onto the space ker T. Similarly, the operator B_{21} maps a (closed) complement of Im S in a one-one way onto a (closed) complement of Im T.

We conclude with a remark about analytical dependence. Assume the entries in the dilation matrices (2.1) depend analytically on a parameter λ for λ in some open subset of $\mathbb{C}$. Then the analytical version of Theorem 1.1 (see the remarks at the end of the previous section) can be used to show that the spectral data (eigenvalues, eigenvectors, generalized eigenvectors, Jordan chains, etc.) of the operator function $T(\lambda)$ can be derived from those of the analytic operator function $S(\lambda)$ (cf. [11]).

I.3 A first example

The usual method of reducing the inversion of an operator $I - F$, F has finite rank, to that of a matrix can be understood and made more precise in the context of matricially coupled operators. To see this, assume $F : X \to X$ is given by

$$F = \sum_{j=1}^{n} \langle \cdot, \varphi_j^* \rangle \psi_j ,$$

where $\psi_1, \cdots, \psi_n$ are given vectors in the Banach space X and $\varphi_1^*, \cdots, \varphi_n^*$ are continuous linear functionals on X. Define $A : X \to \mathbb{C}^n$ and $B : \mathbb{C}^n \to X$ by setting

$$Ax = \mathrm{col}(\langle x, \varphi_i^* \rangle)_{i=1}^{n} , \quad x \in X ,$$

$$B\begin{bmatrix} \alpha_1 \\ \vdots \\ \alpha_n \end{bmatrix} = \sum_{j=1}^{n} \alpha_j \psi_j .$$

Note that $G = AB$ acts on $\mathbb{C}^n$ and its matrix with respect to the standard basis of $\mathbb{C}^n$ is given by

$$\text{(3.1)} \qquad \mathrm{mat}(G) = (\langle \psi_j, \varphi_i^* \rangle)_{i,j=1}^{n} .$$

Since $F = BA$, the operators $I_X - \mu F$ and $I_n - \mu G$ are matricially coupled; in fact (cf., (1.4))

$$\text{(3.2)} \qquad \begin{bmatrix} I_X - \mu F & -\mu B \\ A & I_n \end{bmatrix}^{-1} = \begin{bmatrix} I_X & \mu B \\ -A & I_n - \mu G \end{bmatrix} .$$

Here I_n is the identity operator on $\mathbb{C}^n$. Observe that all entries in (3.2) depend analytically on μ. From (3.2) it follows (cf., formula (2.4)) that

$$(I_X-\mu F)^{-1} = I_X + \mu B(I_n-\mu G)^{-1}A ,$$

whenever $\det(I_n-\mu G) \neq 0$. Further, the non-zero eigenvalues of F and G are the same, the corresponding multiplicities are equal and the relationship between the Jordan chains of F and G (which is not obvious) becomes transparant.

I.4 A test for matricial coupling

The following theorem gives a test for matricial coupling. In the next chapters it is used as a heuristic tool to find the form of certain operators appearing in coupling relations.

THEOREM 4.1 *Let* L *and* K *be invertible* 2×2 *operator matrices, and assume that* L, K *and their inverses are given by*

$$L = \begin{bmatrix} L_1 & U_1R_2 \\ -U_2R_1 & L_2 \end{bmatrix} : N_1 \oplus N_2 \to N_1^\times \oplus N_2^\times ,$$

$$L^{-1} = \begin{bmatrix} L_1^\times & -U_1^\times R_2^\times \\ U_2^\times R_1^\times & L_2^\times \end{bmatrix} : N_1^\times \oplus N_2^\times \to N_1 \oplus N_2 ,$$

$$K = \begin{bmatrix} K_1 & R_1^\times U_1 \\ -R_2^\times U_2 & K_2 \end{bmatrix} : M_1 \oplus M_2 \to M_1^\times \oplus M_2^\times ,$$

$$K^{-1} = \begin{bmatrix} K_1^\times & -R_1U_1^\times \\ R_2U_2^\times & K_2^\times \end{bmatrix} : M_1^\times \oplus M_2^\times \to M_1 \oplus M_2 .$$

Here

$$U_1 : M_2 \to N_1^\times , \quad U_2 : M_1 \to N_2^\times ,$$
$$U_1^\times : M_2^\times \to N_1 , \quad U_2^\times : M_1^\times \to N_2 ,$$
$$R_1 : N_1 \to M_1 , \quad R_2 : N_2 \to M_2 ,$$
$$R_1^\times : N_1^\times \to M_1^\times , \quad R_2^\times : N_2^\times \to M_2^\times .$$

If, in addition,

$$(4.1) \qquad U_1K_2^\times = L_1U_1^\times , \quad U_2K_1^\times = L_2U_2^\times ,$$

$$(4.2) \qquad K_1R_1 = R_1^\times L_1 , \quad K_2R_2 = R_2^\times L_2 ,$$

then the pairs (L_1,K_1), (L_2,K_2), $(L_1^\times,K_1^\times)$ *and* $(L_2^\times,K_2^\times)$ *are pairs of matricially coupled operators. More precisely,*

$$(4.3)\qquad \begin{bmatrix} L_1 & U_1R_2U_2^\times \\ -R_1 & K_1^\times \end{bmatrix}^{-1} = \begin{bmatrix} L_1^\times & -U_1^\times R_2^\times U_2 \\ R_1^\times & K_1 \end{bmatrix},$$

$$(4.4)\qquad \begin{bmatrix} L_2 & -U_2R_1U_1^\times \\ R_2 & K_2^\times \end{bmatrix}^{-1} = \begin{bmatrix} L_2^\times & U_2^\times R_1^\times U_1 \\ -R_2^\times & K_2 \end{bmatrix}.$$

PROOF. First we show that the conditions of the theorem are symmetric in L, K and L^{-1}, K^{-1}. To do this, introduce

$$U = \begin{bmatrix} 0 & U_1 \\ U_2 & 0 \end{bmatrix}, \quad U^\times = \begin{bmatrix} 0 & U_1^\times \\ U_2^\times & 0 \end{bmatrix},$$

$$R = \begin{bmatrix} R_1 & 0 \\ 0 & R_2 \end{bmatrix}, \quad R^\times = \begin{bmatrix} R_1^\times & 0 \\ 0 & R_2^\times \end{bmatrix}.$$

Then (4.1) is equivalent to the statement that $UK^{-1} = LU^\times$, and (4.2) is equivalent to $KR = R^\times L$. It follows that $U^\times K = L^{-1}U$ and $K^{-1}R^\times = RL^{-1}$. So we also have

$$(4.5)\qquad U_1^\times K_2 = L_1^\times U_1, \quad U_2^\times K_1 = L_2^\times U_2,$$

$$(4.6)\qquad K_1^\times R_1^\times = R_1L_1^\times, \quad K_2^\times R_2^\times = R_2L_2^\times.$$

With the identities (4.1), (4.2), (4.5) and (4.6) and with the expressions for L, L^{-1}, K and K^{-1} it is a straightforward matter to check that (4.3) and (4.4) hold true. □

II. INTEGRAL OPERATORS ON A FINITE INTERVAL WITH SEMI-SEPARABLE KERNEL

In this chapter the general principle of matricial coupling is applied to integral operators with semi-separable kernel and their discrete analogues.

II.1 General case

In this chapter we consider the integral operator

$$(1.1)\qquad (K\varphi)(t) = \int_0^\tau k(t,s)\varphi(s)ds, \quad 0 \leqslant t \leqslant \tau,$$

where the kernel k is an $m \times m$ matrix. We assume k to be semi-separable, which means that k admits a representation:

$$(1.2)\qquad k(t,s) = \begin{cases} C(t)(I_n-P)B(s) , & 0 \leq s < t \leq \tau , \\ -C(t)PB(s) , & 0 \leq t < s \leq \tau . \end{cases}$$

Here $B(t)$ and $C(t)$ are matrices of sizes $n \times m$ and $m \times n$, respectively, and the entries of $B(t)$ and $C(t)$ are assumed to be square Lebesque integrable functions on $[0,\tau]$. Further, P is a constant $n \times n$ matrix which is a projection on $\mathbb{C}^n$ and I_n in the $n \times n$ identity matrix. It follows that the integral operator K is a well-defined Hilbert-Schmidt operator on $L_2^m[0,\tau]$.

THEOREM 1.1 *Let* K *be the integral operator* (1.1) *with kernel* (1.2), *and let* $U(t)$ *be the fundamental matrix of the differential equation*

$$(1.3)\qquad \dot{x}(t) = B(t)C(t)x(t) , \quad 0 \leq t \leq \tau .$$

Then $S_\tau = PU(\tau)P : \operatorname{Im} P \to \operatorname{Im} P$ *is an indicator for the operator* $I - K : L_2^m[0,\tau] \to L_2^m[0,\tau]$. *More precisely, the following coupling relation holds true:*

$$(1.4)\qquad \begin{bmatrix} I-K & -R \\ -Q & I_{\operatorname{Im} P} \end{bmatrix}^{-1} = \begin{bmatrix} (I-H)^{-1} & (I-H)^{-1}R \\ Q(I-H)^{-1} & S_\tau \end{bmatrix} .$$

Here

$$H : L_2^m[0,\tau] \to L_2^m[0,\tau] , \quad (H\varphi)(t) = \int_0^t C(t)B(s)\varphi(s)ds ;$$

$$Q : L_2^m[0,\tau] \to \operatorname{Im} P , \quad Q\varphi = P\int_0^\tau B(s)\varphi(s)ds ;$$

$$R : \operatorname{Im} P \to L_2^m[0,\tau] , \quad (Rx)(t) = C(t)Px .$$

PROOF. We shall prove that $I - H$ is invertible and that the following two identities hold true:

$$(1.5)\qquad I - K = I - H + RQ ,$$

$$(1.6)\qquad S_\tau = I_{\operatorname{Im} P} + Q(I-H)^{-1}R .$$

As soon as these two identities are established, one can apply formula (1.4) in Section I.1 to show that the operators $I - K$ and S_τ are matricially coupled with coupling relation given by (1.4). Note that (1.5) follows directly from the definitions of the operators H, Q and R and the description of the kernel k of the integral operator K.

To prove (1.6) we use the identities

$$(1.7)\qquad \frac{d}{dt}U(t) = B(t)C(t)U(t) \quad , \quad 0 \leq t \leq \tau ,$$

$$(1.8)\qquad \frac{d}{dt}U(t)^{-1} = -U(t)^{-1}B(t)C(t) \quad , \quad 0 \leq t \leq \tau ,$$

which are immediate consequences of the fact that $U(t)$ is the fundamental matrix of the differential equation (1.3). Note that the operator H is a Volterra operator. By direct checking, using the identities (1.7) and (1.8), one proves that $I - H$ is invertible and its inverse is given by

$$((I-H)^{-1}f)(t) = f(t) + C(t)U(t)\int_0^t U(s)^{-1}B(s)f(s)ds .$$

Hence for $x \in \operatorname{Im} P$ we have

$$\begin{aligned}((I-H)^{-1}Rx)(t) &= C(t)x + C(t)U(t)\int_0^t U(s)^{-1}B(s)C(s)x\, ds \\ &= C(t)x - C(t)U(t)(U(s)^{-1}\Big|_0^t)x = C(t)U(t)x .\end{aligned}$$

It follows that

$$Q(I-H)^{-1}Rx = P\int_0^\tau B(s)C(s)U(s)x\, ds = P(U(s)\Big|_0^\tau)x = S_\tau x - x$$

for $x \in \operatorname{Im} P$, which proves (1.6). □

COROLLARY 1.2 *Let* K *and* $U(t)$ *be as in the previous theorem, and let* $S_\tau = PU(\tau)P$. *Then*

$$\operatorname{Ker}(I-K) = \{\varphi \mid \varphi(t) = C(t)U(t)Py \ , \ y \in \operatorname{Ker} S_\tau\} \quad ,$$

$$\operatorname{Im}(I-K) = \{f \mid PU(\tau)\int_0^\tau U(s)^{-1}B(s)f(s)ds \in \operatorname{Im} S_\tau\} .$$

Further, a generalized inverse of $I - K$ *is given by the integral operator*

$$((I-K)^{+}f)(t) = f(t) + \int_0^\tau \gamma(t,s)f(s)ds ,$$

with

$$(1.9)\qquad \gamma(t,s) = \begin{cases} C(t)U(t)(I_n - PS_\tau^{+}PU(\tau))U(s)^{-1}B(s) \ , & t > s , \\ -C(t)U(t)PS_\tau^{+}PU(\tau)U(s)^{-1}B(s) \ , & t < s , \end{cases}$$

where S_τ^{+} *is a generalized inverse of* S_τ.

PROOF. According to formula (1.4) and Theorem I.2.1 we have

$$(1.10)\qquad \operatorname{Ker}(I-K) = (I-H)^{-1}R \operatorname{Ker} S_\tau ,$$

$$(1.11)\qquad \operatorname{Im}(I-K) = \{f \mid Q(I-H)^{-1}f \in \operatorname{Im} S_\tau\} .$$

Further,

$$(1.12)\qquad (I-K)^{+} = (I-H)^{-1} - (I-H)^{-1}RS_\tau^{+}Q(I-H)^{-1}$$

is a generalized inverse of $I - K$ if S_τ^+ is a generalized inverse of the indicator $S_\tau = PU(\tau)P$.

To prove the corollary we use the following identities

$$((I-H)^{-1}Rx)(t) = C(t)U(t)x \ , \quad x \in \operatorname{Im} P \ , \tag{1.13}$$

$$Q(I-H)^{-1}f = PU(\tau)\int_0^\tau U(s)^{-1}B(s)f(s)ds \ . \tag{1.14}$$

Formula (1.13) has already been established in the proof of Theorem 1.1. Let us prove (1.14):

$$\begin{aligned} Q(I-H)^{-1}f &= P\int_0^\tau B(s)f(s)ds + P\int_0^\tau B(s)C(s)U(s)(\int_0^s U(\alpha)^{-1}B(\alpha)f(\alpha)d\alpha)ds \\ &= P\int_0^\tau B(s)f(s)ds + P\int_0^\tau \dot{U}(s)(\int_0^s U(\alpha)^{-1}B(\alpha)f(\alpha)d\alpha)ds \\ &= P\int_0^\tau B(s)f(s)ds + PU(\tau)\int_0^\tau U(\alpha)^{-1}B(\alpha)f(\alpha)d\alpha + \\ &\quad - P\int_0^\tau U(s)U(s)^{-1}B(s)f(s)ds \\ &= PU(\tau)\int_0^\tau U(\alpha)^{-1}B(\alpha)f(\alpha)d\alpha \ . \end{aligned}$$

Inserting the formulas for $(I-H)^{-1}R$ and $Q(I-H)^{-1}$ into (1.10), (1.11) and (1.12) yields the description of the kernel and the image of $I - K$ and the formula for a generalized inverse. □

The operator $PS_\tau^+PU(\tau) : \mathbb{C}^n \to \mathbb{C}^n$ appearing in the definition of $\gamma(t,s)$ (see formula (1.9)) is a projection if the generalized inverse S_τ^+ of S_τ has the following additional property: $S_\tau^+ = S_\tau^+ S S_\tau^+$. Indeed, in that case

$$(PS_\tau^+PU(\tau))(PS_\tau^+PU(\tau)) = PS_\tau^+S_\tau S_\tau^+PU(\tau) = PS_\tau^+PU(\tau) \ .$$

It is easily seen that the class of all operators $I - K$, where K is an operator with semi-separable kernel, is an algebra, i.e., the class is closed under forming sums and products. Since the kernel $\gamma(t,s)$ defined by (1.9) is again a semi-separable kernel, it follows that this algebra is also closed under taking inverses (whenever they exist).

II.2 Discrete analogue: semi-separable matrices

The block matrix $[A_{jk}]_{j,k=1}^r$ is said to be in *semi-separable* form whenever

$$A_{jk} = \begin{cases} C_j(I_n-P)B_k \ , & 1 \leqslant k < j \leqslant r \ , \\ -C_jPB_k \ , & 1 \leqslant j \leqslant k \leqslant r \ . \end{cases} \tag{2.1}$$

Here $C_1,\cdots,C_r$ are $m \times n$ matrices, $B_1,\cdots,B_r$ are $n \times m$ matrices, and the $n \times n$ matrix P is a projection on $\mathbb{C}^n$. The symbol I_n denotes the identity operator on $\mathbb{C}^n$, and as usual a $p \times q$ matrix is identified with the operator from $\mathbb{C}^q$ into $\mathbb{C}^p$ given by the canonical action of the matrix with respect to the standard bases in $\mathbb{C}^q$ and $\mathbb{C}^p$.

THEOREM 2.1 *Let* $A = [A_{jk}]_{j,k=1}^r$ *be the operator on* $\mathbb{C}^{mr}$ *with matrix elements* A_{jk} *given by* (2.1). *Then*

$$S = P(I_n+B_rC_r)\cdot\cdots\cdot(I_n+B_1C_1)P : \operatorname{Im} P \to \operatorname{Im} P \tag{2.2}$$

is an indicator for the operator $I - A$. *More precisely the following coupling relation holds true:*

$$\begin{bmatrix} I-A & -R \\ -Q & I_{\operatorname{Im} P} \end{bmatrix}^{-1} = \begin{bmatrix} (I-H)^{-1} & (I-H)^{-1}R \\ Q(I-H)^{-1} & S \end{bmatrix}. \tag{2.3}$$

Here

$$H = \begin{bmatrix} 0 & & & & \\ C_2B_1 & 0 & & & \\ C_3B_1 & C_3B_2 & \ddots & & \\ \vdots & \vdots & \ddots & 0 & \\ C_rB_1 & C_rB_2 & \cdots & C_rB_{r-1} & 0 \end{bmatrix} : \mathbb{C}^{mr} \to \mathbb{C}^{mr} ;$$

$$R = \operatorname{col}(C_jP)_{j=1}^r : \operatorname{Im} P \to \mathbb{C}^{mr} ;$$

$$Q = [PB_1\cdots PB_r] : \mathbb{C}^{mr} \to \operatorname{Im} P .$$

PROOF. The following identities hold true:

$$I - A = I - H + RQ , \tag{2.4}$$

$$S = I_{\operatorname{Im} P} + Q(I-H)^{-1}R . \tag{2.5}$$

Formula (2.4) follows directly from the semi-separable form of A and the definitions of H, R and Q. Clearly, $I - H$ is invertible. To get the block matrix form of its inverse we need the following $n \times n$ matrices U_{jk} $(j = k,\cdots,r+1)$:

$$\begin{cases} U_{j+1,k} = (I_n+B_jC_j)U_{jk} , & j = k,\cdots,r , \\ U_{k,k} = I_n . \end{cases} \tag{2.6}$$

The matrices U_{jk} satisfy two identities, namely:

$$U_{j,k+1}B_kC_k = U_{j,k} - U_{j,k+1}\ , \qquad j = k+1,\cdots,r\ ;$$
$$B_jC_jU_{j,k} = U_{j+1,k} - U_{j,k}\ , \qquad j = k,\cdots,r-1\ .$$

Using these identities one easily checks that

$$(I-H)^{-1} = \begin{bmatrix} I & 0 & & & \\ C_2U_{22}B_1 & I & & & \\ C_3U_{32}B_1 & C_3U_{33}B_2 & \ddots & & \\ \vdots & \vdots & \ddots & I & \\ C_rU_{r2}B_1 & C_rU_{r3}B_3 & \cdots & C_rU_{rr}B_{r-1} & I \end{bmatrix} .$$

Next, we consider $Q(I-H)^{-1}R$. From the block matrix representation for $(I-H)^{-1}$ it is not difficult to derive the following formulas:

(2.7) $\qquad (I-H)^{-1}R = \mathrm{col}(C_jU_{j1}P)_{j=1}^{r}$

(2.8) $\qquad Q(I-H)^{-1} = [PU_{r+1,2}B_1 \cdots PU_{r+1,r+1}B_r]\ .$

In particular, one sees that

$$\begin{aligned} I_{\mathrm{Im}\,P} + Q(I-H)^{-1}R &= I_{\mathrm{Im}\,P} + \sum_{j=1}^{r} PB_jC_jU_{j1}P \\ &= I_{\mathrm{Im}\,P} + \sum_{j=1}^{r} P(U_{j+1,1}-U_{j,1})P \\ &= I_{\mathrm{Im}\,P} + PU_{r+1,1}P - P = S\ . \end{aligned}$$

This proves formula (2.5).

From the identities (2.4) and (2.5) it is clear that (2.3) holds true (cf. formula (1.4) in Section I.1). Hence $I - A$ and S are matricially coupled and the theorem is proved. □

COROLLARY 2.2 *Let* $A = [A_{jk}]_{j,k=1}^{r}$ *be the operator on* $\mathbb{C}^{mr}$ *with matrix entries given by* (2.1). *For* $1 \leq k \leq j \leq r+1$ *let* U_{jk} *be the matrix defined by*

$$\begin{cases} U_{j+1,k} = (I_n+B_jC_j)U_{j,k}\ , \\ U_{k,k} = I_n\ . \end{cases}$$

Put $S = PU_{r+1,1}P$. *Then*

$$\mathrm{Ker}(I-A) = \{\mathrm{col}(C_jU_{j1}y)_{j=1}^{r} \mid y \in \mathrm{Ker}\,S\}\ ,$$

$$\mathrm{Im}(I-A) = \{\mathrm{col}(x_j)_{j=1}^{r} \mid \sum_{j=1}^{r} PU_{r+1,j+1}B_jx_j \in \mathrm{Im}\,S\}\ .$$

Further, given a generalized inverse S^+ *of* S, *the operator* $I + A^\times$ *with* $A^\times = [A^\times_{jk}]_{j,k=1}^r$ *and*

$$A^\times_{jk} = \begin{cases} C_j(U_{j,k+1} - U_{j1}PS^+PU_{r+1,k+1})B_k \,, & j > k \,, \\ -C_jU_{j1}PS^+PU_{r+1,k+1}B_k \,, & j \leq k \,, \end{cases}$$

is a generalized inverse of $I - A$.

PROOF. Note that $S = PU_{r+1,1}P$ is equal to the operator defined by (2.2). So S is an indicator of $I - A$, and we can apply Theorem I.2.1. This yields

$$\mathrm{Ker}(I-A) = (I-H)^{-1}R\ \mathrm{Ker}\ S \,,$$

$$\mathrm{Im}\ (I-A) = \{x \mid Q(I-H)^{-1}x \in \mathrm{Im}\ S\} \,.$$

Further, if S^+ is a generalized inverse of S, then

$$(I-A)^+ = (I-H)^{-1} - (I-H)^{-1}RS^+Q(I-H)^{-1}$$

is a generalized inverse of $I - A$. Now use the identities (2.7) and (2.8) to complete the proof. □

II.3 Spectrum and indicator

In this section we return to the integral operator

$$(3.1)\qquad (K\varphi)(t) = \int_0^\tau k(t,s)\varphi(s)ds \,, \quad 0 \leq t \leq \tau \,.$$

As in Section 1 we assume that k is the semi-separable kernel given by (1.2). Our aim is to describe the characteristic values of K in terms of the indicator of $I - \mu K$. Recall that μ_0 $(\neq 0)$ is said to be a *characteristic value* of K if μ_0^{-1} is an eigenvalue of K, and in that case the *multiplicity* of μ_0 is equal to the algebraic multiplicity of μ_0^{-1} as an eigenvalue of K.

Let $\Omega(t;\mu)$ be the fundamental matrix of the differential equation

$$(3.2)\qquad \dot{x}(t) = \mu B(t)C(t)x(t) \,, \quad 0 \leq t \leq \tau \,.$$

Put $S_\tau(\mu) = P\Omega(t;\mu)P$. We consider $S_\tau(\mu)$ as an operator acting on the finite dimensional space $\mathrm{Im}\ P$. From the theory of differential equations we know that the fundamental matrix of (3.2) depends analytically on μ. Thus $S_\tau(\mu)$ is an entire operator function. Since $S_\tau(0) = I_{\mathrm{Im}\ P}$ it follows that $\det S_\tau(\mu)$ does not vanish identically. We shall refer to $S_\tau(\cdot)$ as the *indicator* of the integral operator K. Note that this terminology is justified by Theorem 1.1.

THEOREM 3.1 *The characteristic values of the integral operator* K *with semi-separable kernel* k *given by* (1.2) *are precisely equal to the zeros of the determinant of its indicator* S_τ *(multiplicities taken into account).*

PROOF. We use the notation and terminology of Section 1. Since $S_\tau(\mu)$ is an indicator for the operator $I - \mu K$ we know (see Theorem I.1.1) that

$$(3.3)\qquad \begin{bmatrix} I-\mu K & 0 \\ 0 & I_{\operatorname{Im} P} \end{bmatrix} = E(\mu)\begin{bmatrix} S_\tau(\mu) & 0 \\ 0 & I \end{bmatrix} F(\mu)\ ,$$

where $E(\mu)$ and $F(\mu)$ are certain invertible 2×2 operator matrices which depend analytically on μ when μ varies over $\mathbb{C}$. From the equivalence relation (3.3) it is immediately clear (cf. the remark made at the end of Section I.2) that the characteristic values of K are precisely equal to the zeros of the function $\det S_\tau(\mu)$ and the multiplicity of μ_0 as a characteristic value of K is equal to the multiplicity of μ_0 as a zero of $\det S_\tau(\mu)$. □

From the equivalence relation (3.3) it also follows that

$$(3.4)\qquad \dim \ker(I-\mu K) \leq \operatorname{rank} P\ .$$

In particular, if $\operatorname{rank} P = 1$, then each non-zero eigenvalue of K has precisely one Jordan block. Since the kernels of the operators $I - \mu K$ and $I - \bar{\mu}K^*$ have the same dimension, the analogue of (3.4) for the adjoint operator yields

$$\dim \ker(I-\mu K) \leq \operatorname{rank}(I-P)\ .$$

From Theorem 3.1 it is clear that $\det S_\tau(\mu)$ has properties of a (generalized) Fredholm determinant. The next corollary makes this statement more precise.

COROLLARY 3.2 *If the integral operator* K *is a trace class operator, then* $\det(I-\mu K) = \det S_\tau(\mu)$.

PROOF. Using the notation of Section 1 we may write $K = H - RQ$. Here R and Q are both finite rank operators. Thus H is a trace class operator too. Since $I - \mu H$ is invertible for all μ, $\det(I-\mu H) = 1$. It follows that

$$\det(I-\mu K) = \det((I-\mu H)(I+\mu(I-\mu H)^{-1}RQ)) = \det(I+\mu(I-\mu H)^{-1}RQ)$$

$$= \det(I_{\operatorname{Im} P} + \mu Q(I-\mu H)^{-1}R) = \det S_\tau(\mu) \ . \quad \square$$

We conclude with a remark about the discrete case. Let $A = [A_{jk}]_{j,k=1}^{r}$ be a block matrix in semi-separable form. More precisely assume that the matrix elements A_{jk} are given by (2.1). Then (cf. Theorem 2.1) the finite dimensional operator

$$S(\mu) = P(I_n+\mu B_r C_r) \cdots (I_n+\mu B_1 C_1)P : \operatorname{Im} P \to \operatorname{Im} P$$

is an indicator for $I - \mu A$. Note that in this case the indicator $S(\mu)$ is a polynomial in μ with matrix coefficients of order n, where n may be much smaller than the size of A.

II.4 Semi-separable kernels on disconnected intervals

The method of matricial coupling can also be applied to integral operators with semi-separable kernels on disconnected intervals. For instance, consider the integral operator

$$(4.1) \qquad (K\varphi)(t) = \int_{a_1}^{b_1} k_1(t,s)\varphi(s)ds + \int_{a_2}^{b_2} k_2(t,s)\varphi(s)ds \ , \qquad t \in [a_1,b_1] \cup [a_2,b_2].$$

Here $-\infty < a_1 < b_1 \leq a_2 < b_2$, both k_1 and k_2 are assumed to be semi-separable $m \times m$ matrix kernels, and K is considered as an operator on $L_2^m(\Lambda)$, where $\Lambda = [a_1,b_1] \cup [a_2,b_2]$.

First assume that k_1 and k_2 have the same semi-separable representation. Later on we shall see that the general case may be reduced to this one. So assume

$$(4.2) \qquad k_1(t,s) = k(t,s) \ , \quad t \in \Lambda \ , \quad a_1 \leq s \leq b_1 \ ,$$

$$(4.3) \qquad k_2(t,s) = k(t,s) \ , \quad t \in \Lambda \ , \quad a_2 \leq s \leq b_2 \ ,$$

and for $(t,s) \in \Lambda \times \Lambda$

$$(4.4) \qquad k(t,s) = \begin{cases} C(t)(I_n-P)B(s) \ , & t > s \ , \\ -C(t)PB(s) \ , & t < s \ . \end{cases}$$

THEOREM 4.1 *Let* K *be the integral operator* (4.1) *and assume both* k_1 *and* k_2 *have the same semi-separable representation* (4.4). *Let* $\Omega(t)$ *be the unique solution on* $\Lambda = [a_1,b_1] \cup [a_2,b_2]$ *of the* $n \times n$ *matrix differential equation*

$$\begin{cases} \dot{\Omega}(t) = B(t)C(t)\Omega(t) , & t \in \Lambda , \\ \Omega(a_1) = I_n , \quad \Omega(b_1) = \Omega(a_2) . \end{cases}$$

Then $S = P\Omega(b_2)P : \operatorname{Im} P \to \operatorname{Im} P$ *is an indicator for the operator* $I - K$. *More precisely the following coupling relation holds true:*

$$(4.5) \qquad \begin{bmatrix} I-K & -R \\ -Q & I_{\operatorname{Im} P} \end{bmatrix}^{-1} = \begin{bmatrix} (I-H)^{-1} & (I-H)^{-1}R \\ Q(I-H)^{-1} & S \end{bmatrix},$$

where

$$H : L_2^m(\Lambda) \to L_2^m(\Lambda) , \quad (H\varphi)(t) = \int_{\Lambda\cap[a_1,t]} C(t)B(s)\varphi(s)ds ;$$

$$Q : L_2^m(\Lambda) \to \operatorname{Im} P , \quad Q\varphi = P\int_\Lambda B(s)\varphi(s)ds ;$$

$$R : \operatorname{Im} P \to L_2^m(\Lambda) , \quad (Rx)(t) = C(t)Px .$$

Note that for $b_1 = a_2$ the above theorem reduces to Theorem 1.1. For $b_1 < a_2$ the proof of Theorem 4.1 is similar to that of Theorem 1.1. We omit the details.

The coupling relation (4.5) may be used to derive information about the invertibility properties of $I - K$ in terms of those of the indicator S. We give here only the formula for a generalized inverse of $I - K$.

COROLLARY 4.2 *Let* K *and* $\Omega(t)$ *be as in the previous theorem. A generalized inverse of* $I - K$ *is given by the integral operator*

$$((I-K)^+f)(t) = f(t) + \int_\Lambda \gamma(t,s)f(s)ds , \quad t \in \Lambda .$$

Here $\Lambda = [a_1,b_1] \cup [a_2,b_2]$ *and*

$$\gamma(t,s) = \begin{cases} C(t)\Omega(t)(I_n-PS^+P\Omega(b_2))\Omega(s)^{-1}B(s) , & t > s , \\ -C(t)\Omega(t)PS^+P\Omega(b_2)\Omega(s)^{-1}B(s) , & t < s , \end{cases}$$

where S^+ *is a generalized inverse of the indicator* $S = P\Omega(b_2)P$.

We omit the proof of Corollary 4.2; it is similar to that of Corollary 1.2.

The general case may be reduced to the case that k_1 and k_2 have the same semi-separable representation. To see this, assume that for $\nu = 1,2$

$$k_\nu(t,s) = \begin{cases} C_\nu(t)(I_{n_\nu}-P_\nu)B_\nu(s) , & t > s , \\ -C_\nu(t)P_\nu B(s) , & t < s . \end{cases}$$

Now introduce

$$C(t) = [C_1(t)C_2(t)] , \quad t \in \Lambda = [a_1,b_1] \cup [a_2,b_2] ;$$

$$P = \begin{bmatrix} P_1 & 0 \\ 0 & P_2 \end{bmatrix}, \quad n = n_1 + n_2 ;$$

$$B(t) = \begin{bmatrix} B_1(t) \\ 0 \end{bmatrix}, \quad a_1 \leq t \leq b_1 , \quad B(t) = \begin{bmatrix} 0 \\ B_2(t) \end{bmatrix}, \quad a_2 \leq t \leq b_2 ,$$

and consider

$$(4.6) \qquad k(t,s) = \begin{cases} C(t)(I_n-P)B(s) , & t > s , \\ -C(t)PB(s) , & t < s . \end{cases}$$

Then (4.2) and (4.3) hold true with k given by (4.6). So we can apply Theorem 4.1 to find an indicator. Note that in this way the indicator becomes an operator acting on $\text{Im } P_1 \oplus \text{Im } P_2$.

II.5 Kernels with exponential representation

In this section we specify the results of this chapter for kernels of the form

$$(5.1) \qquad k(t,s) = \begin{cases} iCe^{-itA}(I_n-P)e^{isA}B , & t > s , \\ -iCe^{-itA}Pe^{isA}B , & t < s . \end{cases}$$

Here A, B and C are constant matrices, B and C are (possibly non-square) matrices of sizes $n \times m$ and $m \times n$, respectively, and A is a square matrix of order n. The symbol P stands for a projection of $\mathbb{C}^n$. Let $K : L_2^m[0,\tau] \to L_2^m[0,\tau]$ be the integral operator with kernel k, i.e.,

$$(5.2) \qquad (K\varphi)(t) = \int_0^\tau k(t,s)\varphi(s)ds , \quad 0 \leq t \leq \tau ,$$

where k is given by (5.1).

To find the indicator of $I - \mu K$ we have to determine the fundamental matrix of the differential equation

$$(5.3) \qquad \dot{x}(t) = ie^{itA}BCe^{-itA} , \quad 0 \leq t \leq \tau .$$

To do this put $A^\times = A - BC$. Note that $A^\times$ not only depends on A but also on B and C. The $n \times n$ matrix function $U(t) = e^{itA}e^{-itA^\times}$ is the fundamental matrix of (5.3). Indeed, $U(0) = I_n$ and

$$\dot{U}(t) = e^{itA}(iA-iA^{\times})e^{-itA^{\times}} = ie^{itA}BCe^{-itA^{\times}} .$$

It follows that $S_\tau = Pe^{itA}e^{-itA^{\times}}P : \operatorname{Im} P \to \operatorname{Im} P$ is an indicator of the operator $I - K$. The next theorem is Corollary 1.2 specified for kernels of the form (5.1)

THEOREM 5.1 *Let* K *be the integral operator* (5.2) *with kernel* (5.1). *Put* $A^{\times} = A - BC$, *and let* $S_\tau = Pe^{i\tau A}e^{-i\tau A^{\times}}P$. *Then*

$$\operatorname{Ker}(I-K) = \{\varphi \mid \varphi(t) = Ce^{-itA^{\times}}Py \ , \ y \in \operatorname{Ker} S_\tau\} ,$$

$$\operatorname{Im}(I-K) = \{f \mid Pe^{i\tau A}e^{-i\tau A^{\times}}\int_0^\tau e^{isA^{\times}}Bf(s)ds \in \operatorname{Im} S_\tau\} .$$

Further, a generalized inverse of $I - K$ *is given by the integral operator*

$$((I-K)^{+}f)(t) = f(t) + \int_0^\tau \gamma(t,s)f(s)ds ,$$

with

$$\gamma(t,s) = \begin{cases} iCe^{-itA^{\times}}(I_n-Q)e^{isA^{\times}}B , & t > s , \\ -iCe^{-itA^{\times}}Qe^{isA^{\times}}B , & t < s . \end{cases}$$

Here $Q = PS_\tau^{+}Pe^{i\tau A}e^{-i\tau A^{\times}} : \mathbb{C}^n \to \mathbb{C}^n$, *where* S_τ^{+} *is a generalized inverse of* S_τ.

Put $A^{\times}(\mu) = A - \mu BC$. Then

$$S_\tau(\mu) = Pe^{itA}e^{-itA^{\times}(\mu)}P : \operatorname{Im} P \to \operatorname{Im} P$$

is an indicator for $I - \mu K$. We know (see Theorem 3.1) that the zeros of $\det S_\tau(\mu)$ are the characteristic values of the operator K (multiplicities taken into account).

THEOREM 5.2 *Let* K *be the integral operator* (5.2) *with kernel* (5.1). *Then* K *is a trace class operator if and only if* $CB = 0$, *and in that case* $\det(I-\mu K) = \det S_\tau(\mu)$.

PROOF. We may write $K = H - RQ$, where

$$(H\varphi)(t) = \int_0^t iCe^{-i(t-s)A}B\varphi(s)ds , \quad 0 \leqslant t \leqslant \tau ,$$

and R and Q are operators of finite rank. Now H is a trace class operator if and only if $CB = 0$. This statement has been proved in Section I.6 of [5]. It follows that K is a trace class operator if and only if $CB = 0$. To finish to proof we apply Corollary 3.2. □

In [5] Theorems 5.1 and 5.2 have been proved for the case when the projection P appearing in (5.1) commutes with A. The extra condition "P commutes with A" implies that k is a difference kernel. This means that $k(t,s) = h(t-s)$. In our case

$$(5.4)\qquad h(t) = \begin{cases} iCe^{-itA}(I_n-P)B\,, & 0 < t \leq \tau\,, \\ -iCe^{-itA}PB\,, & -\tau \leq t < 0\,. \end{cases}$$

The method described here is applicable to difference kernels $h(t-s)$ if, for instance, h has an extension to an $m \times m$ matrix L_1-function $\tilde{h}$ on the full real line such that the Fourier transform of $\tilde{h}$ is a rational function. A function h which admits such an extension can always be written in the form (5.4) where now A is a square matrix with no eigenvalues on the real line and P is the Riesz projection corresponding to the eigenvalues of A in the upper half plane. The converse of the latter statement is also true. For more details we refer to Section I.1 of [5] (see also Section IV.1 in this paper).

The discrete analogues of the integral operators considered in this section are block matrices $T = [A_{jk}]_{j,k=1}^{r}$ with $m \times m$ matrix entries given by

$$(5.5)\qquad A_{jk} = \begin{cases} -CA^{j-1}(I_n-P)A^{-k}B\,, & j > k\,, \\ CA^{j-1}PA^{-k}B\,, & j \leq k\,. \end{cases}$$

Here A is assumed to be an invertible matrix of order n. The matrices B and C have sizes $n \times m$ and $m \times n$, respectively, and as usual P is a projection of $\mathbb{C}^n$. Further, $T = [A_{jk}]_{j,k=1}^{r}$ is considered as an operator on $\mathbb{C}^{mr}$. Using Theorem 2.1 one easily computes that

$$(5.6)\qquad S = PA^{-r}(A-\mu BC)^{r}P : \operatorname{Im} P \to \operatorname{Im} P$$

is an indicator for the operator $I - \mu T$. By applying Corollary 2.2 one finds that a generalized inverse of $I - T$ is given by $I + T^{\times}$, where $T^{\times} = [A^{\times}_{jk}]_{j,k=1}^{r}$ and

$$A^{\times}_{jk} = \begin{cases} -C(A^{\times})^{j-k-1}(I_n-(A^{\times})^{k}PS^{+}PA^{-r}(A^{\times})^{r-k})B\,, & j > k\,, \\ C(A^{\times})^{j-1}PS^{+}PA^{-r}(A^{\times})^{r-k}B\,, & j \leq k\,. \end{cases}$$

Here S^{+} is a generalized inverse of the indicator (5.6) and $A^{\times} = A - BC$. If in (5.5) the projection P commutes with A, then $T = [A_{jk}]_{j,k=1}^{r}$ is a finite block Toeplitz matrix and one gets back (in a somewhat improved form) the results of [5], Section I.9.

III. SINGULAR INTEGRAL EQUATIONS

In this and the next chapter the principle of matricial coupling is applied to certain classes of non-compact operators. Singular integral operators and Toeplitz operators with (operator-valued) analytical symbols are treated in the present chapter.

III.1 Preliminaries

The main topic of the present chapter is the singular integral equation

$$A(\lambda)\varphi(\lambda) + B(\lambda)\Big(\frac{1}{\pi i}\int_\Gamma \frac{\varphi(\lambda)}{\mu-\lambda}\,d\mu\Big) = f(\lambda)\ , \qquad \lambda \in \Gamma\ . \tag{1.1}$$

We shall think about (1.1) as a system of equations. So A and B are matrix or operator functions defined on Γ and the given function f and the unknown function φ are vector functions. The contour Γ consists of a finite number of disjoint smooth simple Jordan curves. The inner domain of Γ will be denoted by F_+, the outer domain by F_-, and it is assumed that $\infty \in F_-$.

As usual (see e.g. [9]) we rewrite (1.1) in the form

$$(M_A + M_B S_\Gamma)\varphi = f\ , \tag{1.2}$$

where M_A and M_B are the operators of multiplication by A and B, respectively, and S_Γ is the basic singular integral operator. The operators S_Γ, M_A and M_B are assumed to act as bounded linear operators on a suitable function space E. In what follows E will always be the space $H_\alpha(\Gamma;Y)$ of all functions from Γ into a given complex Banach space Y that are Hölder continuous with Hölder exponent α, $0 < \alpha < 1$ (cf. [12]), but with only minor modifications our results hold for other choices for E as well.

Put $P_\Gamma = \frac{1}{2}(I+S_\Gamma)$ and $Q_\Gamma = \frac{1}{2}(I-S_\Gamma)$, where $I = I_E$ is the identity operator on E. The operators P_Γ and Q_Γ are complementary projections, which can be used to rewrite the operator $M_A + M_B S_\Gamma$. Assume that the values of the function $A - B$ are invertible operators on Y and that the operator $M_{(A-B)^{-1}}$ of multiplication by $[A(\lambda)-B(\lambda)]^{-1}$ is a well-defined bounded linear operator on E. Then

$$M_A + M_B S_\Gamma = M_{A-B}(P_\Gamma M_W P_\Gamma + Q_\Gamma)(I + Q_\Gamma M_Q P_\Gamma) ,$$

where M_W is the operator of multiplication by $W(\lambda) = [A(\lambda)-B(\lambda)]^{-1}[A(\lambda)+B(\lambda)]$. Observe that M_{A-B} and $I + Q_\Gamma M_W P_\Gamma$ are both invertible operators. It follows that the invertibility properties of $M_A + M_B S_\Gamma$ are completely determined by those of the operator

$$T_W = P_\Gamma M_W P_\Gamma : E^+ \to E^+ .$$

Here $E^+ = \text{Im } P_\Gamma$ is the space consisting of all functions in E that have an extension which is analytic on the inner domain F_+ of Γ and continuous on the closure $F_+ \cup \Gamma$ of F_+. We shall write E^- for $\text{Im } Q_\Gamma$, and thus $E = E^+ \oplus E^-$. The operator T_W is called the *Toeplitz operator with symbol* W. The action of T_W on E^+ is given by

(1.3) $$(T_W\varphi)(\lambda) = \frac{1}{2} W(\lambda)\varphi(\lambda) + \frac{1}{2\pi i}\int_\Gamma \frac{W(\mu)\varphi(\mu)}{\mu-\lambda} d\mu , \quad \lambda \in \Gamma .$$

Note that the symbol W is a function on Γ whose values are in $L(Y)$, the space of all bounded linear operators on the Banach space Y.

III.2 The indicator

Continuing the discussion of the previous section we now assume that the symbol $W : \Gamma \to L(Y)$ admits an analytic extension to a neighbourhood of the given contour Γ. The following result (taken from [3], Ch. II) will play a fundamental role in our analysis of the Toeplitz operator T_W.

THEOREM 2.1 *The operator function* $W : \Gamma \to L(Y)$ *admits an analytic continuation to a neighbourhood of the contour* Γ *if and only if* W *can be written in the form*

(2.1) $$W(\lambda) = I_Y + C(\lambda I_X - A)^{-1}B , \quad \lambda \in \Gamma \subset \rho(A) ,$$

where X *is a complex Banach space and* $A : X \to X$, $B : Y \to X$, $C : X \to Y$ *are bounded linear operators.*

Here $\rho(A)$ denotes the resolvent set of A.

A representation of the form (2.1) is called a *realization* of W. This terminology is taken from system theory where it is connected with the notion of the transfer function of a linear dynamical system. It is well-known (cf. [1,16,3]) that in case Y is finite dimensional, W has a realization (2.1) with finite dimensional "state space" X if and only if W is (or can be viewed as) a rational matrix function having the value I_Y at ∞.

Another condition that we shall impose is that W takes invertible values on Γ. In terms of the realization (2.1) this means that the operator $A - BC$ has no spectrum on Γ and

$$W(\lambda)^{-1} = I_Y - C(\lambda I_X - A + BC)^{-1}B \,, \qquad \lambda \in \Gamma \cap \rho(A-BC)$$

(cf. [3], Ch. II). Instead of $A - BC$ we often write $A^{\times}$. Note that $A^{\times}$ does not depend on A only, but also on the operators B and C appearing in (2.1).

THEOREM 2.2 *Consider the Toeplitz operator* $T_W : E^+ \to E^+$ *given by* (1.3) *and assume that* $W : \Gamma \to L(y)$ *has the realization* (2.1), *where* A *and* $A^{\times}$ *have no spectrum on the contour* Γ. *Let* $P, P^{\times}$ *be the Riesz projections given by*

$$P = \frac{1}{2\pi i}\int_{\Gamma}(\lambda-A)^{-1}d\lambda \,, \qquad P^{\times} = \frac{1}{2\pi i}\int_{\Gamma}(\lambda-A^{\times})^{-1}d\lambda \,.$$

Then the operator

$$S = P^{\times}|_{\operatorname{Im} P} : \operatorname{Im} P \to \operatorname{Im} P^{\times}$$

is an indicator for the Toeplitz operator T_W. *More precisely, the following coupling relation holds:*

$$(2.2) \qquad \begin{bmatrix} T_W & U \\ R & Q \end{bmatrix}^{-1} = \begin{bmatrix} T_{W^{-1}} & U^{\times} \\ R^{\times} & S \end{bmatrix} .$$

Here $W^{-1} : \Gamma \to L(y)$ *is given by* $W^{-1}(\lambda) = W(\lambda)^{-1}$ *and*

$$U : \operatorname{Im} P^{\times} \to E^+ \,, \qquad [Ux](\lambda) = C(\lambda-A)^{-1}(I-P)x \,,$$

$$U^{\times} : \operatorname{Im} P \to E^+ \,, \qquad [U^{\times}x](\lambda) = C(\lambda-A^{\times})^{-1}(I-P^{\times})x \,,$$

$$R : E^+ \to \operatorname{Im} P \,, \qquad R\varphi = \frac{1}{2\pi i}\int_{\Gamma} P(\lambda-A)^{-1}B\varphi(\lambda)d\lambda \,,$$

$$R^{\times} : E^+ \to \operatorname{Im} P^{\times} \,, \qquad R^{\times}\varphi = \frac{-1}{2\pi i}\int_{\Gamma} P^{\times}(\lambda-A^{\times})^{-1}B\varphi(\lambda)d\lambda \,,$$

$$Q : \operatorname{Im} P^{\times} \to \operatorname{Im} P \,, \qquad Qx = Px \,.$$

The form of the operators appearing in the theorem may be guessed from the "test for matricial coupling" (Theorem I.4.1) by taking $L = M_W : E^+ \oplus E^- \to E^+ \oplus E^-$ and $K = I_Y : \operatorname{Im} P \oplus \operatorname{Ker} P \to \operatorname{Im} P^{\times} \oplus \operatorname{Ker} P^{\times}$. By Cauchy's theorem $R\varphi$ and $R^{\times}\varphi$ can also be written as

$$R\varphi = \frac{1}{2\pi i}\int_{\Gamma}(\lambda-A)^{-1}B\varphi(\lambda)d\lambda \,, \qquad R^{\times}\varphi = \frac{-1}{2\pi i}\int_{\Gamma}(\lambda-A^{\times})^{-1}B\varphi(\lambda)d\lambda \,.$$

PROOF. Proving (2.2) comes down to verifying eight identities.

Here we shall establish four of them, namely

(2.3) $T_W T_{W^{-1}} + UR^\times = I_{E^+}$,

(2.4) $RT_{W^{-1}} + QR^\times = 0$,

(2.5) $T_W U^\times + US = 0$,

(2.6) $RU^\times + QS = I_{\mathrm{Im}\,P}$.

The other four can be obtained similarly or by interchanging the roles of W and W^{-1}.

Take $\varphi \in E^+$. Then

$$[T_{W^{-1}}\varphi](\lambda) = \frac{1}{2} W(\lambda)^{-1}\varphi(\lambda) + \frac{1}{2\pi i}\int_\Gamma \frac{W(\mu)^{-1}\varphi(\mu)}{\mu-\lambda}\,d\mu \ .$$

Applying the multiplication operator M_W yields

$$[M_W T_{W^{-1}}\varphi](\lambda) = \frac{1}{2}\varphi(\lambda) + \frac{1}{2\pi i}\int_\Gamma \frac{W(\lambda)W(\mu)^{-1}\varphi(\mu)}{\mu-\lambda}\,d\mu \ .$$

Using that $W(\mu)^{-1} = I_Y - C(\lambda I_X - A^\times)^{-1}B$, we get

$$W(\lambda)W(\mu)^{-1} = I_Y + (\mu-\lambda)C(\lambda-A)^{-1}(\mu-A^\times)^{-1}B \ ,$$

and so

$$\begin{aligned}
[M_W T_{W^{-1}}\varphi](\lambda) &= \frac{1}{2}\varphi(\lambda) + \frac{1}{2\pi i}\int_\Gamma \frac{\varphi(\mu)}{\mu-\lambda}\,d\mu + C(\lambda-A)^{-1} \\
&\quad \times \left(\frac{1}{2\pi i}\int_\Gamma (\mu-A^\times)^{-1}B\varphi(\mu)\,d\mu\right) \\
&= [P_\Gamma\varphi](\lambda) - C(\lambda-A)^{-1}\left(\frac{-1}{2\pi i}\int_\Gamma (\mu-A^\times)^{-1}B\varphi(\mu)\,d\mu\right) \\
&= \varphi(\lambda) - C(\lambda-A)^{-1}R^\times\varphi \\
&= \varphi(\lambda) - C(\lambda-A)^{-1}(I-P)R^\times\varphi - C(\lambda-A)^{-1}PR^\times\varphi \\
&= \varphi(\lambda) - [UR^\times\varphi](\lambda) - C(\lambda-A)^{-1}PR^\times\varphi \ .
\end{aligned}$$

The function $C(\lambda-A)^{-1}PR^\times\varphi$ belongs to $E^- = \mathrm{Ker}\,P_\Gamma$. It follows that

$$T_W T_{W^{-1}}\varphi = P_\Gamma M_W T_{W^{-1}}\varphi = \varphi - UR^\times\varphi \ ,$$

and (2.3) is proved.

In order to establish (2.4) we argue as follows. Put $\psi = Q_\Gamma M_{W^{-1}}\varphi$. Then $\psi \in E^-$ and the function $P(\lambda-A)^{-1}B\psi(\lambda)$ has a zero of order 2 at infinity. So

$$\frac{1}{2\pi i}\int_\Gamma P(\lambda-A)^{-1}B\psi(\lambda)d\lambda = 0 .$$

Note that $T_{W^{-1}}\varphi = P_\Gamma M_{W^{-1}}\varphi = M_{W^{-1}}\varphi - \psi$. Hence

$$RT_{W^{-1}}\varphi = \frac{1}{2\pi i}\int_\Gamma P(\lambda-A)^{-1}B([M_{W^{-1}}\varphi](\lambda)-\psi(\lambda))d\lambda$$

$$= \frac{1}{2\pi i}\int_\Gamma P(\lambda-A)^{-1}BW(\lambda)^{-1}\varphi(\lambda)d\lambda .$$

Now $(\lambda-A)^{-1}BW(\lambda)^{-1} = (\lambda-A)^{-1}B[I_Y-C(\lambda-A^\times)^{-1}B] = (\lambda-A^\times)^{-1}B$. Thus

$$RT_{W^{-1}}\varphi = P\left(\frac{1}{2\pi i}\int_\Gamma(\lambda-A^\times)^{-1}B\varphi(\lambda)d\lambda\right) = -PR^\times\varphi = -QR^\times\varphi .$$

Next take $x \in \operatorname{Im} P$. Then

$$[M_W U^\times x](\lambda) = W(\lambda)[U^\times x](\lambda)$$

$$= [I_Y+C(\lambda I_X-A)^{-1}B]C(\lambda-A^\times)^{-1}(I-P^\times)x$$

$$= C(\lambda-A)^{-1}(I-P^\times)x$$

$$= C(\lambda-A)^{-1}(I-P)(I-P^\times)x + C(\lambda-A)^{-1}P(I-P^\times)x$$

$$= -C(\lambda-A)^{-1}(I-P)P^\times x + C(\lambda-A)^{-1}P(I-P^\times)x$$

$$= -[USx](\lambda) + C(\lambda-A)^{-1}P(I-P^\times)x .$$

The function $C(\lambda-A)^{-1}P(I-P^\times)x$ belongs to $E^- = \operatorname{Ker} P_\Gamma$. So $T_W U^\times x = -USx$ and we have established (2.5).

Finally, identity (2.6) is clear from the following computation with $x \in \operatorname{Im} P$:

$$RU^\times x = \frac{1}{2\pi i}\int_\Gamma P(\lambda-A)^{-1}BC(\lambda-A^\times)^{-1}(I-P^\times)x\, d\lambda$$

$$= \frac{1}{2\pi i}\int_\Gamma P(\lambda-A)^{-1}[(\lambda-A^\times)-(\lambda-A)](\lambda-A^\times)^{-1}(I-P^\times)x\, d\lambda$$

$$= \frac{1}{2\pi i}\int_\Gamma P(\lambda-A)^{-1}(I-P^\times)x\, d\lambda - \frac{1}{2\pi i}\int_\Gamma P(\lambda-A^\times)^{-1}(I-P^\times)x\, d\lambda$$

$$= P(I-P^\times)x - PP^\times(I-P^\times)x = x - QSx .$$

This proves the theorem. □

Combining Theorem 2.2 and Theorem I.1.1, we obtain the following result.

COROLLARY 2.3 *Let* T_W, P *and* $P^\times$ *be as in* Theorem 2.2. *Then*

$$\dim \operatorname{Ker} T_W = \dim(\operatorname{Im} P \cap \operatorname{Ker} P^\times) ,$$

$$\dim(E^+/\operatorname{Im} T_W) = \dim(\operatorname{Im} P^\times/\operatorname{Im} P^\times P) = \dim(X/\operatorname{Im} P + \operatorname{Ker} P^\times) .$$

In particular, T_W *is a Fredholm operator if and only if*

$$\dim(\operatorname{Im} P \cap \operatorname{Ker} P^\times) < \infty , \quad \dim(X/\operatorname{Im} P + \operatorname{Ker} P^\times) < \infty ,$$

and in that case

$$\operatorname{ind}(T_W) = \dim(\operatorname{Im} P \cap \operatorname{Ker} P^\times) - \dim(X/\operatorname{Im} P + \operatorname{Ker} P^\times).$$

Suppose Y is finite dimensional and W is rational having the value I_Y at ∞. We then may assume that the state space X in the realization (2.1) is finite dimensional. In this situation T_W is a Fredholm operator and the expression for the index of T_W given above can be rewritten as

$$\operatorname{ind}(T_W) = \operatorname{rank} P - \operatorname{rank} P^\times .$$

This opens the way for still another description of the index of T_W. With each pole μ of W one can associate a positive integer $\delta(W;\mu)$ called the *local degree* or *pole multiplicity* of W at μ. For the definition we refer to [3]. It is convenient to put $\delta(W;\mu) = 0$ if W is analytic at μ. The number $\sum_{\mu\in\mathbb{C}} (W;\mu)$ is an important characteristic of the rational matrix function W. It is called the *McMillan degree* of W and plays a role in the theory of electrical networks. By definition, a *zero* of W is a pole of W^{-1} and the number $\delta(W^{-1};\nu)$ is called the *zero multiplicity* of W at ν. Taking for (2.1) a minimal realization of W and using the results of [3], Ch. IV, one sees that

$$\operatorname{ind}(T_W) = \sum_{\mu\in F_+} \delta(W;\mu) - \sum_{\nu\in F_+} \delta(W^{-1};\nu) .$$

Here, as before, F_+ denotes the inner domain of Γ. Recall that a realization is *minimal* if its state space has the least possible dimension (which happens to be equal to the McMillan degree of W).

Returning to the general case, a more explicit description of $\operatorname{Ker} T_W$ and $\operatorname{Im} T_W$ can be obtained by combining Theorem 2.2 and the first part of Theorem I.2.1.

COROLLARY 2.4 *Let* T_W, P *and* $P^\times$ *be as in* Theorem 2.2. *Then*

$$\operatorname{Ker} T_W = \{\varphi : \Gamma \to Y \mid \varphi(\lambda) = C(\lambda-A^\times)^{-1}x , \ x \in \operatorname{Im} P \cap \operatorname{Ker} P^\times\} ,$$

$$\operatorname{Im} T_W = \{\varphi \in E^+ \mid \int_\Gamma P^\times(\lambda-A^\times)^{-1}B\varphi(\lambda)d\lambda \in \operatorname{Im} P + \operatorname{Ker} P^\times\} .$$

From the first identity we see that if $\varphi \in \operatorname{Ker} T_W$, then there exists $x \in \operatorname{Im} P \cap \operatorname{Ker} P^\times$ such that $\varphi(\lambda) = C(\lambda-A^\times)^{-1}x$. It is easily verified that the vector x is uniquely determined by φ.

The kernel of the Toeplitz operator T_W gives the solutions of the *Hilbert boundary value problem*

$$W(\lambda)\psi_+(\lambda) = \psi_-(\lambda)\ , \qquad \lambda \in \Gamma\ . \tag{2.7}$$

Recall that a pair of functions (ψ_+,ψ_-) is said to be a solution of (2.7) if $\psi_+ \in E^+$, $\psi_- \in E^-$ and the identity (2.7) holds true. Note that this implies that $T_W\psi_+ = P_\Gamma M_W\psi_+ = 0$. Conversely, $\psi_+ \in \operatorname{Ker} T_W$ and $\psi_- = M_{W^{-1}}\psi_+$, then (ψ_+,ψ_-) is a solution of (2.7). This yields the following corollary.

COROLLARY 2.5 *Let* $W : \Gamma \to L(Y)$ *have a realization* (2.1), *where* A *and* $A^\times$ *have no spectrum on* Γ. *Let* P *and* $P^\times$ *be the Riesz projections*

$$P = \frac{1}{2\pi i}\int_\Gamma (\lambda-A)^{-1}d\lambda\ , \qquad P^\times = \frac{1}{2\pi i}\int_\Gamma (\lambda-A^\times)^{-1}d\lambda\ .$$

Then the general solution of the Hilbert boundary value problem (2.7) *is given by*

$$\psi_+(\lambda) = C(\lambda-A^\times)^{-1}x\ , \qquad \psi_-(\lambda) = C(\lambda-A)^{-1}x\ ,$$

where x *is an arbitrary vector of* $\operatorname{Im} P \cap \operatorname{Ker} P^\times$. *Moreover, the vector* x *is uniquely determined by the solution* (ψ_+,ψ_-).

The next corollary gives a generalized inverse of the Toeplitz operator T_W in terms of its indicator.

COROLLARY 2.6 *Let* T_W, P *and* $P^\times$ *be as in* Theorem 2.2. *Then* T_W *has a generalized inverse if and only if its indicator* $S = P^\times \mid \operatorname{Im} P : \operatorname{Im} P \to \operatorname{Im} P^\times$ *has a generalized inverse. If* S^+ *is a generalized inverse of* S, *a generalized inverse* T_W^+ *of* T_W *is given by*

$$[T_W^+\varphi](\lambda) = W(\lambda)^{-1}\varphi(\lambda) + \frac{1}{2\pi i}\int_\Gamma C(\lambda-A^\times)^{-1}\Pi^+(\mu-A^\times)^{-1}B\varphi(\mu)d\mu\ , \tag{2.8}$$

where Π^+ *is the projection of* X *along* $\operatorname{Ker} P^\times$ *defined by*

$$\Pi^+ = P^\times + (I-P^\times)S^+P^\times\ . \tag{2.9}$$

The corollary remains correct when "generalized inverse" is replaced by "left inverse", "right inverse" or "(two-sided) inverse". Note that T_W is invertible if and only if $X = \operatorname{Im} P \oplus \operatorname{Ker} P^\times$ and this, in turn, corresponds to the case when Π^+ is the projection of X along $\operatorname{Ker} P^\times$ onto $\operatorname{Im} P$.

It is also possible to express a generalized inverse of S in terms of one of T_W; we omit the details.

Proof of Corollary 2.6. From Theorem 2.2 and Theorem I.2.1 we know that T_W has a generalized inverse if and only if this is the case for S. The same theorems imply that $T_W^+ = T_{W^{-1}} - U^\times S^+ R^\times$ is a generalized inverse of T_W, provided S^+ is one for S. For $\varphi \in E^+$ and $\lambda \in \Gamma$ we now compute $[T_W^+\varphi](\lambda) = [T_{W^{-1}}\varphi](\lambda) - [U^\times S^+ R^\times \varphi](\lambda)$.

Note first that $T_{W^{-1}}\varphi = P_\Gamma M_W \varphi = M_{W^{-1}}\varphi - Q_\Gamma M_{W^{-1}}\varphi$. Since $Q_\Gamma = \frac{1}{2}(I - S_\Gamma)$ and $Q_\Gamma \varphi = 0$, we have $Q_\Gamma M_{W^{-1}}\varphi = \frac{1}{2}(M_{W^{-1}} S_\Gamma - S_\Gamma M_{W^{-1}})\varphi$, and so

$$\begin{aligned}
[Q_\Gamma M_{W^{-1}}\varphi](\lambda) &= \frac{1}{2\pi i}\int_\Gamma \frac{[W(\lambda)^{-1} - W(\mu)^{-1}]\varphi(\mu)}{\mu-\lambda}\,d\mu \\
&= \frac{1}{2\pi i}\int_\Gamma \frac{C[(\mu-A^\times)^{-1} - (\lambda-A^\times)^{-1}]B\varphi(\mu)}{\mu-\lambda}\,d\mu \\
&= \frac{-1}{2\pi i}\int_\Gamma C(\lambda-A^\times)^{-1}(\mu-A^\times)^{-1}B\varphi(\mu)\,d\mu \\
&= \frac{-1}{2\pi i}\int_\Gamma C(\lambda-A^\times)^{-1}P^\times(\mu-A^\times)^{-1}B\varphi(\mu)\,d\mu \;.
\end{aligned}$$

Also it is clear from the definition of $U^\times$ and $R^\times$ that

$$[U^\times S^+ R^\times](\lambda) = \frac{-1}{2\pi i}\int_\Gamma C(\lambda-A^\times)^{-1}(I-P^\times)S^+P^\times(\mu-A^\times)^{-1}B\varphi(\mu)\,d\mu \;.$$

It follows that $[T_W^+\varphi](\lambda)$ is indeed given by (2.8), where Π^+ is defined by (2.9). One verifies without difficulty that Π^+ is a projection of X along $\operatorname{Ker} P^\times$, and the proof is complete. □

We conclude this section with some remarks. The analysis presented above is heavily based on the fact that W can be written in the form (2.1), where A and $A^\times = A - BC$ have no spectrum on the contour Γ. Such realizations are not unique. However, it is clear from Corollary 2.2 that the quantities

$$\dim(\operatorname{Im} P \cap \operatorname{Ker} P^\times)\;, \quad \dim(X/\operatorname{Im} P + \operatorname{Ker} P^\times)$$

do not depend on the particular choice of the realization (2.1). More generally this is true for the quantities

$$\dim(\operatorname{Im} P \cap \operatorname{Ker} P^\times \cap \operatorname{Ker} C \cap \operatorname{Ker} CA \cap \cdots \cap \operatorname{Ker} CA^{j-1})\;,$$

$$\dim(X/\operatorname{Im} P + \operatorname{Ker} P^\times + \operatorname{Im} B + \operatorname{Im} AB + \cdots + \operatorname{Im} A^{j-1}B)\;.$$

Here $j = 0,1,2,\cdots$. One way to see this is as follows. For $j = 0,1,2,\cdots$, let $\mathrm{Ker}_j T_W$ be the set of all $\varphi : \Gamma \to Y$ such that the function $\varphi(\lambda), \lambda\varphi(\lambda), \cdots, \lambda^j\varphi(\lambda)$ all belong to $\mathrm{Ker}\, T_W$. Then $\mathrm{Ker}_j T_W$ is a subspace of $\mathrm{Ker}\, T_W$ and the operator $U^\times$ maps $\mathrm{Im}\, P \cap \mathrm{Ker}\, P^\times \cap \mathrm{Ker}\, C \cap \cdots \cap \mathrm{Ker}\, CA^{j-1}$ one-to-one onto $\mathrm{Ker}_j T_W$. Similarly, let $\mathrm{Im}_j T_W$ be the set of all functions $\psi \in E^+$ such that for some Y-valued polynomial p of degree not exceeding $j - 1$ the function $\psi(\lambda) + p(\lambda)$ belongs to $\mathrm{Im}\, T_W$. Then $\mathrm{Im}_j T_W$ corresponds to $\mathrm{Im}\, P + \mathrm{Ker}\, P^\times + \mathrm{Im}\, B + \cdots + \mathrm{Im}\, A^{j-1}B$ in the same way as $\mathrm{Im}\, T_W (= \mathrm{Im}_0 T_W)$ corresponds to $\mathrm{Im}\, P + \mathrm{Ker}\, P^\times$. A more direct proof (not involving the singular integral operator) is given in [6] in the context of a detailed analysis of Wiener-Hopf factorization of analytic operator functions. For a concise summary of [6], see [7].

IV. WIENER-HOPF EQUATIONS

In this chapter the principle of matricial coupling of operators is applied to Wiener-Hopf integral equations and their discrete analogues.

IV.1 Wiener-Hopf integral equations

In this section we recast the results obtained in [8] (see also [5]) in the framework described in Chapter I.

The operator we wish to investigate is the convolution integral operator $K : L_p([0,\infty),Y) \to L_p([0,\infty),Y)$ given by

$$[K\varphi](t) = \int_0^\infty k(t-s)\varphi(s)ds \ . \tag{1.1}$$

Here Y is a (non-trivial) complex Banach space, $1 \leq p \leq \infty$ and k is a Bochner integrable kernel whose values are bounded linear operators on Y, i.e., $k \in L_1((-\infty,\infty),\mathcal{L}(Y))$. The familiar Wiener-Hopf integral equation

$$\varphi(t) - \int_0^\infty k(t-s)\varphi(s)ds = f(t) \ , \quad 0 \leqslant t < \infty \tag{1.2}$$

(with $\varphi, f \in L_p([0,\infty),Y)$) can be written as $(I-K)\varphi = f$.

By the *symbol* of the operator K or of the equation (1.2) we mean the function defined for real λ by

$$W(\lambda) = I_Y - \int_{-\infty}^\infty e^{i\lambda t}k(t)dt \ .$$

Putting $W(\infty) = I_Y$, we have that W is continuous on the extended real line $\mathbb{R}_\infty$. We shall assume that W admits an analytic continuation to a neighbourhood (in the Riemann sphere $\mathbb{C}_\infty$) of $\mathbb{R}_\infty$. The following result (cf. [3], Chapter II) is essential for the analysis of the operator $I - K$ to be presented below.

THEOREM 1.1 *The symbol* W *of the Wiener-Hopf integral equation* (1.2) *admits an analytic continuation to a neighbourhood (in* $\mathbb{C}_\infty$*) of the extended real line if and only if* W *can be written in the form*

$$(1.3) \qquad W(\lambda) = I_Y + C(\lambda I_X - A)^{-1}B \;, \quad \lambda \in \mathbb{R} \subset \rho(A) \;,$$

where X *is a complex Banach space and* $A : X \to X$, $B : Y \to X$, $C : X \to Y$ *are bounded linear operators.*

In case Y is finite dimensional, W has a realization (1.3) with finite dimensional state space X if and only if W is a rational matrix function.

Another condition that we shall impose is that W takes invertible values on the real line. In terms of the realization (1.3) this means that the spectrum $\sigma(A^\times)$ of the operator $A^\times = A - BC$ lies off the real line.

THEOREM 1.2 *Consider the Wiener-Hopf integral operator* $K : L_p([0,\infty),Y) \to L_p([0,\infty),Y)$ *given by* (1.2), *and assume that its symbol* $W : \mathbb{R}_\infty \to L(Y)$ *has the realization* (1.3), *where* A *and* $A^\times$ *have no spectrum on the real line. Let* P *[respectively* $P^\times$*] be the Riesz projection corresponding to the part of* $\sigma(A)$ *[respectively* $\sigma(A^\times)$*] lying in the upper half plane. Then the operator*

$$S = P^\times|_{\operatorname{Im} P} : \operatorname{Im} P \to \operatorname{Im} P^\times$$

is an indicator for $I - K$. *More precisely, the following coupling relation holds:*

$$\begin{bmatrix} I-K & U \\ R & Q \end{bmatrix}^{-1} = \begin{bmatrix} I-K^\times & U^\times \\ R^\times & S \end{bmatrix} .$$

Here

$$\begin{aligned}
U &: \operatorname{Im} P^\times \to L_p([0,\infty),Y) \,, & [Ux](t) &= iCe^{-itA}(I-P)x \,, \\
U^\times &: \operatorname{Im} P \to L_p([0,\infty),Y) \,, & [U^\times x](t) &= iCe^{-itA^\times}(I-P^\times)x \,, \\
R &: L_p([0,\infty),Y) \to \operatorname{Im} P \,, & R\varphi &= -\int_0^\infty Pe^{isA}B\varphi(s)ds \,, \\
R^\times &: L_p([0,\infty),Y) \to \operatorname{Im} P^\times \,, & R^\times\varphi &= \int_0^\infty P^\times e^{isA^\times}B\varphi(s)ds \,, \\
Q &: \operatorname{Im} P^\times \to \operatorname{Im} P \,, & Qx &= Px \,.
\end{aligned}$$

$$K^\times : L_p([0,\infty),Y) \to L_p([0,\infty),Y) , \quad [K^\times\varphi](t) = \int_0^\infty k^\times(t-s)\varphi(s)ds ,$$

where $k^\times \in L_1((-\infty,\infty),L(Y))$ *is given by*

$$(1.4) \qquad k^\times(t) = \begin{cases} -iCe^{-itA^\times}(I-P^\times)B , & t > 0 , \\ iCe^{-itA^\times}P^\times B , & t < 0 . \end{cases}$$

Again the form of the operators appearing in the theorem can be guessed from the "test for matricial coupling" (Theorem I.4.1). On the other hand, the eight identities needed to prove Theorem 1.2 were already established in Section 2 of [8]. The proof is based on the fact that the kernel k can be written in the form

$$(1.5) \qquad k(t) = \begin{cases} iCe^{-itA}(I-P)B , & t > 0 , \\ -iCe^{-itA}PB , & t < 0 . \end{cases}$$

This fact also plays an important role in [4,5]. The expression (1.5) is called a (*spectral*) *exponential representation* of k. This notion is related to (or may even be considered as a special case of) the concept of balanced realization appearing in system theory (cf. [10,2]). In [8] the property of having a (spectral) exponential representation is characterized in terms of growth conditions on k and its derivatives. Note that (1.4) is a spectral exponential representation of $k^\times$.

It is now clear that we can apply Theorems I.1.1 and I.2.1 to obtain the analogues of Corollaries III.2.3, III.2.4 and III.2.6. Since the resulting theorems are already contained in [8], it is not necessary to present the details here.

IV.2 Block Toeplitz equations

In this section we discuss the discrete analogue of the Wiener-Hopf integral equation (1.2), that is the Toeplitz equation

$$(2.1) \qquad \sum_{k=0}^{\infty} T_{j-k}\xi_k = \eta_j , \quad j = 0,1,2,\cdots .$$

More precisely, we study the operator $T : \ell_p(\mathbb{N},Y) \to \ell_p(\mathbb{N},Y)$ given by

$$(2.2) \qquad (T\xi)_j = \sum_{k=0}^{\infty} T_{j-k}\xi_k , \quad j = 0,1,2,\cdots .$$

Here Y is a (non-trivial) complex Banach space, $1 \leq p \leq \infty$ and $\cdots,T_{-1},T_0,T_1,\cdots$ is a (two-sided) ℓ_1-sequence of bounded linear operators on

Y.

In fact we shall impose the stronger convergence condition that

$$\sum_{m=-\infty}^{\infty} q^{|m|} \, ||T_m|| < \infty$$

for some $q > 1$. This amounts to the same as saying that the symbol of the operator T (or of the equation (2.1)) admits an analytic continuation to an annulus around the unit circle $\Gamma = \{\lambda \in \mathbb{C} \mid |\lambda| = 1\}$. The *symbol* is the operator function $W : \Gamma \to L(Y)$ defined by

$$W(\lambda) = \sum_{m=-\infty}^{\infty} \lambda^m T_m .$$

Our analysis will be based on Theorem III.2.1 which guarantees that under the hypothesis formulated above W can be written in the form

$$W(\lambda) = I_Y + C(\lambda I_X - A)^{-1}B , \quad \lambda \in \Gamma \subset \rho(A) , \tag{2.3}$$

where X is a complex Banach space and $A : X \to X$, $B : Y \to X$, $C : X \to Y$ are bounded linear operators and A has no spectrum on Γ. Note that this implies that the operators T_n can be written as

$$T_n = \begin{cases} CA^{-n-1}PB , & n = -1,-2,\cdots , \\ I_Y - CA^{-1}(I-P)B , & n = 0 , \\ -CA^{-n-1}(I-P)B , & n = 1,2,\cdots , \end{cases} \tag{2.4}$$

where P is the Riesz projection corresponding to the part of $\sigma(A)$ inside the unit circle. Since the restriction of A to $\mathrm{Im}(I-P)$ has its spectrum outside the unit circle, this operator is invertible.

We shall also assume that W takes invertible values on Γ. In terms of the realization (2.3) this means that $A^\times = A - BC$ has no spectrum on Γ.

THEOREM 2.1 *Consider the operator* $T : \ell_p(\mathbb{N},Y) \to \ell_p(\mathbb{N},Y)$ *given by* (2.2) *and assume that its symbol* $W : \Gamma \to L(Y)$ *has the realization* (2.3), *where* A *and* $A^\times$ *have no spectrum on the unit circle* Γ. *Let* P, $P^\times$ *be the Riesz projections given by*

$$P = \frac{1}{2\pi i} \int_\Gamma (\lambda - A)^{-1} d\lambda , \quad P^\times = \frac{1}{2\pi i} \int_\Gamma (\lambda - A^\times)^{-1} d\lambda .$$

Then the operator

$$S = P^\times|_{\mathrm{Im}\, P} : \mathrm{Im}\, P \to \mathrm{Im}\, P^\times$$

is an indicator for T. *More precisely, the following coupling relation holds:*

$$\begin{bmatrix} T & U \\ R & Q \end{bmatrix}^{-1} = \begin{bmatrix} T^{\times} & U^{\times} \\ R^{\times} & S \end{bmatrix}.$$

Here

$$\begin{aligned}
U &: \operatorname{Im} P^{\times} \to \ell_p(\mathbb{N},Y), & (Ux)_m &= -CA^{-m-1}(I-P)x, \\
U^{\times} &: \operatorname{Im} P \to \ell_p(\mathbb{N},Y), & (U^{\times}x)_m &= -C(A^{\times})^{-m-1}(I-P^{\times})x, \\
R &: \ell_p(\mathbb{N},Y) \to \operatorname{Im} P, & R\xi &= \sum_{k=0}^{\infty} PA^{k}B\xi_k, \\
R^{\times} &: \ell_p(\mathbb{N},Y) \to \operatorname{Im} P^{\times}, & R^{\times}\xi &= -\sum_{k=0}^{\infty} P^{\times}(A^{\times})^{k}B\xi_k, \\
Q &: \operatorname{Im} P^{\times} \to \operatorname{Im} P, & Qx &= Px, \\
T^{\times} &: \ell_p(\mathbb{N},Y) \to \ell_p(\mathbb{N},Y), & (T^{\times}\xi)_j &= \sum_{k=0}^{\infty} T^{\times}_{j-k}\xi_k,
\end{aligned}$$

where the (two-sided) sequence $\cdots T^{\times}_{-1}, T^{\times}_{0}, T^{\times}_{1}\cdots$ *of bounded linear operators on* Y *is given by*

$$(2.5) \qquad T^{\times}_n = \begin{cases} -C(A^{\times})^{-n-1}P^{\times}B, & n = -1,-2,\cdots, \\ I_Y + C(A^{\times})^{-1}(I-P^{\times})B, & n = 0, \\ C(A^{\times})^{-n-1}(I-P^{\times})B, & n = 1,2,\cdots. \end{cases}$$

The operators appearing in the theorem are well-defined, linear and bounded. In this context, recall that the restriction of A to $\operatorname{Im}(I-P) = \operatorname{Ker} P$ is invertible. Also, $A^{\times}|\operatorname{Ker} P^{\times}$ is invertible.

Note that the operator T may be viewed as the discrete version of the Toeplitz operator T_W on the unit circle. In fact, all operators appearing in Theorem 2.1 are discretizations of the corresponding operators in Theorem III.2.2. Hence Theorem 2.1 is just an analogue of Theorem III.3.2. "Power representations" of the form (2.4) and (2.5) play an important role in [4]. The details of the proof of Theorem 2.1 are omitted.

We can now complete our discussion of the operator T along the lines suggested by Section III.2. Since the modifications to be made are straightforward there is no need to present all the details. Here we only mention the following. In the situation of Theorem 2.1 the kernel and image of T are given by

$$\operatorname{Ker} T = \{\xi : \mathbb{N} \to Y \mid \xi_m = C(A^\times)^{-m-1}x \ , \ x \in \operatorname{Im} P \cap \operatorname{Ker} P^\times\} \ ;$$

$$\operatorname{Im} \ T = \{\eta \in \ell_p(\mathbb{N},Y) \mid \sum_{k=0}^{\infty} P^\times(A^\times)^k B\eta_k \in \operatorname{Im} P + \operatorname{Ker} P^\times\} \ .$$

Moreover, if S^+ is a generalized inverse of the indicator S, a generalized inverse T^+ of T is defined by

$$(T^+\xi)_j = \sum_{k=0}^{\infty} T^+_{jk}\xi_k \ , \quad j = 0,1,2,\cdots \ ,$$

$$T^+_{jk} = \begin{cases} C(A^\times)^{-j-1}[I-\Pi^+](A^\times)^k B & , \quad k = 0,\cdots,j-1 \ , \\ I_Y + C(A^\times)^{-j-1}[I-\Pi^+](A^\times)^j B \ , & \quad k = j \ , \\ -C(A^\times)^{-j-1}\Pi^+(A^\times)^k B & , \quad k = j+1,j+2,\cdots \ , \end{cases}$$

where $\Pi^+ = P^\times + (I-P^\times)S^+P^\times$ is a projection of X along $\operatorname{Ker} P^\times$. The expression for T^+_{jk} with $k > j$ has to be handled with a little bit of care. It makes sense when $A^\times$ is invertible (cf. [4], Theorem II.1.2); otherwise it should be read as $T_{jk} = -C(A^\times)^{k-j-1}P^\times B - C(A^\times)^{-j-1}(I-P^\times)S^+P^\times(A^\times)^k B$. The operator T is invertible if and only if $X = \operatorname{Im} P \oplus \operatorname{Ker} P^\times$ and this in turn corresponds to the case when Π^+ is the projection of X along $\operatorname{Ker} P^\times$ onto $\operatorname{Im} P$.

The basis of the analysis presented above is the realization (2.3) of the symbol W. We could instead have used a realization of the function $W(\lambda^{-1})$ which has also an analytic continuation to an annulus around the unit circle. In fact such a realization was employed in [4], Chapter II. This explains the differences in the formulas derived in [4] and those obtained here.

REFERENCES

1. Barnett, S.: Introduction to mathematical control theory. Oxford, Clarendon Press, 1975.

2. Barras, J.S. and Brockett, R.W.: H^2-functions and infinite-dimensional realization theory. SIAM J. Control Optimization 13 (1975), 221-241.

3. Bart, H., Gohberg, I. and Kaashoek, M.A.: Minimal factorization of matrix and operator functions. Operator Theory: Advances and Applications, Vol. 1, Basel-Boston-Stuttgart, Birkhäuser Verlag, 1979.

4. Bart, H., Gohberg, I. and Kaashoek, M.A.: Wiener-Hopf integral equations, Toeplitz matrices and linear systems. In: Toeplitz Centennial (ed. I. Gohberg), Operator Theory: Advances and Applications, Vol. 4, Basel-Boston-Stuttgart, Birkhäuser Verlag, 1982, 85-135.

5. Bart, H., Gohberg, I. and Kaashoek, M.A.: Convolution equations and linear systems. Integral Equations and Operator Theory, 5/3 (1982), 283-340.

6. Bart, H., Gohberg, I. and Kaashoek, M.A.: Wiener-Hopf factorization of analytic operator functions and realization. Wiskundig Seminarium der Vrije Universiteit, Amsterdam, Rapport nr. 231, 1983.

7. Bart, H., Gohberg, I. and Kaashoek, M.A.: Wiener-Hopf factorization and realization. To appear in Proceedings MTNS, Beer-Sheva, 1983.

8. Bart, H. and Kroon, L.S.: An indicator for Wiener-Hopf integral equations with invertible analytic symbol. Integral Equations and Operator Theory, 6/1 (1983), 1-20.
See also the addendum to this paper to appear in Integral Equations and Operator Theory, Vol. 6/6 (1983).

9. Clancey, K. and Gohberg, I.: Factorization of matrix functions and singular integral operators. Operator Theory: Advances and Applications, Vol. 3, Basel-Boston-Stuttgart, Birkhäuser Verlag, 1981.

10. Fuhrmann, P.: Linear systems and operators in Hilbert space. New-York, McGraw-Hill, 1981.

11. Gohberg, I., Kaashoek, M.A. and Lay, D.C.: Equivalence, linearization and decomposition of holomorphic operator functions. J. Funct. Anal. 28 (1978), 102-144.

12. Gohberg, I. and Leiterer, J.: Factorization of operator functions relative to a contour. I. Finitely meromorphic operator functions. Math. Nachr. 52 (1972), 259-282 [Russian].

13. Gohberg, I. and Lerer, L.: Fredholm properties of Wiener-Hopf operators. Private communication (1980).

14. Kaashoek, M.A., Van der Mee, C.V.M. and Rodman, L.: Analytic operator functions with compact spectrum, I. Spectral nodes, linearization and equivalence. Integral Equations and Operator Theory 4 (1981), 504-547.

15. Kaashoek, M.A., Van der Mee, C.V.M. and Rodman, L.: Analytic operator functions with compact spectrum, III. Hilbert space case: inverse problem and applications. J. of Operator Theory (to appear).

16. Kalman, R.E., Fall, P.F. and Arbib, M.A.: Topics in mathematical systems theory. New-York, McGraw-Hill, 1969.

Operator Theory:
Advances and Applications, Vol. 12

THE EXPANSION THEOREM FOR HILBERT SPACES OF ANALYTIC FUNCTIONS

Louis de Branges

Some improvements are made in the spectral theory of transformations which are nearly selfadjoint. Consider a densely defined transformation T, with domain and range in a Hilbert space $\mathcal{H}$, such that the adjoint T* of T has the same domain as T and T - T* is (the restriction of) a completely continuous transformation. The existence of invariant subspaces is not known for (the resolvents of) T if no further hypothesis is made on T - T*. But if T - T* belongs to the class of completely continuous operators introduced by Macaev [1], then invariant subspaces exist which cleave the spectrum of the transformation. The hypotheses of the Brodskii expansion [2] are satisfied. A generalization of the Fourier transformation results when the expansion is formulated in terms of the nodal Hilbert spaces of analytic functions of a Livšič-Brodskii node [3].

Some coefficient space $\mathcal{C}$ is assumed given. A vector is always an element of this space. An operator is a bounded linear transformation of vectors into vectors. The absolute value symbol is used for the norm of a vector and for the operator norm of an operator. Complex numbers are regarded as multiplication operators. A bar is used to denote the adjoint of an operator. If b is a vector, b^- denotes the linear functional on vectors such that the inner product of any vector a with b is $b^- a$. If a and b are vectors, $a b^-$ is the operator defined by $(a b^-)c = a (b^- c)$ for every vector c. The notation I is used for a unitary operator such that $I^- = - I$.

Some basic examples of Hilbert spaces of analytic functions appear in the perturbation theory of selfadjoint transformations [4]. An axiomatic characterization of the relevant spaces will be stated.

Assume that $\mathcal{L}$ is a given Hilbert space, whose elements are vector-valued functions of z, defined and analytic separately in the upper and lower half-planes. Assume that $[F(z) - F(w)]/(z - w)$ belongs to the space whenever $F(z)$ belongs to the space, for every nonreal number w, and that the identity

$$0 = \langle [F(z) - F(\beta)]/(z - \beta), G(z) \rangle_{\mathcal{L}} - \langle F(z), [G(z) - G(\alpha)]/(z - \alpha) \rangle_{\mathcal{L}} - (\beta - \bar{\alpha}) \langle [F(z) - F(\beta)]/(z - \beta), [G(z) - G(\alpha)]/(z - \alpha) \rangle_{\mathcal{L}}$$

holds for all nonreal numbers α and β. Assume that $\lim z F(z)$ exists as z goes to infinity through imaginary values, for every element $F(z)$ of the space, and that a completely continuous transformation of the space into $\mathcal{C}$ is defined by taking $F(z)$ into $\lim z F(z)$.

Then a unique operator-valued analytic function $\varphi(z)$ exists, which is defined and analytic in the upper and lower half-planes, such that $\varphi(z)\, c$ belongs to the space for every vector c and such that the identity

$$-\pi i c^- \lim z F(z) = \langle F(z), \varphi(z)\, c \rangle_{\mathcal{L}}$$

holds for every element $F(z)$ of the space. The function

$$[\varphi(z) + \bar{\varphi}(w)]\, c/[\pi i(\bar{w} - z)]$$

belongs to the space for every vector c when w is not real. The identity

$$c^- f(w) = \langle f(z), [\varphi(z) + \bar{\varphi}(w)]\, c/[\pi i(\bar{w} - z)] \rangle_{\mathcal{L}}$$

holds for every element $f(z)$ of the space. The space $\mathcal{L} = \mathcal{L}(\varphi)$ is uniquely determined by a knowledge of $\varphi(z)$.

Assume given an operator-valued function $\varphi(z)$ which is defined and analytic separately in the upper and lower half-planes. A necessary and sufficient condition that a space $\mathcal{L}(\varphi)$ exists is that the identity

$$\varphi^*(z) + \varphi(z) = 0$$

holds for nonreal values of z, the operator inequality

$$[\varphi(z) + \bar{\varphi}(z)]/[i(\bar{z} - z)] \geq 0$$

holds for nonreal values of z, and

$$m = \lim 2 y\, \varphi(iy)$$

exists in the limit of large $|y|$ and is a completely continuous operator. The notation $\varphi^*(z)$ is used for $\bar{\varphi}(\bar{z})$.

If $\mathcal{L}(\varphi)$ is a given space, then a selfadjoint transformation H exists whose resolvent $(H - w)^{-1}$ takes $F(z)$ into $[F(z) - F(w)]/(z - w)$ for nonreal values of w. The integral representation of a selfadjoint transformation is equivalent to the Poisson representation

$$\pi\, i\, \varphi(z) = \int_{-\infty}^{+\infty} (t - z)^{-1}\, d\mu(t)$$

which holds for a bounded, nondecreasing, operator-valued function $\mu(t)$ of real t, with completely continuous increments, such that

$$\tfrac{1}{2}\, \pi\, m = \mu(+\infty) - \mu(-\infty).$$

An interval (a,b) of the real axis fails to contain a point of the spectrum of H if, and only if, $\varphi(z)$ has an analytic extension from the lower half-plane to the upper half-plane through the interval (a,b). An equivalent condition is that $\mu(t)$ is constant in (a,b). Then every element of $\mathcal{L}(\varphi)$ has an analytic continuation from the lower half-plane to the upper half-plane through the interval (a,b).

Thus the spectral properties of the selfadjoint transformation H are expressed in analyticity properties of the function $\varphi(z)$. This useful relation is perpetuated in the Livšič-Brodskii model of nonselfadjoint transformations.

Assume that $\varphi(z)$ is an operator-valued analytic function of z such that a corresponding space $\mathcal{L}(\varphi)$ exists. Then an operator-valued analytic function M(z) of z is defined by

$$M(z) = [1 - \varphi(z)\, i\, I]/[1 + \varphi(z)\, i\, I]$$

on the set of elements z of the domain of φ such that $1 + \varphi(z)\, i\, I$ is invertible. This set contains all but isolated points of the domain of φ and it contains all points z of the complex plane such that $|m| < |z - \bar{z}|$.

A Hilbert space, denoted $\mathcal{V}(M)$, exists, whose elements are vector-valued functions, defined and analytic in the domain of M(z), such that multiplication by $\frac{1}{2}[1 + M(z)]$ is an isometry of $\mathcal{L}(\varphi)$ onto $\mathcal{V}(M)$. A densely defined transformation T exists in $\mathcal{V}(M)$ such that $(T - w)^{-1}$ is an everywhere defined and bounded transformation which takes F(z) into $[F(z) - F(w)]/(z - w)$ whenever M(w) is defined. The adjoint T* of T has the same domain as T and $T - T^*$ has a completely continuous extension. The transformation $(T^* - w)^{-1}$ takes $F(z)$ into

$$[F(z) + M(z)\ I\ M^*(w)\ I\ F(w)]/(z - w)$$

whenever $M^*(w)$ is defined. The function

$$[M(z)\ I\ \bar{M}(w) - I]\ c/[2\pi(z - \bar{w})]$$

belongs to $\mathcal{V}(M)$ for every vector c when $M(w)$ is defined. The identity

$$c^- F(w) = \langle F(z), [M(z)\ I\ \bar{M}(w) - I]\ c/[2\pi(z - \bar{w})]\rangle_{\mathcal{V}(M)}$$

holds for every element $F(z)$ of $\mathcal{V}(M)$. The function $M(z)\ c - c$ belongs to $\mathcal{V}(M)$ for every vector c. The identity

$$2\pi\ c^-\ I \lim_{w = \infty} w\ F(w) = \langle F(z), M(z)\ c - c\rangle_{\mathcal{V}(M)}$$

holds for every vector c.

Assume that $\mathcal{V}$ is a given Hilbert space, whose elements are vector-valued functions, which are defined and analytic, with isolated exceptions, at nonreal points of the complex plane, such that a completely continuous transformation of $\mathcal{V}$ into $\mathcal{C}$ is defined by taking $F(z)$ into $\lim w\ F(w)$. Assume that $[F(z) - F(w)]/(z - w)$ belongs to the space whenever $F(z)$ belongs to the space for all except isolated nonreal numbers w. Then $\mathcal{V}$ is isometrically equal to a space $\mathcal{V}(M)$ if, and only if, the identity

$$\begin{aligned} -2\pi\ \bar{G}(\alpha)\ I\ F(\beta) &= \langle [F(z) - F(\beta)]/(z - \beta), G(z)\rangle_{\mathcal{V}} \\ &\quad - \langle F(z), [G(z) - G(\alpha)]/(z - \alpha)\rangle_{\mathcal{V}} \\ &\quad - (\beta - \bar{\alpha})\ \langle [F(z) - F(\beta)]/(z - \beta), [G(z) - G(\alpha)]/(z - \alpha)\rangle_{\mathcal{V}} \end{aligned}$$

holds for all nonreal numbers α and β for which the identity is meaningful. The function $M(z)$ is then uniquely determined by the space $\mathcal{V}$.

The relation of the space $\mathcal{V}(M)$ to the space $\mathcal{L}(\varphi)$ is a calculation in perturbation theory. The selfadjoint transformation H in $\mathcal{L}(\varphi)$ is unitarily equivalent to the selfadjoint part $\frac{1}{2}\ (T + T^*)$ of the transformation T in $\mathcal{V}(M)$. The unitary equivalence is given by multiplication by $\frac{1}{2}\ [1 + M(z)]$ as a transformation of $\mathcal{L}(\varphi)$ into $\mathcal{V}(M)$.

The invariant subspaces of the underlying transformation in a space $\mathcal{V}(M)$ are related to the factorizations of $M(z)$. If

$\mathcal{H}(M(a))$ and $\mathcal{H}(M(a,b))$ are given spaces, then a space $\mathcal{H}(M(b))$ exists such that

$$M(b,z) = M(a,z)\ M(a,b,z)\ .$$

The elements of $\mathcal{H}(M(b))$ are the functions $H(z)$ of the form

$$H(z) = F(z) + M(a,z)\ G(z)$$

with $F(z)$ in $\mathcal{H}(M(a))$ and $G(z)$ in $\mathcal{H}(M(a,b))$. The inequality

$$\|F(z)\|^2_{\mathcal{H}(M(b))} \le \|F(z)\|^2_{\mathcal{H}(M(a))} + \|G(z)\|^2_{\mathcal{H}(M(a,b))}$$

is always satisfied. Every element $H(z)$ of $\mathcal{H}(M(b))$ admits a decomposition for which equality holds.

The minimal decomposition is unique. The element $F(z)$ is obtained from $H(z)$ under the adjoint of the inclusion of $\mathcal{H}(M(a))$ in $\mathcal{H}(M(b))$. The element $G(z)$ is obtained from $H(z)$ under the adjoint of multiplication by $M(a,z)$ as a transformation of $\mathcal{H}(M(a,b))$ into $\mathcal{H}(M(b))$.

The overlapping space of the space $\mathcal{H}(M(a))$ with respect to the space $\mathcal{H}(M(a,b))$ is the set of elements $F(z)$ of $\mathcal{H}(M(a,b))$ such that $M(a,z)\ F(z)$ belongs to $\mathcal{H}(M(a))$. The overlapping space is a Hilbert space $\mathcal{L}$ in the overlapping norm,

$$\|F(z)\|^2_{\mathcal{L}} = \|F(z)\|^2_{\mathcal{H}(M(a,b))} + \|M(a,z)\ F(z)\|^2_{\mathcal{H}(M(a))}\ .$$

The space is isometrically equal to a space $\mathcal{L}(\varphi)$.

The space $\mathcal{H}(M(a))$ is contained isometrically in the space $\mathcal{H}(M(b))$ if, and only if, it contains no nonzero element of the form $M(a,z)\ F(z)$ with $F(z)$ in $\mathcal{H}(M(a,b))$.

Conditions which imply the triviality of overlapping spaces are relevant in applications. A vertical strip $s < \text{Re}\ z < t$ contains no point of the spectrum of the underlying selfadjoint transformation in a space $\mathcal{L}(\varphi)$ if, and only if, $\varphi(z)$ has an analytic extension to the strip. Then every element of $\mathcal{L}(\varphi)$ has an analytic extension to the strip.

Assume that $\mathcal{H}(M(a))$ and $\mathcal{H}(M(a,b))$ are given spaces such that that $M(a,z)$ has an analytic extension to the half-plane $\text{Re}\ z > t$ and $M(a,b,z)$ has an analytic extension to the half-plane $\text{Re}\ z < t$. Then $(z-t)\ F(z)$ is a constant whenever $F(z)$ belongs to $\mathcal{L}(\varphi)$. Although the inclusion of $\mathcal{H}(M(a))$ in $\mathcal{H}(M(b))$ need not be isometric, related factorizations can be found which determine isometric inclusions. But usually refactorization is unnecessary.

Factorizations are useful when the inclusions are not isometric.

The existence of factorizations which cleave the spectrum can be obtained from an estimate, due to Macaev, of the bound of a Volterra transformation from a knowledge of the eigenvalues of its imaginary part. A generalization of Macaev's estimate applies to transformations with imaginary spectrum [3]. A correction is now made of an error which was noted in the proof by Gohberg and Krein [1].

THEOREM 1. Assume that T is a densely defined transformation with domain and range in a Hilbert space $\mathcal{K}$ such that T^* has the same domain as T and such that $T - T^* \subset 2i\, b\, b^-$ for an element b of $\mathcal{K}$. If T has imaginary spectrum, then T is bounded and

$$T + T^* = 2 \Sigma \operatorname{sgn}(n)\, a_n\, a_n^-$$

for an orthogonal set of elements a_n of $\mathcal{K}$, indexed by the odd integers, such that $\|a_{n+2}\|_{\mathcal{K}} \leq \|a_n\|_{\mathcal{K}}$ when n is positive, such that $\|a_{n-2}\|_{\mathcal{K}} \leq \|a_n\|_{\mathcal{K}}$ when n is negative, and such that the inequality

$$\pi\, |n|\, \|a_n\|_{\mathcal{K}}^2 \leq \|T - T^*\|$$

holds for every index n.

A duality principle is used to obtain estimates of more general transformations with imaginary spectrum.

THEOREM 2. Assume that S and T are linear transformations of an r-dimensional Hilbert space $\mathcal{K}$ into itself which have their spectra restricted to the imaginary axis. Assume that orthogonal projections $0 = P_0 \leq P_1 \leq \cdots \leq P_r = 1$ exist into subspaces which are invariant under S and T, such that $P_n - P_{n-1}$ has one-dimensional range for $n = 1, \ldots, r$. Then

$$\operatorname{spur}\, [(S + S^*)(T - T^*)] + \operatorname{spur}\, [(T + T^*)(S - S^*)] = 0 .$$

The Macaev estimate remains valid under the weakened hypothesis of imaginary spectrum.

THEOREM 3. Assume that T is a densely defined transformation of a Hilbert space $\mathcal{K}$ into itself such that T^* has the same domain as T and

$$T - T^* \subset 2i \Sigma \operatorname{sgn}(n)\, c_n \bar{c}_n$$

for an orthogonal set of elements c_n of $\mathscr{H}$, indexed by the odd integers, such that $\|c_{n+2}\|_{\mathscr{H}} \le \|c_n\|_{\mathscr{H}}$ when n is positive, such that $\|c_{n-2}\|_{\mathscr{H}} \le \|c_n\|_{\mathscr{H}}$ when n is negative, and such that

$$\delta = \Sigma\, |n|^{-1}\, \|c_n\|^2_{\mathscr{H}}$$

is finite. If the spectrum of T is imaginary, then T is bounded and the spectrum of $\frac{1}{2}(T + T^*)$ is contained in the interval $[-2\delta/\pi,\ 2\delta/\pi]$. If the spectrum of T is contained in the half-plane Re $z \le t$, then the spectrum of $\frac{1}{2}(T + T^*)$ is contained in the half line $(-\infty,\ t + 2\delta/\pi]$. If the spectrum of T is contained in the half-plane Re $z \ge t$, then the spectrum of $\frac{1}{2}(T + T^*)$ is contained in the half-line $[t - 2\delta/\pi,\ \infty)$.

The Macaev existence theorem for invariant subspaces is now obtained in the form of a factorization theorem. A nonnegative operator m is said to be of Macaev class if

$$m = \Sigma\, a_n \bar{a}_n$$

for an orthogonal set of vectors a_n, indexed by the positive integers, such that $|a_{n+1}| \le |a_n|$ for every n and such that $\Sigma\, n^{-1}\, |a_n|^2$ is finite.

THEOREM 4. If $\mathscr{H}(M(b))$ is a given space such that m(b) is of Macaev class and if t is a given real number, then spaces $\mathscr{H}(M(a))$ and $\mathscr{H}(M(a,b))$ exist such that

$$M(b,z) = M(a,z)\ M(a,b,z)$$

with M(a,z) analytic in the half-plane Re $z > t$ and M(a,b,z) analytic in the half-plane Re $z < t$.

An admissible family of Hilbert spaces of analytic functions is a family of spaces $\mathscr{H}(M(t))$, parametrized by real numbers t, which has the following properties with respect to a measurable real-valued function $\lambda(t)$ of real t. For every finite interval (a,b), a space $\mathscr{H}(M(a,b))$ exists such that

$$M(b,z) = M(a,z)\ M(a,b,z)\ .$$

The limit of m(a,b) is zero as a increases to b for any given b or as b decreases to a for any given a. A completely continuous transformation with imaginary spectrum exists in $\mathscr{H}(M(a,b))$ which takes $F(a,b,z)$ into

$$z\, F(a,b,z) - \lim_{w=\infty} w\, F(a,b,w) - \int_a^b \lambda(t)\, dF(a,t,z)$$

where $F(a,t,z)$ is the element of $\mathcal{H}(M(a,t))$ obtained from $F(a,b,z)$ under the adjoint of the inclusion of $\mathcal{H}(M(a,t))$ in $\mathcal{H}(M(a,b))$. (The transformation is defined by continuity when the spectral integral does not converge.)

The Macaev factorization allows the construction of an admissible family of factors when the underlying transformation has real spectrum.

THEOREM 5. If $\mathcal{H}(M)$ is a given space such that m is of Macaev class and such that $M(z)$ has only real singularities, then an admissible family of spaces $\mathcal{H}(M(t))$ exists, with respect to a continuous nondecreasing function $\lambda(t)$ of real t, such that $M(t,z)$ has limit 1 at $t = -\infty$ and has limit $M(z)$ at $t = +\infty$.

Assume given an admissible family of spaces $\mathcal{H}(M(t))$ with respect to a measurable real-valued function $\lambda(t)$ of real t. A parameter function $m(t)$ is an operator-valued function of t such that the identity

$$m(a,b) = m(b) - m(a)$$

holds when $a < b$. It is normalized to have limit 0 at $t = -\infty$ if $M(t,z)$ has limit 1 there. A related Hilbert space $L^2(m)$ is used in the formulation of the expansion.

A continuous vector-valued function $f(t)$ of real t is said to determine an element of $L^2(m)$ if the Stieltjes sums

$$\sum_{n=1}^{r} \bar{f}(t_{n-1})\, [m(t_n) - m(t_{n-1})]\, f(t_{n-1})$$

taken over all finite sets $t_0 < t_1 < \cdots < t_r$ of real numbers have a finite limit as the weighted mesh

$$\max_{n=1,\ldots,r} |m(t_n) - m(t_{n-1})|$$

goes to zero. The limit is then written

$$\|f\|^2_{L^2(m)} = \int_{-\infty}^{+\infty} \bar{f}(t)\, dm(t)\, f(t) .$$

The set of continuous vector-valued functions which determine elements of $L^2(m)$ form a vector space over the complex numbers.

Elements f and g of the space are considered equivalent if g - f has zero norm in $L^2(m)$. The Hilbert space $L^2(m)$ is defined as the completion of the resulting inner product space.

A functional representation of arbitrary elements of $L^2(m)$ is not needed for the expansion theorem. But it is necessary to explain how certain vector-valued functions which are not continuous represent elements of the space. The required functions are products h(t) f(t) of a measurable complex-valued function h(t) and a continuous vector-valued function f(t) of real t which determines an element of $L^2(m)$.

If h(t) is a bounded continuous function of t, the continuous function h(t) f(t) determines an element of $L^2(m)$ such that

$$\int_{-\infty}^{+\infty} \bar{h}(t)\ \bar{f}(t)\ dm(t)\ f(t)\ h(t) \leq C^2 \int_{-\infty}^{+\infty} \bar{f}(t)\ dm(t)\ f(t)$$

where C is a bound of $|h(t)|$. Bounded approximation is used to define the element of $L^2(m)$ represented by h f when h is any bounded measurable function. The same inequality then holds.

If h(t) is an unbounded function of t, define $h_n(t) = h(t)$ when $|h(t)| \leq n$ and $h_n(t) = 0$ otherwise. The element of $L^2(m)$ represented by h f is taken to be the limit of the elements of $L^2(m)$ represented by h_n f when the limit exists in the metric of $L^2(m)$.

Examples of continuous functions which represent elements of $L^2(m)$ are obtained by an argument which is due to Brodskii [2]. A similar application of the construction has previously been given in the theory of Hilbert spaces of entire functions [5].

THEOREM 6. Assume given an admissible family of spaces $\mathfrak{H}(M(t))$ with respect to a measurable real-valued function $\lambda(t)$ of real t. Assume that an element F(t,z) of $\mathfrak{H}(M(t))$ is given for every index t such that the identity

$$F(b,z) = F(a,z) + M(a,z)\ F(a,b,z)$$

holds for every finite interval (a,b) for an element F(a,b,z) of $\mathfrak{H}(M(a,b))$ such that

$$\|F(b,z)\|^2_{\mathfrak{H}(M(b))} = \|F(a,z)\|^2_{\mathfrak{H}(M(a))} + \|F(a,b,z)\|^2_{\mathfrak{H}(M(a,b))} .$$

Then the expression

$$g(t) = \lim_{w = \infty} w\, F(t,w)$$

is a continuous function of real t which satisfies the identity

$$\|G(a,z) + [M(a,z) - 1]\, c\|^2_{\mathcal{H}(M(a))}$$
$$= 2\pi \int_{-\infty}^{a} [g(t) - c]^{-}\, \bar{I}\, dm(t)\, I\, [g(t) - c]$$

for every vector c, where

$$G(a,z) = z\, F(a,z) - M(a,z)\, g(a) - \int_{-\infty}^{a} \lambda(t)\, dF(t,z) .$$

The formulation of the expansion theorem for Volterra transformations as a generalization of the Fourier transformation is a theme of the theory of Hilbert spaces of entire functions [6] which has been pursued in a related context by Dym and Gohberg [7]. A similar generalization of the Fourier transformation applies to any admissible family of spaces $\mathcal{H}(M(t))$.

THEOREM 7. Assume given an admissible family of spaces $\mathcal{H}(M(t))$ with respect to a measurable real-valued function $\lambda(t)$ of real t such that $M(t,z)$ is 1 in the limit $t = -\infty$. For every element f of $L^2(m)$ a corresponding element $F(a,z)$ of $\mathcal{H}(M(a))$ is given by

$$2\pi\, F(a,z) = \int_{-\infty}^{a} \frac{M(t,z)}{z - \lambda(t)}\, dm(t)\, f(t)$$

when $|m(a)| < |z - \bar{z}|$. The inequality

$$2\pi\, \|F(a,z)\|^2_{\mathcal{H}(M(a))} \leq \int_{-\infty}^{a} \bar{f}(t)\, dm(t)\, f(t)$$

is satisfied. Every element of $\mathcal{H}(M(a))$ is of the form $F(a,z)$ for some such element f of $L^2(m)$ for which equality holds. If f is an element of $L^2(m)$ for which equality holds, then a continuous vector-valued function $g(x)$ of real x is defined by

$$g(x) = \int_{x}^{a} I\, dm(t)\, f(t)$$

when $x < a$ and $g(x) = 0$ otherwise. The function determines an element of $L^2(m)$. The corresponding element of $\mathcal{H}(M(a))$ is

$$G(a,z) = z\, F(a,z) - M(a,z) \lim_{w = \infty} w\, F(a,w) - \int_{-\infty}^{a} \lambda(t)\, dF(t,z)$$

and the identity

$$2\pi \ \|G(a,z)\|^2_{\mathcal{H}(M(a))} = \int_{-\infty}^{a} \bar{g}(t) \ dm(t) \ g(t)$$

is satisfied.

The integrals appearing in the statement of theorem are interpreted as inner products in the metric of $L^2(m)$. When w is a complex number such that $|m(a)| < |w - \bar{w}|$ for a real number a, $\bar{M}(t,w)\, c$ is a continuous vector-valued function of t in $(-\infty, a]$ for every vector c and it determines an element of $L^2(m)$. The notation $\chi_a(t)$ will be used for a real-valued function of real t which is 1 when $t < a$ and which is 0 otherwise. Since the complex-valued function

$$\chi_a(t) \ [\bar{w} - \lambda(t)]^{-1}$$

is a bounded and measurable function of t, the product

$$\chi_a(t) \ [\bar{w} - \lambda(t)]^{-1} \ \bar{M}(t,w) \ c$$

determines an element of $L^2(m)$. The interpretation of the generalized Fourier integral is the $L^2(m)$ inner product

$$2\pi \ \bar{c} \ F(a,w) = \langle \chi_a(\cdot)[\bar{w} - \lambda(\cdot)]^{-1}\bar{M}(\cdot,w) \ c, \ f\rangle_{L^2(m)} \ .$$

The functional notation $f(t)$ appearing in the statement of the theorem is a convenient abuse of language which should not be interpreted as a restriction on the element f of $L^2(m)$.

The definition of g is interpreted in a similar way as

$$\bar{c} \ g(x) = \langle [\chi_x(\cdot) - \chi_a(\cdot)] \ c, \ f\rangle_{L^2(m)}$$

for every vector c.

The identity

$$\int_{-\infty}^{a} \lambda(t) \ dF(t,z) = \int_{-\infty}^{a} \lambda(t) \ \frac{M(t,z)}{z - \lambda(t)} \ dm(t) \ f(t)$$

should also be noted in connection with the statement of the theorem. The left side belongs to $\mathcal{H}(M(a))$ whenever the product $\chi_a(t) \ \lambda(t) \ f(t)$ is a function of t which determines an element of $L^2(m)$ which is orthogonal to the kernel of the Fourier transformation. The identity then holds with a similar interpretation as an $L^2(m)$ inner product when $|m(a)| < |z - \bar{z}|$.

A uniqueness theorem is a consequence of properties of the generalized Fourier transformation.

THEOREM 8. Assume given admissible families of spaces $\mathcal{H}(M_+(t))$ and $\mathcal{H}(M_-(t))$ with respect to a measurable real-valued function $\lambda(t)$ of real t, such that $M_+(t,z)$ and $M_-(t,z)$ are 1 in the limit $t = -\infty$. If the families have the same parameter function $m(t)$, then the identity

$$M_+(t,z) = M_-(t,z)$$

holds for all real numbers t.

The expansion theorem is the spectral analysis of a, generally speaking, unbounded and partially defined transformation in $L^2(m)$ which takes f into g whenever they are elements of $L^2(m)$ formally related by

$$g(x) = \lambda(x)\, f(x) - \int_{-\infty}^{x} I\, dm(t)\, f(t)$$

for all real x. The integral on the right is interpreted as an inner product in $L^2(m)$ whereas the formal multiplication by λ denotes the unique selfadjoint extension of a transformation in $L^2(m)$ which is determined by multiplication by λ on continuous vector-valued functions which represent elements of $L^2(m)$.

The spectral theory of such transformations has been previously studied by Lubin [7], Kriete [8], and Ball [9] using methods which are related to the present work. An interesting feature of such a transformation, from the point of view of the model theory, is that it may have a nontrivial selfadjoint part, a phenomenon which is reflected in a nonzero kernel in the generalized Fourier transformation. No explicit computation of the elements of the kernel is known. Yet the spectral theory of the selfadjoint part of the transformation requires no new conceptions.

If an admissible family of spaces $\mathcal{H}(M(t))$ is given and if (a,b) and (b,c) are adjacent finite intervals, define $\mathcal{L}(\varphi(a,b,c))$ to be the overlapping space of $\mathcal{H}(M(a,b))$ with respect to $\mathcal{H}(M(b,c))$.

An expansion theorem for overlapping spaces is a corollary of the previous expansion theorem.

THEOREM 9. Assume given an admissible family of spaces $\mathcal{H}(M(t))$ with respect to a measurable real-valued function $\lambda(t)$ of real t. Let (a,b) and (b,c) be adjacent finite intervals. If f is an element of $L^2(M)$ such that

$$\int_a^c \frac{M(a,t,z)}{z-\lambda(t)} \, dm(t) \, f(t)$$

vanishes identically, then a corresponding element F(z) of $\mathcal{L}(\varphi(a,b,c))$ is given by

$$2\pi \, F(z) = \int_b^c \frac{M(b,t,z)}{z-\lambda(t)} \, dm(t) \, f(t)$$

and the inequality

$$2\pi \, \|F(z)\|^2_{\mathcal{L}(\varphi(a,b,c))} \leq \int_a^c \bar{f}(t) \, dm(t) \, f(t)$$

is satisfied. Every element of $\mathcal{L}(\varphi(a,b,c))$ is of the form F(z) for some such element f of $L^2(m)$ for which equality holds. If f is an element of $L^2(m)$ for which equality holds and if w is a nonreal number, then an element g of $L^2(m)$ exists such that

$$f(x) = [\lambda(x) - w] \, g(x) + \int_x^c I \, dm(t) \, g(t)$$

for $a < x < c$, such that

$$\int_a^c \frac{M(a,t,z)}{z-\lambda(t)} \, dm(t) \, g(t)$$

vanishes identically, such that

$$2\pi \, \frac{F(z)-F(w)}{z-w} = \int_b^c \frac{M(b,t,z)}{z-\lambda(t)} \, dm(t) \, g(t)$$

and such that

$$2\pi \, \left\| \frac{F(z)-F(w)}{z-w} \right\|^2_{\mathcal{L}(\varphi(a,b,c))} = \int_a^c \bar{g}(t) \, dm(t) \, g(t) \, .$$

An underlying concept of invariant subspace theory is complementation [11]. If $\mathcal{H}_+$ is a Hilbert space which is contained contractively in a Hilbert space $\mathcal{H}$, then a unique Hilbert space $\mathcal{H}_-$ exists, which is contained contractively in $\mathcal{H}$, such that the inequality

$$\|h\|^2_{\mathcal{H}} \leq \|f\|^2_{\mathcal{H}_+} + \|g\|^2_{\mathcal{H}_-}$$

holds when $h = f + g$ with f in $\mathcal{H}_+$ and g in $\mathcal{H}_-$, and such that every element h of $\mathcal{H}$ admits a decomposition for which equality holds. The space $\mathcal{H}_-$ is called the complementary space to $\mathcal{H}_+$ in $\mathcal{H}$. The space $\mathcal{H}_+$ is recovered as the complementary space to $\mathcal{H}_-$ in $\mathcal{H}$.

A space $\mathcal{L}(\varphi_+)$ is contained contractively in a space $\mathcal{L}(\varphi)$ if, and only if, a space $\mathcal{L}(\varphi_-)$ exists such that

$$\varphi(z) = \varphi_+(z) + \varphi_-(z) \ .$$

The space $\mathcal{L}(\varphi_-)$ is then isometrically equal to the complementary space to $\mathcal{L}(\varphi_+)$ in $\mathcal{L}(\varphi)$.

A theorem on the propagation of overlapping spaces is an application of complementation theory [12].

THEOREM 10. Assume given an admissible family of spaces $\mathcal{H}(M(t))$ with respect to a measurable real-valued function $\lambda(t)$ of real t. Let (a,b), (b,c), and (c,d) be adjacent finite intervals. Then the adjoint of multiplication by M(b,c,z) as a transformation of $\mathcal{H}(M(c,d))$ into $\mathcal{H}(M(b,d))$ is a contractive transformation of $\mathcal{H}(M(b,d))$ into $\mathcal{H}(M(c,d))$ which commutes with the transformation which takes F(z) into z F(z) - lim w F(w). The transformation induces a partial isometry of $\mathcal{L}(\varphi(a,c,d))$ onto the complementary space to $\mathcal{L}(\varphi(b,c,d))$ in $\mathcal{L}(\varphi(a,c,d))$.

The theorem allows a global picture of the selfajoint part of the transformation which takes f(x) into

$$\lambda(x)\ f(x) - \int_{-\infty}^{x} I\ dm(t)\ f(t)$$

in $L^2(m)$ to be composed from the local expansions given by Theorem 9. A complete analysis of the transformation is obtained in the limit of small subdivisions.

It would be interesting to construct examples in which the kernel of the partial isometry is nontrivial.

The results of the paper are the outcome of a visit to the Weizmann Institute of Science during the summer of 1982. The author thanks Professors Harry Dym and Israel Gohberg for their support of the work. He is also indebted to Professor Shmuel Kantorovitz for arranging the visit to Israel during which the project was initiated.

PROOF OF THEOREM 1. The proof makes use of the Livšič-Brodskii model theory for a transformation T, which is densely defined in a Hilbert space, such that T* has the same domain as T and T - T* has a completely continuous extension. Only nodes whose state space is a Hilbert space are considered. The input and output spaces are always chosen to be the given coefficient space $\mathcal{C}$.

A node is a system (A,B,C,D) of linear transformations with these properties: The transformation A has domain and range in the state space $\mathcal{H}$. The transformation B maps the coefficient space $\mathcal{C}$ into the state space $\mathcal{H}$. The transformation C maps the state space $\mathcal{H}$ into the coefficient space $\mathcal{C}$. The transformation D maps the coefficient space $\mathcal{C}$ into itself. The transfer function M(z) of the node is

$$M(z) = D + C\,(z - A)^{-1}\,B\ .$$

The nodes which are now used have additional properties: The transformation A is densely defined, its adjoint A* has the same domain as A, and $A - A^* \subset B\,C$. The transformations B and C are related by the identity $B = 2\pi\,C^*\,I$. The transformation D is the identity operator. A space $\mathcal{H}(M)$ exists corresponding to the transfer function M(z). The transformation which takes f into $C\,(z - A)^{-1}\,f$ is a partial isometry of the state space $\mathcal{H}$ onto the space $\mathcal{H}(M)$. The kernel of the partial isometry is a reducing subspace for A to which the restriction of A is selfadjoint.

The Livšič-Brodskii model theory is now applied with $\mathcal{C}$ equal to the complex numbers, considered as a Hilbert space with absolute value as norm. The transformation A is taken equal to T. The transformation C is equal to $\pi^{-\frac{1}{2}}\,\bar{b}$. The operator I is i. The transfer function of the node is

$$M(z) = 1 + 2i\ b\,(z - T)^{-1}\,\bar{b}\ .$$

The identity

$$M(z) = [1 + \varphi(z)]/[1 - \varphi(z)]$$

holds for a complex-valued function φ(z), which is analytic in the upper and lower half-planes, satisfies the identity $\varphi^*(z) = -\,\varphi(z)$, and has positive real part in the upper half-plane. These conditions imply that $M(z)$ is analytic in the lower half-plane, satisfies the identity

$$M^*(z) = 1/M(z),$$

and is bounded by one in the lower half-plane. Since the spectrum of T is imaginary, the singularities of $M(z)$ lie on the imaginary axis. The identity

$$M(z) = \exp(2ia/z) \prod \frac{z + i\, t_n}{z - i\, t_n}$$

holds for a positive number a and a summable sequence of positive numbers t_n. Since

$$\exp(2ia/z) = \lim \frac{(z + i\, a/n)^n}{(z - i\, a/n)^n},$$

it is sufficient to consider the case $a = 0$.

For real numbers x, $M(x) = \exp(i\theta(x))$ where

$$\theta(x) = 2 \sum \arctan(t_n/x)$$

and

$$\sum t_n = |b|^2 .$$

Since the inequality

$$\arctan(x) \leq x$$

holds when x is positive, the inequality

$$x\, \theta(x) \leq 2\, |b|^2$$

holds for all real numbers x.

The nonzero eigenvalues of $\frac{1}{2}(T + T^*)$ are simple and are the nonzero real numbers x such that $\theta(x)$ is congruent to π modulo 2π. If x_n denotes the solution of the equation $\theta(x_n) = n\,\pi$, when n is an odd integer, then the theorem follows from the inequality

$$\pi\, n\, x_n \leq 2\, |b|^2 .$$

PROOF OF THEOREM 2. An orthonormal basis for $\mathcal{K}$ is obtained on choosing an element c_n of norm one in the range of P_n which is orthogonal to the range of P_{n-1} for every $n = 1,\ldots,r$. Then $S = \sum S_{ij}\, e_i\, \bar{e}_j$ and $T = \sum T_{ij}\, e_i\, \bar{e}_j$ where $S_{ij} = \bar{e}_i\, S\, e_j$ and $T_{ij} = \bar{e}_i\, T\, e_j$ vanish when $i > j$ and are imaginary when $i = j$. Since the composed transformation has the representation $S\, T = \sum A_{ij}\, e_i\, \bar{e}_j$ where $A_{ij} = 0$ when $i > j$ and $A_{ij} = S_{ij}\, T_{ij}$

is real when $i = j$, spur(ST) is real. The desired identity follows.

PROOF OF THEOREM 3. An argument which is due to Macaev [1] for Volterra transformations will be adapted to transformations with imaginary spectrum. The use of transformations with imaginary spectrum simplifies the argument by allowing an immediate reduction of the problem to a finite dimensional situation.

Consider first the case in which the space $\mathfrak{N}$ has finite dimension r. Then orthogonal projections

$$0 = P_0 \leq P_1 \leq \cdots \leq P_r = 1$$

exist into invariant subspaces for T such that $P_n - P_{n-1}$ has one-dimensional range for $n = 1,\ldots,r$. Choose an element e_n of norm one in the range of P_n which is orthogonal to the range of P_{n-1}. Then $T = \Sigma\, T_{ij}\, e_i\, \overline{e}_j$ where $T_{ij} = \overline{e}_i\, T\, e_j$ vanishes when $i > j$. The spectrum of T is the set of numbers T_{nn} for $n = 1,\ldots,r$. Since real numbers can be added to the numbers T_{nn}, the estimation problem reduces to the case in which the numbers T_{nn} are imaginary.

If c is a given element of $\mathfrak{N}$, then a unique transformation S exists, which has the range of P_n as an invariant subspace for every $n = 1,\ldots,r$, such that $S - S^* = 2i\, c\, \overline{c}$. In the formulation of Gohberg and Krein [1], the transformation S is given as a sum

$$S = \sum_{n=1}^{r} P_n\, i\, c\, \overline{c}\, (P_n - P_{n-1}) \, .$$

Another computation of S is to write $S = \Sigma\, S_{ij}\, e_i\, \overline{e}_j$ where $S_{ij} = \overline{e}_i\, S\, e_j$ vanishes for $i > j$ and is imaginary for $i = j$. The entries S_{ij} are then the unique solutions of the equations

$$S_{ij} - \overline{S}_{ji} = 2i\, (\overline{e}_i\, c)(\overline{c}\, e_j) \, .$$

By Theorem 2, the identity

$$2\, \overline{c}\, (T + T^*)\, c = i \text{ spur } [(S + S^*)(T - T^*)]$$

is satisfied. It follows that

$$2\, |\overline{c}\, (T + T^*)\, c| \leq 4\, \Sigma\, \|a_n\|_{\mathfrak{N}}^2\, \|c_n\|_{\mathfrak{N}}^2$$

where

$$S + S^* = 2 \sum \operatorname{sgn}(n)\ a_n\ \bar{a}_n$$

and

$$\pi\ |n|\ \|a_n\|_{\mathscr{H}}^2 \le 2\ \|c\|_{\mathscr{H}}^2$$

by Theorem 1. It follows that

$$\|\tfrac{1}{2}\ (T + T^*)\|_{\mathscr{H}} \le 2\delta/\pi\ .$$

This completes the proof of the theorem when the space $\mathscr{H}$ has finite dimension.

In the general case, use is made of the Livšič-Brodskii model theory, described in the proof of Theorem 1, to approximate by finite-dimensional spaces. The estimation problem reduces to the case in which the space $\mathscr{H}$ is a space $\mathscr{H}(M)$ and T is the underlying transformation in the space. Then

$$M(z) = [1 - \varphi(z)\ i\ I]/[1 + \varphi(z)\ i\ I]$$

for a space $\mathscr{L}(\varphi)$ such that multiplication by $\frac{1}{2}\ [1 + M(z)]$ is an isometry of $\mathscr{L}(\varphi)$ onto $\mathscr{H}(M)$. By the Poisson representation of $\varphi(z)$, a sequence of finite-dimensional spaces $\mathscr{L}(\varphi_n)$ can be chosen so that $m_n \le m$ for every n and so that

$$\varphi(z) = \lim \varphi_n(z)\ .$$

Because of the complete continuity of the operator m, convergence is obtained in the operator norm. If

$$M_n(z) = [1 - \varphi_n(z)\ i\ I]/[1 + \varphi_n(z)\ i\ I]\ ,$$

then a space $\mathscr{H}(M_n)$ exists for every index n and

$$M(z) = \lim M_n(z)$$

in the operator norm when $|m| \le |z - \bar{z}|$.

If a real number t exists such that each function $M_n(z)$ is analytic in the half-plane Re $z > t$, then each function $\varphi_n(z)$ is analytic in the half-plane Re $z > t + 2\delta/\pi$. It then follows from the Poisson representation that the function $\varphi(z)$ is analytic in the half-plane Re $z > t + 2\delta/\pi$.

If a real number t exists such that each function $M_n(z)$ is analytic in the half-plane Re $z < t$, then each function $\varphi_n(z)$ is analytic in the half-plane Re $z < t - 2\delta/\pi$. It then follows from the Poisson representation that the function $\varphi(z)$ is analytic in the half-plane Re $z < t - 2\delta/\pi$.

The theorem now follows provided that suitable approximations can be made. If $M(z)$ is analytic in the half-plane $\operatorname{Re} z > t$, then it must be shown that $M_n(z)$ can be chosen analytic in the same half-plane for every index n. If $M(z)$ is analytic in the half-plane $\operatorname{Re} z < t$, then it must be shown that $M_n(z)$ can be chosen analytic in the half-plane for every n.

The approximation will be made explicitly only in the case that $M(z)$ is analytic in a half-plane $\operatorname{Re} z > t$. An analogous argument can be given for a half-plane $\operatorname{Re} z < t$.

Write $M(z) = M(b,z)$ and choose $M_n(b,z)$ as above so that

$$M(b,z) = \lim M_n(b,z) .$$

Let ϵ be a given positive number. Since each space $\mathfrak{H}(M_n(b))$ has finite dimension, a space $\mathfrak{H}(M_n(a))$ exists, which is contained isometrically in $\mathfrak{H}(M_n(b))$, such that $M_n(a,z)$ is analytic in the half-plane $\operatorname{Re} z > t+\epsilon$ and such that

$$M_n(b,z) = M_n(a,z)\, M_n(a,b,z)$$

for a space $\mathfrak{H}(M_n(a,b))$ such that $M_n(a,b,z)$ is analytic in the half-plane $\operatorname{Re} z < t+\epsilon$. Since $m_n(a,b) \le m$ for every n, the approximations can be made (by passing to a subsequence) in such a way that the limit

$$M(a,b,z) = \lim M_n(a,b,z)$$

exists in the operator norm when $|m| < |z-\bar{z}|$. Then a space $\mathfrak{H}(M(a))$ exists such that

$$M(b,z) = M(a,z)\, M(a,b,z) .$$

A consequence of the estimates obtained in finite dimensional spaces is that $M(a,b,z)$ is analytic in the half-plane $\operatorname{Re} z < t+\epsilon$. Since $M(b,z)$ is analytic in the half-plane $\operatorname{Re} z > t$, $M(a,b,z)$ is analytic in the same half-plane.

Since $M(a,b,z)$ is analytic in the complex plane, it is identically one. Since $M_n(a,b,z)$ has limit one in the operator norm when $|m| < |z-\bar{z}|$,

$$M(b,z) = \lim M_n(a,z)$$

in the operator norm for these values of z. The desired approximation is now possible by the arbitrariness of ϵ.

PROOF OF THEOREM 4. The theorem is proved by an approximation procedure which has already been described in the proof of Theorem 3. It will now be shown that half-planes of analyticity are preserved in approximation. This supplies a verification which was omitted in the proof of Theorem 3.

It has been shown that a sequence of finite-dimensional spaces $\mathcal{H}(M_n(b))$ can be constructed so that

$$M(b,z) = \lim M_n(b,z)$$

exists when $|m| < |z - \bar{z}|$, where $M(b,z) = M(z)$, and so that $m_n \leq m$ for every index n. Spaces $\mathcal{H}(M_n(a))$ and $\mathcal{H}(M_n(a,b))$ exist for every index n such that

$$M_n(b,z) = M_n(a,z)\ M_n(a,b,z)$$

with $M_n(a,z)$ analytic in the half-plane Re $z > t$ and $M_n(a,b,z)$ analytic in the half-plane Re $z < t$. Furthermore the spaces can be chosen so that

$$M(a,b,z) = \lim M_n(a,b,z)$$

exists for $|m| < |z - \bar{z}|$, where $\mathcal{H}(M(a,b))$ is a space such that

$$M(b,z) = M(a,z)\ M(a,b,z)$$

for a space $\mathcal{H}(M(a))$. The identity

$$M(a,z) = \lim M_n(a,z)$$

holds when $|m| < |z - \bar{z}|$. By Theorem 3, $\varphi(a,z)$ is analytic in a half-plane of the form Re $z > t + 2\delta/\pi$ and $\varphi(a,b,z)$ is analytic in a half-plane of the form Re $z < t - 2\delta/\pi$. It follows that $M(a,z)$ is analytic in the half-plane Re $z > t + 2\delta/\pi$ and $M(a,b,z)$ is analytic in the half-plane Re $z < t - 2\delta/\pi$.

But if $\mathcal{H}(M)$ is a given space and if ϵ is a given positive number, then spaces $\mathcal{H}(M_1),\ldots,\mathcal{H}(M_r)$ can be chosen for a positive integer r so that

$$M(z) = M_1(z) \cdots M_r(z)$$

and so that the corresponding numbers δ_n, defined as in Theorem 3, satisfy the inequality $\delta_n < \epsilon$ for every $n = 1,\ldots,r$. From this it follows that $M(a,z)$ is analytic in the half-plane Re $z > t + 2\epsilon/\pi$ and that $M(a,b,z)$ is analytic in the half-plane Re $z < t - 2\epsilon/\pi$.

By the arbitrariness of ϵ, $M(a,z)$ is analytic in the half-plane Re $z > t$ and $M(a,b,z)$ is analytic in the half-plane

Re $z < t$.

PROOF OF THEOREM 5. For the construction of an admissible family, define a partial ordering of spaces $\mathcal{H}(M)$ by taking $\mathcal{H}(M(a))$ less than or equal to $\mathcal{H}(M(b))$ if

$$M(b,z) = M(a,z)\ M(a,b,z)$$

for a space $\mathcal{H}(M(a,b))$ such that $M(a,b,z)$ is analytic in a half-plane Re $z < t$ for a real number t such that $M(a,z)$ is analytic in the half-plane Re $z > t$.

By Zorn's lemma, a maximal totally ordered family of spaces $\mathcal{H}(M(t))$ exists which contains the given space $\mathcal{H}(M)$. By the definition of the partial ordering, a nondecreasing function $\lambda(t)$ of real t exists such that $M(a,z)$ is analytic in the half-plane Re $z > \lambda(a)$ and such that $M(a,b,z)$ is analytic in the half-plane Re $z < \lambda(a)$ whenever $a < b$.

The parametrization of the family can be made in such a way that $\lambda(t)$ is a continuous function of t since it will be permitted for different parameters to apply to the same space.

Since $M(z)$ has only real singularities by hypothesis, the functions $M(a,z)$ and $M(a,b,z)$ have only real singularities. By Theorem 4 and the factorization of functions whose singularities are concentrated at a real point, $m(a,b)$ has limit zero as a increases to b for any real number b or as b decreases to a for any real number a. By the maximal choice of the family, $M(t,z)$ is 1 in the limit $t = -\infty$ and is $M(z)$ in the limit $t = +\infty$.

The singularities of $M(a,b,z)$ are contained in the interval $[\lambda(a),\lambda(b)]$ for every finite interval (a,b). It will be shown that the family of spaces $\mathcal{H}(M(t))$ is admissible. Some preliminary remarks on spectral integration will be made.

Define $P(t)$ to be the adjoint of the inclusion of $\mathcal{H}(M(t))$ in $\mathcal{H}(M)$ for every real number t. Then $P(t)$ is a nonnegative and contractive transformation in $\mathcal{H}(M)$ which is weakly continuous and nondecreasing when considered as a function of t. The Stieltjes integral $\int \lambda(t)\ dP(t)$ defines a selfadjoint transformation in $\mathcal{H}(M)$. The action of the transformation on $F(z)$ to produce $G(z)$ is written

$$G(z) = \int \lambda(t) \; dF(t,z)$$

where $F(t,z)$ is the action of $P(t)$ on $F(z)$.

For every finite interval (a,b), the minimal decomposition of $F(b,z)$ as an element of $\mathcal{H}(M(b))$ is obtained with $F(a,z)$ as the element of $\mathcal{H}(M(a))$ and with, say, $F(a,b,z)$ as the element of $\mathcal{H}(M(a,b))$. Then $F(a,t,z)$ is the element of $\mathcal{H}(M(a,t))$ obtained from $\mathcal{H}(M(a,b))$ under the adjoint of the inclusion of $\mathcal{H}(M(a,t))$ in $\mathcal{H}(M(a,b))$ when $a < t < b$.

The notation $F(a,b,z)$ will also be used for an element of $\mathcal{H}(M(a,b))$ which is not derived from an element of $\mathcal{H}(M(b))$. But $F(a,t,z)$ will still denote the element of $\mathcal{H}(M(a,t))$ obtained from $F(a,b,z)$ under the adjoint of the inclusion of $\mathcal{H}(M(a,t))$ in $\mathcal{H}(M(a,b))$. In this notation a bounded linear transformation $T(a,b)$ of the space $\mathcal{H}(M(a,b))$ into itself is defined by taking $F(a,b,z)$ into

$$z \, F(a,b,z) - \lim_{w=\infty} w \, F(a,b,w) - \int_a^b \lambda(t) \; dF(a,t,z) \; .$$

The transformations so defined are consistent in the following sense: Assume that (a,b) and (b,c) are adjacent finite intervals. Assume that the minimal decomposition of an element $F(a,c,z)$ of $\mathcal{H}(M(a,c))$ is obtained with $F(a,b,z)$ as the element of $\mathcal{H}(M(a,b))$ and with $F(b,c,z)$ as the element of $\mathcal{H}(M(b,c))$. If $T(a,c)$ takes $F(a,c,z)$ into $G(a,c,z)$, if $T(a,b)$ takes $F(a,b,z)$ into $G(a,b,z)$, and if $T(b,c)$ takes $F(b,c,z)$ into $G(b,c,z)$, then the minimal decomposition of $G(a,c,z)$ as an element of $\mathcal{H}(M(a,c))$ is obtained with

$$G(a,b,z) + [M(a,b,z) - 1] \lim_{w=\infty} w \; F(b,c,w)$$

as the element of $\mathcal{H}(M(a,b))$ and with $G(b,c,z)$ as the element of $\mathcal{H}(M(b,c))$.

Let ϵ be a given positive number. If (a,b) is a given finite interval and if a partition

$$a = t_0 < t_1 < \cdots < t_r = b$$

of the interval exists such that $\lambda(t_n) - \lambda(t_{n-1}) < \epsilon$ for every $n = 1,\ldots,r$, then the spectrum of $T(t_{n-1},t_n)$ is contained in the interval $[-2\,\epsilon,\; 2\,\epsilon]$ for every n. Since the spectrum of $T(a,b)$ is contained in the union of the spectra of the transformations $T(t_{n-1}, t_n)$, it is contained in the same interval. By the arbitrariness of ϵ, the origin is the only point of the spectrum of $T(a,b)$.

By the arbitrariness of a and b, the transformation which takes $F(z)$ into

$$z\, F(z) - \lim_{w=\infty} w\, F(w) - \int_{-\infty}^{+\infty} \lambda(t)\, dF(t,z)$$

in $\mathcal{H}(M)$ has the origin as the only point of its spectrum. It is a consequence of the Livšič-Brodskii model theory that a transformation is completely continuous if it has the origin as the only point of its spectrum and if the difference between the transformation and its adjoint is completely continuous. For it can be assumed that the given transformation is the underlying transformation in a space $\mathcal{H}(M)$. The construction of the related space $\mathcal{L}(\varphi)$ then shows that the sum of the transformation and its adjoint is completely continuous.

PROOF OF THEOREM 6. The proof adapts an argument due to Brodskii [2] to a context in which the inclusions of spaces need not be isometric. Such a generalization of the argument has previously been made in the theory of Hilbert spaces of entire functions [5].

By the definition of an admissible family, a completely continuous transformation $T(a,b)$ with imaginary spectrum exists which takes $F(a,b,z)$ into

$$z\, F(a,b,z) - \lim w\, F(a,b,w) - \int_a^b \lambda(t)\, dF(a,t,z)$$

in $\mathcal{H}(M(a,b))$. The adjoint $T^*(a,b)$ takes $F(a,b,z)$ into

$$z\, F(a,b,z) - M(a,b,z) \lim w\, F(a,b,w) - \int_a^b \lambda(t)\, dF(a,t,z) .$$

By the Livšič-Brodskii model theory, the bound $\kappa(a,b)$ of the transformation has limit zero as a increases to b for any b or as b decreases to a for any a. It follows that the expression

$$\lim w\, F(a,b,w)$$

has limit zero as a increases to b or as b decreases to a for any given a. This verifies that the function $g(t)$ given in the statement of the theorem is a continuous function of t.

If (a,b) is a given finite interval and if ε is a given positive number, then a partition

$$a = t_0 < t_1 < \cdots < t_r = b$$

of the interval exists for some positive integer r such that $\kappa(t_{n-1}, t_n) < \varepsilon$ for every $n = 1, \ldots, r$.

Assume that an element $F(a,b,z)$ of $\mathcal{H}(M(a,b))$ is given a minimal decomposition

$$F(a,b,z) = \sum_{n=1}^{r} M(a, t_{n-1}, z)\ F(t_{n-1}, t_n, z)$$

with $F(t_{n-1}, t_n, z)$ in $\mathcal{H}(M(t_{n-1}, t_n))$ for every $n = 1, \ldots, r$. If $T^*(a,b)$ takes $F(a,b,z)$ into $G(a,b,z)$, then $G(a,b,z)$ has a minimal decomposition of the form

$$G(a,b,z) = \sum_{n=1}^{r} M(a, t_{n-1}, z)\ \{G(t_{n-1}, t_n, z) - [M(t_{n-1}, t_n, z) - 1] \lim w\, F(a, t_{n-1}, w)\}$$

with

$$G(t_{n-1}, t_n, z) - [M(t_{n-1}, t_n, z) - 1] \lim w\, F(a, t_{n-1}, w)$$

in $\mathcal{H}(M(t_{n-1}, t_n))$ for every $n = 1, \ldots, r$. Since the inequality

$$\|G(t_{n-1}, t_n, z)\|_{\mathcal{H}(M(t_{n-1}, t_n))} \le \varepsilon\ \|F(t_{n-1}, t_n, z)\|_{\mathcal{H}(M(t_{n-1}, t_n))}$$

holds for every $n = 1, \ldots, r$, the inequality

$$\begin{aligned}
\Big\| \sum_{n=1}^{r} M(a, t_{n-1}, z)\ G(t_{n-1}, t_n, z) \Big\|^2_{\mathcal{H}(M(a,b))}
&\le \sum_{n=1}^{r} \|G(t_{n-1}, t_n, z)\|^2_{\mathcal{H}(M(t_{n-1}, t_n))} \\
&\le \varepsilon^2 \sum_{n=1}^{r} \|F(t_{n-1}, t_n, z)\|^2_{\mathcal{H}(M(t_{n-1}, t_n))} \\
&\le \varepsilon^2\ \|F(a,b,z)\|^2_{\mathcal{H}(M(a,b))}
\end{aligned}$$

is satisfied.

The identity

$$\begin{aligned}
&\|G(a,b,z)\|^2_{\mathcal{H}(M(a,b))} \\
&= \lim \sum_{n=1}^{r} \|[M(t_{n-1}, t_n, z) - 1] \lim w\, F(a, t_{n-1}, w)\|^2_{\mathcal{H}(M(t_{n-1}, t_n))} \\
&= 2\pi \lim \sum_{n=1}^{r} [\lim w\, F(a, t_{n-1}, w)]^- \bar{I} \\
&\qquad m(t_{n-1}, t_n)\ I\ [\lim w\, F(a, t_{n-1}, w)]
\end{aligned}$$

is obtained in the limit as the weighted mesh of the partition goes to zero. By the definition of the integral in $L^2(m)$, the identity can be written

$$\|G(a,b,z)\|^2_{\mathcal{H}(M(a,b))}$$
$$= 2\pi \int_a^b [\lim w\, F(a,t,w)]^- \bar{I}\, dm(t)\, I\, [\lim w\, F(a,t,w)] .$$

If c is a vector, then $[M(a,b,z) - 1]\, c$ has a minimal decomposition of the form

$$[M(a,b,z) - 1]\, c = \sum_{n=1}^{r} M(a,t_{n-1},z)[M(t_{n-1},t_n,z) - 1]\, c$$

with

$$[M(t_{n-1},t_n,z) - 1]\, c$$

in $\mathcal{H}(M(t_{n-1},t_n))$ for every $n = 1,\ldots,r$. The identity

$$\langle G(a,b,z),\, [M(a,b,z) - 1]\, c\rangle_{\mathcal{H}(M(a,b))}$$
$$= \lim \sum_{n=1}^{r} - \langle [M(t_{n-1},t_n,z) - 1] \lim w\, F(a,t_{n-1},w),$$
$$[M(t_{n-1},t_n,z) - 1]\, c\rangle_{\mathcal{H}(M(t_{n-1},t_n))}$$
$$= 2\pi \lim \sum_{n=1}^{r} \bar{c}\, I\, m(t_{n-1},t_n)\, I \lim w\, F(a,t_{n-1},w)$$
$$= 2\pi \int_a^b \bar{c}\, I\, dm(t)\, I \lim w\, F(a,t,w)$$

is satisfied.

The identity

$$\|G(a,b,z) + [M(a,b,z) - 1]\, c\|^2_{\mathcal{H}(M(a,b))}$$
$$= 2\pi \int_a^b [\lim w\, F(a,t,w) - c]^- \bar{I}\, dm(t)\, [\lim w\, F(a,t,w) - c]$$

is obtained by a similar argument.

The results given in the statement of the theorem now follow by the arbitrariness of a.

PROOF OF THEOREM 7. Assume that the choice of an element $F(a,z)$ of $\mathcal{H}(M(a))$ is made for every real number a such that, when $a < b$, $F(a,z)$ is always the element of $\mathcal{H}(M(a))$ obtained from $F(b,z)$ under the adjoint of the inclusion of $\mathcal{H}(M(a))$ in $\mathcal{H}(M(b))$. For each real number a, consider the element $G(a,z)$ of $\mathcal{H}(M(a))$ defined by

$$G(a,z) = z\, F(a,z) - M(a,z) \lim w\, F(a,w) - \int_{-\infty}^{a} \lambda(t)\, dF(t,z) .$$

When $a < b$, $G(a,z)$ is the element of $\mathcal{H}(M(a))$ obtained from $G(b,z)$ under the adjoint of the inclusion of $\mathcal{H}(M(a))$ in $\mathcal{H}(M(b))$. Consider the vector-valued function $g(t)$ of real t defined by

$$g(t) = -2\pi \; I \lim w \; F(t,w) \; .$$

By Theorem 6 (with a change of variable), $g(t)$ is a continuous function of t which satisfies the identity

$$2\pi \; \|G(a,z) + [M(a,z) - 1] \; c/(2\pi)\|^2_{\mathcal{H}(M(a))} = \int_{-\infty}^{a} [g(t) + I \; c]^- \; dm(t) \; [g(t) + I \; c]$$

for every vector c when a is real.

A bounded linear transformation f into $F(a,z)$ of $L^2(m)$ into $\mathcal{H}(M(a))$ is defined so as to take $\chi_a(g + I \; c)$ into

$$G(a,z) + [M(a,z) - 1] \; c/(2\pi)$$

for every such family of function $F(t,z)$ and for every vector c. The transformation is defined by continuity on the closure in $L^2(m)$ of such functions $\chi_a \; (g + I \; c)$. It is defined to be zero on the orthogonal complement of such functions. It will be shown that the transformation of $L^2(m)$ into $\mathcal{H}(M(a))$ so defined is the desired generalization of the Fourier transformation.

When f is the element of $L^2(m)$ defined by $f(t) = \chi_a(t) \; I \; c$ for a vector c, then the corresponding element $F(a,z)$ of $\mathcal{H}(M(a))$ is given by

$$2\pi \; F(a,z) = [M(a,z) - 1] \; c$$

and

$$2\pi \; \|F(a,z)\|^2_{\mathcal{H}(M(a))} = \int_{-\infty}^{a} \bar{f}(t) \; dm(t) \; f(t) \; .$$

It follows that the identity

$$\langle F(a,z), \; [M(a,z) - 1] \; c \rangle_{\mathcal{H}(M(a))} = \int_{-\infty}^{a} \bar{c} \; \bar{I} \; dm(t) \; f(t)$$

holds whenever f is an element of $L^2(m)$ and $F(a,z)$ is the corresponding element of $\mathcal{H}(M(a))$. The identity can be written

$$2\pi \lim_{w=\infty} w \; F(a,w) = \int_{-\infty}^{a} dm(t) \; f(t) \; .$$

Assume that f is an element of $L^2(m)$ and that $F(a,z)$ is the corresponding element of $\mathcal{H}(M(a))$ for every real number a. If the identity

$$2\pi \; \|F(a,z)\|^2_{\mathcal{H}(M(a))} = \int_{-\infty}^{a} \bar{f}(t) \; dm(t) \; f(t)$$

holds when $a = b$, then it holds when $a < b$ and $F(a,z)$ is the element of $\mathcal{H}(M(a))$ obtained from $F(b,z)$ under the adjoint of the inclusion of $\mathcal{H}(M(a))$ in $\mathcal{H}(M(b))$. An element of $L^2(m)$ is represented by the continuous function g defined by

$$g(x) = \int_x^b I\, dm(t)\, f(t)$$

when $x < b$ and by $g(x) = 0$ otherwise. The corresponding element $G(b,z)$ of $\mathcal{H}(M(b))$ is

$$G(b,z) = z\, F(b,z) - M(b,z) \lim w\, F(b,w) - \int_{-\infty}^b \lambda(t)\, dF(t,z)$$

and the identity

$$2\pi\, \|G(b,z)\|^2_{\mathcal{H}(M(b))} = \int_{-\infty}^b \bar{g}(t)\, dm(t)\, g(t)$$

is satisfied.

When $F(b,z)$ is the element of $\mathcal{H}(M(b))$ defined by

$$F(b,z) = [M(b,z)\, I\, \bar{M}(b,w) - I]\, c/[2\pi(z - \bar{w})]$$

for a vector c and a complex number w such that $|m(b)| < |w - \bar{w}|$, and when $a < b$, then the element $F(a,z)$ of $\mathcal{H}(M(a))$ obtained from $\mathcal{H}(M(b))$ under the adjoint of the inclusion of $\mathcal{H}(M(a))$ in $\mathcal{H}(M(b))$ is

$$F(a,z) = [M(a,z)\, I\, \bar{M}(a,w) - I]\, c/[2\pi(z - \bar{w})]$$

where

$$-\, 2\pi\, I \lim_{z=\infty} z\, F(a,z) = [\bar{M}(a,w) - 1]\, c.$$

It follows than an element of $L^2(m)$ is determined by the continuous function

$$f(t) = \chi_a(t)\, [\bar{M}(t,w) - 1]\, c\ .$$

The corresponding element of $\mathcal{H}(M(a))$ is

$$z\, F(a,z) - M(a,z) \lim w\, F(a,w) - \int_{-\infty}^a \lambda(t)\, dF(t,z)\ .$$

The identity

$$2\pi\, \|z\, F(a,z) - M(a,z) \lim w\, F(a,w) - \int_{-\infty}^a \lambda(t)\, dF(t,z)\|^2_{\mathcal{H}(M(a))}$$

$$= \int_{-\infty}^a \bar{f}(t)\, dm(t)\, f(t)$$

is satisfied.

If f is any element of $L^2(m)$ and if $F(a,z)$ is the corresponding element of $\mathcal{H}(M(a))$, then the identity

$$2\pi\ \bar{c}\ [z\ F(a,z) - \lim\ w\ F(a,w) - \int_{-\infty}^{a} \lambda(t)\ dF(t,z)]$$
$$= \int_{-\infty}^{a} \bar{c}\ [M(t,w) - 1]\ dm(t)\ f(t)$$

holds for every vector c when $|m(a)| < |w - \bar{w}|$.

This information will be used to show that the range of the transformation of $L^2(m)$ into $\mathcal{H}(M(b))$ contains every element of $\mathcal{H}(M(b))$. It is sufficient to show that no nonzero element $F(b,z)$ of $\mathcal{H}(M(b))$ is orthogonal to the range of the transformation. When $a < b$, let $F(a,z)$ be the element of $\mathcal{H}(M(a))$ obtained from $F(b,z)$ under the adjoint of the inclusion of $\mathcal{H}(M(a))$ in $\mathcal{H}(M(b))$. Since the identity

$$0 = z\ F(b,z) - M(b,z)\ \lim\ w\ F(b,w) - \int_{-\infty}^{b} \lambda(t)\ dF(t,z)$$

is satisfied, the identity

$$0 = z\ F(a,z) - M(a,z)\ \lim\ w\ F(a,w) - \int_{-\infty}^{a} \lambda(t)\ dF(t,z)$$

is satisfied when $a < b$. Since the identity

$$0 = z\ F(b,z) - \lim\ w\ F(b,w) - \int_{-\infty}^{b} \lambda(t)\ dF(t,z)$$

is satisfied, the identity

$$0 = z\ F(a,z) - \lim\ w\ F(a,w) - \int_{-\infty}^{a} \lambda(t)\ dF(t,z)$$

is satisfied when $a < b$. It follows that

$$\lim\ w\ F(a,w) = 0$$

when $a \le b$.

Consider the set $\mathcal{M}(b)$ of elements $F(b,z)$ of $\mathcal{H}(M(b))$ such that the last identity holds when $a \le b$, where $F(a,z)$ is the element of $\mathcal{H}(M(a))$ obtained from $F(b,z)$ under the adjoint of the inclusion of $\mathcal{H}(M(a))$ in $\mathcal{H}(M(b))$. Then $\mathcal{M}(b)$ is a reducing subspace for the transformation which takes $F(z)$ into

$$z\ F(b,z) - \lim\ w\ F(b,w) - \int_{-\infty}^{b} \lambda(t)\ dF(t,w)$$

and hence for the transformation which takes $F(z)$ into

$$z\ F(b,z) - \lim\ w\ F(b,w)$$

in $\mathcal{H}(M(b))$. It follows that $[F(z) - F(w)]/(z - w)$ belongs to $\mathcal{M}(b)$ whenever $F(z)$ belongs to $\mathcal{M}(b)$ when $|m(b)| < |w - \bar{w}|$. An application of the definition of the space $\mathcal{M}(b)$ now shows that $F(w) = 0$. By the arbitrariness of w, $\mathcal{M}(b)$ contains no nonzero element.

Let c be a given vector and let w be a given number, $|m(b)| < |w - \bar{w}|$. Since the expression

$$[M(b,z) \, I \, \bar{M}(b,w) - I] \, c/[2\pi(z - \bar{w})]$$

belongs to $\mathcal{H}(M(b))$, it is the generalized Fourier transform of an element k of $\mathcal{H}(M(b))$ such that the identity

$$2\pi \, \bar{c} \, F(b,w) = \int_{-\infty}^{b} \bar{k}(t) \, dm(t) \, f(t)$$

holds whenever f is an element of $L^2(m)$ and F(b,z) is the corresponding element of $\mathcal{H}(M(b))$. The identity

$$2\pi \, \bar{c} \, [w \, F(b,w) - \lim_{z=\infty} z \, F(b,z) - \int_{-\infty}^{b} \lambda(t) \, dF(t,w)]$$

$$= \int_{-\infty}^{b} [(w - \lambda(t))\bar{k}(t) - c] \, dm(t) \, f(t)$$

then holds for every element f of $L^2(m)$. It follows that

$$(\bar{w} - \lambda(t)) \, k(t) - c = \bar{M}(t,w) \, c - c$$

and that

$$k(t) = \bar{M}(t,w) \, c/[\bar{w} - \lambda(t)] \, .$$

The stated form of the generalized Fourier transformation follows.

PROOF OF THEOREM 8. For each element f of $L^2(m)$, a corresponding element $F_+(a,z)$ of $\mathcal{H}(M_+(a))$ and a corresponding element $F_-(a,z)$ of $\mathcal{H}(M_-(a))$ have been constructed by Theorem 7. By the proof of the theorem, the identities

$$2\pi \lim_{z=\infty} z \, F_+(a,z) = \int_{-\infty}^{a} dm(t) \, f(t) = 2\pi \lim_{z=\infty} z \, F_-(a,z)$$

are satisfied. Assume that w is a complex number such that $|m(a)| < |w - \bar{w}|$. By Theorem 7, an element g of $L^2(m)$ exists such that the corresponding element $G_+(a,z)$ of $\mathcal{H}(M_+(a))$ and the corresponding element $G_-(a,z)$ of $\mathcal{H}(M_-(a))$ are

$$G_+(a,z) = [F_+(a,z) - F_+(a,w)]/(z - w)$$

and

$$G_-(a,z) = [F_-(a,z) - F_-(a,w)]/(z - w) \, .$$

It follows from the previous identity that $F_+(a,w) = F_-(a,w)$. By the arbitrariness of w, $F_+(a,z)$ and $F_-(a,z)$ are identically equal. The desired identity

$$M_+(a,z) = M_-(a,z)$$

follows because the spaces $\mathcal{H}(M_+(a))$ and $\mathcal{H}(M_-(a))$ are isometrically equal by Theorem 7.

PROOF OF THEOREM 9. The stated expansion follows from Theorem 7 on applying the definition of overlapping space.

PROOF OF THEOREM 10. The theorem is a variant of Theorem 7 of [11] and has a similar proof. Underlying both theorems is a general decomposition which will now be stated and proved.

Assume that $\mathcal{H}(a,b)$, $\mathcal{H}(b,c)$, $\mathcal{H}(c,d)$, $\mathcal{H}(b,d)$, and $\mathcal{H}(a,d)$ are Hilbert spaces such that $\mathcal{H}(a,b)$ is contained contractively in $\mathcal{H}(a,d)$, and $\mathcal{H}(b,d)$ is the complementary space to $\mathcal{H}(a,b)$ in $\mathcal{H}(a,d)$, and such that $\mathcal{H}(b,c)$ is contained contractively in $\mathcal{H}(b,d)$, and $\mathcal{H}(c,d)$ is the complementary space to $\mathcal{H}(b,c)$ in $\mathcal{H}(b,d)$.

Define $\mathcal{H}(a,b,d)$ to be the Hilbert space which is obtained as the intersection of $\mathcal{H}(a,b)$ and $\mathcal{H}(b,d)$ in the norm

$$\|f\|^2_{\mathcal{H}(a,b,d)} = \|f\|^2_{\mathcal{H}(a,b)} + \|f\|^2_{\mathcal{H}(b,d)} .$$

Define $\mathcal{H}(b,c,d)$ to be the Hilbert space which is obtained as the intersection of $\mathcal{H}(b,c)$ and $\mathcal{H}(c,d)$ in the norm

$$\|f\|^2_{\mathcal{H}(b,c,d)} = \|f\|^2_{\mathcal{H}(b,c)} + \|f\|^2_{\mathcal{H}(c,d)} .$$

Define $\mathcal{H}(a,c,d)$ to be the Hilbert space which is obtained as the intersection of $\mathcal{H}(a,c)$ and $\mathcal{H}(c,d)$ in the norm

$$\|f\|^2_{\mathcal{H}(a,c,d)} = \|f\|^2_{\mathcal{H}(a,c)} + \|f\|^2_{\mathcal{H}(c,d)} .$$

The space $\mathcal{H}(b,c,d)$ is contained contractively in the space $\mathcal{H}(a,c,d)$. It will be shown that the adjoint of the inclusion of $\mathcal{H}(c,d)$ in $\mathcal{H}(b,d)$ induces a partial isometry of $\mathcal{H}(a,b,d)$ onto the complementary space to $\mathcal{H}(b,c,d)$ in $\mathcal{H}(a,c,d)$.

The adjoint of the inclusion of $\mathcal{H}(c,d)$ in $\mathcal{H}(b,d)$ is a contractive transformation P of $\mathcal{H}(b,d)$ in $\mathcal{H}(c,d)$.

Assume that f is an element of $\mathcal{H}(a,b,d)$. Then f is an element of $\mathcal{H}(a,b)$ which belongs to $\mathcal{H}(b,d)$. So P f is an element of $\mathcal{H}(c,d)$. Since f belongs to $\mathcal{H}(a,b)$, it belongs to $\mathcal{H}(a,c)$. Since $(1-P)f$ belongs to $\mathcal{H}(b,c)$, it belongs to $\mathcal{H}(a,c)$. It follows that P f belongs to $\mathcal{H}(a,c)$ and hence to $\mathcal{H}(a,c,d)$.

The inequality

$$\begin{aligned}\|P\ f\|^2_{\mathcal{H}(a,c,d)} &= \|f-(1-P)\ f\|^2_{\mathcal{H}(a,c)} + \|P\ f\|^2_{\mathcal{H}(c,d)} \\ &\le \|f\|^2_{\mathcal{H}(a,b)} + \|(1-P)\ f\|^2_{\mathcal{H}(b,c)} + \|P\ f\|^2_{\mathcal{H}(c,d)} \\ &\le \|f\|^2_{\mathcal{H}(a,b)} + \|f\|^2_{\mathcal{H}(b,d)} \\ &\le \|f\|^2_{\mathcal{H}(a,b,d)}\end{aligned}$$

is satisfied.

If g is any element of $\mathcal{H}(b,c,d)$, then the inequality

$$\begin{aligned}&\|P\ f+g\|^2_{\mathcal{H}(a,c,d)} - \|g\|^2_{\mathcal{H}(b,c,d)} \\ &= \|f-(1-P)\ f+g\|^2_{\mathcal{H}(a,c)} + \|P\ f+g\|^2_{\mathcal{H}(c,d)} \\ &\qquad - \|g\|^2_{\mathcal{H}(b,c)} - \|g\|^2_{\mathcal{H}(c,d)} \\ &\le \|f\|^2_{\mathcal{H}(a,b)} + \|(1-P)\ f-g\|^2_{\mathcal{H}(b,c)} + \|P\ f+g\|^2_{\mathcal{H}(c,d)} \\ &\qquad - \|g\|^2_{\mathcal{H}(b,c)} - \|g\|^2_{\mathcal{H}(c,d)} \\ &\le \|f\|^2_{\mathcal{H}(a,b)} + \|(1-P)\ f\|^2_{\mathcal{H}(b,c)} + \|P\ f\|^2_{\mathcal{H}(c,d)} \\ &\qquad \le \|f\|^2_{\mathcal{H}(a,b)} + \|f\|^2_{\mathcal{H}(b,d)} \\ &\qquad \le \|f\|^2_{\mathcal{H}(a,b,d)}\end{aligned}$$

is satisfied. This verifies that P acts as a contractive transformation of $\mathcal{H}(a,b,d)$ into the complementary space to $\mathcal{H}(b,c,d)$ in $\mathcal{H}(a,c,d)$

A similar argument shows that the adjoint Q of the inclusion of $\mathcal{H}(a,b)$ in $\mathcal{H}(a,c)$ acts as a contractive transformation of $\mathcal{H}(a,c,d)$ in $\mathcal{H}(a,b,d)$. It will be shown that $1-P\ Q$ maps $\mathcal{H}(a,c,d)$ into $\mathcal{H}(b,c,d)$

Assume that f is an element of $\mathcal{H}(a,c,d)$. Since P maps $\mathcal{H}(b,d)$ into $\mathcal{H}(c,d)$ and since f belongs to $\mathcal{H}(a,d)$, $(1-P\ Q)\ f$ belongs to $\mathcal{H}(c,d)$. Since $1-P$ maps $\mathcal{H}(b,d)$ into $\mathcal{H}(b,c)$ and since $(1-Q)\ f$ belongs to $\mathcal{H}(b,c)$, $(1-P\ Q)\ f$ belongs to $\mathcal{H}(b,c)$. It follows that $(1-P\ Q)\ f$ belongs to $\mathcal{H}(b,c,d)$.

It will now be verified that the transformation $1-P\ Q$ is the adjoint of the inclusion of $\mathcal{H}(b,c,d)$ in $\mathcal{H}(a,c,d)$. Assume that f belongs to $\mathcal{H}(a,c,d)$ and that g belongs to $\mathcal{H}(b,c,d)$. Then the identity

$$\begin{aligned}
&\langle (1 - P\,Q)\, f,\ g\rangle_{\mathcal{H}(b,c,d)} \\
&\quad = \langle (1 - P\,Q)\, f,\ g\rangle_{\mathcal{H}(b,c)} + \langle (1 - P\,Q)\, f,\ g\rangle_{\mathcal{H}(c,d)} \\
&\quad = \langle (1 - Q)\, f,\ g\rangle_{\mathcal{H}(b,c)} + \langle (1 - P)\, Q\, f,\ g\rangle_{\mathcal{H}(b,c)} \\
&\qquad + \langle f,\ g\rangle_{\mathcal{H}(c,d)} - \langle P\,Q\, f,\ g\rangle_{\mathcal{H}(c,d)} \\
&\quad = \langle f,\ g\rangle_{\mathcal{H}(a,c)} + \langle Q\, f,\ g\rangle_{\mathcal{H}(b,d)} \\
&\qquad + \langle f,\ g\rangle_{\mathcal{H}(c,d)} - \langle Q\, f,\ g\rangle_{\mathcal{H}(b,d)} \\
&\quad = \langle f,\ g\rangle_{\mathcal{H}(a,c,d)}
\end{aligned}$$

is the required verification. It follows that the transformation $P\,Q$ is the adjoint of the inclusion of $\mathcal{M}$ in $\mathcal{H}(a,c,d)$, where $\mathcal{M}$ is the complementary space to $\mathcal{H}(b,c,d)$ in $\mathcal{H}(a,c,d)$.

It will now be shown that P acts as an isometry from the norm of $\mathcal{H}(a,b,d)$ into the norm of $\mathcal{M}$ on elements in the Q-image of $\mathcal{H}(a,c,d)$. If f belongs to $\mathcal{H}(a,c,d)$, then

$$\begin{aligned}
\|P\,Q\, f\|^2_{\mathcal{M}} &= \|f\|^2_{\mathcal{H}(a,c,d)} - \|(1 - P\,Q)\, f\|^2_{\mathcal{H}(b,c,d)} \\
&= \|f\|^2_{\mathcal{H}(a,c)} + \|f\|^2_{\mathcal{H}(c,d)} \\
&\quad - \|(1 - P\,Q)\, f\|^2_{\mathcal{H}(b,c)} - \|(1 - P\,Q)\, f\|^2_{\mathcal{H}(c,d)} \\
&= \|Q\, f\|^2_{\mathcal{H}(a,b)} + \|(1 - Q)\, f\|^2_{\mathcal{H}(b,c)} + \|f\|^2_{\mathcal{H}(c,d)} \\
&\quad - \|(1 - P\,Q)\, f\|^2_{\mathcal{H}(b,c)} - \|(1 - P\,Q)\, f\|^2_{\mathcal{H}(c,d)} \\
&= \|Q\, f\|^2_{\mathcal{H}(a,b)} - \|(1 - P)\, Q\, f\|^2_{\mathcal{H}(b,c)} \\
&\quad - \|P\,Q\, f\|^2_{\mathcal{H}(c,d)} + 2\, \|Q\, f\|^2_{\mathcal{H}(b,d)} \\
&= \|Q\, f\|^2_{\mathcal{H}(a,b)} + \|Q\, f\|^2_{\mathcal{H}(b,d)} \\
&= \|Q\, f\|^2_{\mathcal{H}(a,b,d)}\ .
\end{aligned}$$

It follows that P acts as a partial isometry of $\mathcal{H}(a,b,d)$ onto $\mathcal{M}$. The kernel of the partial isometry is the orthogonal complement in $\mathcal{H}(a,b,d)$ of the Q-image of $\mathcal{H}(a,c,d)$.

REFERENCES

1. Gohberg, I. C., and Krein, M. G.: "Theory and Applications of Volterra Operators in Hilbert Space," Translations of Mathematical Monographs, vol. 24, American Mathematical Society, Providence, 1970.

2. Brodskii, M.S.: "Triangular and Jordan Representations of Linear Operators," Translations of Mathematical Monographs, vol. 32, American Mathematical Society, Providence, 1970.

3. de Branges, L.: Some Hilbert spaces of analytic functions II, J. Math. Anal. Appl. 11 (1965), 44-72, and 12 (1965), 149-186.

4. de Branges, L. Perturbations of selfadjoint transformations, Amer. J. Math. 84 (1962), 543-560.

5. de Branges, L.: The comparison theorem for Hilbert spaces of entire functions, Integral Equations and Operator Theory, to appear.

6. de Branges, L.: The expansion theorem for Hilbert spaces of entire functions, in "Entire Functions and Related Parts of Analysis," American Mathematical Society, Providence, 1968, pp. 79-148.

7. Dym, H., and Gohberg, I.: On an extension problem, generalized Fourier analysis and an entropy problem, Integral Equations and Operator Theory 3 (1980), 143-215.

8. Lubin, A.: Concrete model theory for a class of operators with unitary part, J. Functional Anal. 17 (1974), 388-394.

9. Kriete, T.: Canonical models and the nonselfadjoint parts of dissipative operators, J. Functional Anal. 23 (1976), 39-84.

10. Ball, J.A.: Factorization and model theory for contractive operators with unitary part, Memoirs of the American Mathematical Society, Number 198, 1978.

11. de Branges, L.: "Square Summable Power Series," Addison-Wesley, to appear.

12. de Branges, L.: Factorization and invariant subspaces, J. Math. Anal. Appl. 29 (1970), 163-200.

L. de Branges
Department of Mathematics
Purdue University
Lafayette, Indiana 47907
U.S.A.

Operator Theory:
Advances and Applications, Vol. 12

THE LOSSLESS INVERSE SCATTERING PROBLEM IN THE NETWORK-THEORY CONTEXT

Patrick M. Dewilde

We present the theory leading to a theorem that describes all the rational solutions of the lossless inverse scattering problem (LIS-problem) for lossless networks. They are parametrized by a set of points in the closed unit disc of the complex plane. Quite a few classical problems in estimation theory and network theory may be viewed as a special case of the LIS problem. We present a global method to construct LIS solutions using reproducing kernel Hilbert space methods. Finally, we give connections with applications and with some classical interpolation problems and relate the results to maximum entropy approximation theory.

ACKNOWLEDGEMENT

Most of the results presented in this paper were obtained as a result of a collaboration with H. Dym of the Weizmann Institute. His help is hereby gratefully acknowledged.

1. INTRODUCTION

A lossless inverse scattering problem is concerned with identifying an unknown, linear, causal time-invariant, passive system, given an "incident wave" (input signal) $a(t)$ and a "reflected wave" (output signal) $b(t)$ from the system. In practice a and b will be vector valued and t may be either discrete or continuous. For the sake of simplicity of exposition, a and b are taken to be scalar valued and t is discrete. The main goal is the description of all the solutions of the scalar rational LIS problem. It turns out that these solutions may be generated from a fundamental set of solutions. These may in turn be parametrized by a set of points in the closed unit disc of the complex plane.

In the course of the history of this problem particular sets of solutions have been derived. In this paper solutions which have not been studied before are given, and a complete set of solutions is given. Finally, the relation between solutions of the LIS problem and linear least squares prediction theory and spectral estimation theory is given.

What seems to be the first set of solutions of the LIS problem has been derived by Schur in a classical paper [1]. Anticipating the results, it is possible to parametrize solutions with points on the closed unit disc. These points will be called 'transmission zeros' [2] following a usage well established in network theory [3].
In Schur's case, all TZ's are at the origin. The emphasis of the earlier papers is not on construction of solutions of a LIS problem but on some related mathematical problem, e.g. the problem of interpolating values at certain points in the unit disc of the complex plane by means of a p.r. function or contractive analytic function. This interpolation problem at other points than at the origin was considered by Nevanlinna-Pick [4] and by Caratheodory-Fejer [5]. In performing such an interpolation, there is an implicit construction of the solution of an LIS problem. Further connections e.g. with the Szegö theory of polynomials orthogonal on the unit circle are easily made. Hence many publications in that area [6],[7],[8] are relevant to our theory. Other connections are with linear least squares estimation problem problem [9],[10] and with classical network theory [11],[12],[13],[14],[15], [16],[17].

The solutions of the LIS problem may all be described in terms of a 'scattering matrix' or of a 'chain scattering matrix' (CSM). These matrices have the property that they are mesomorphic in the unit disc of the complex plane, and that they are contractive with respect to a certain, possibly non definite, metric described by a Gaussian matrix J. A complete description and parametrization of all possible J-contractive matrices has been given by Potapov [18]. Our problem hence is to select solutions of the LIS problem from this large collection, when given an incident and the corresponding reflected wave.

The lossless inverse scattering problem has many well-understood and interesting mathematical connections. We mention a few with the "lifting theorem", see Sarason [19], Nagy-Foias [20],[21], Adamyan-Arov and Krein [22],[23],[24],[25] and Helton [26],[27],[28], with the Nevanlinna-Pick interpolation problem, see e.g. Delsarte, Genin and Kamp [29],[30],[31] and with factorization theory as discussed in Gohberg [32] and Bart, Gohberg and Kaashoek [33]. Finally, the work of Kailath and his colloborators [34],[35],

[36] and [37] has been instrumental in establishing the connection between various topics in estimation theory and scattering theory.

Usage

$\mathbb{R}$	the set of real numbers
$\mathbb{C}$	the set of complex numbers
$\bar{z}$	the complex conjugate of z
$\bar{A}$	the closure of the set A
$\mathbb{D}$	the set $\{z \mid \|z\| < 1\}$
$\mathbb{T}$	the set $\{z \mid \|z\| = 1\}$
$\mathbb{E}$	the set $\{z \mid \|z\| > 1\}$
$\tilde{A}$	the Hermitian conjugate of the matrix A

$A_*(z) = \tilde{A}(1/\bar{z})$

H^p the Hardy space of class p over $\mathbb{D}$

Preliminaries

Let $a(t) \in \ell_2$, i.e. $\sum_{t=-\infty}^{\infty} |a(t)|^2 < \infty$, we define the Fourier transform $A(e^{i\theta})$ of a as:

$$A(e^{i\theta}) = \sum_{t=-\infty}^{\infty} a(t)e^{it\theta} \tag{1.1}$$

If $a(t) = 0$ for $t < 0$, then $A \in H^2(\mathbb{D})$ and $A(z)$, the analytic continuation of A to $\mathbb{D}$ will be called the 'z-transform' of A.

An n x n matrix valued function $S(z)$, with $z \in \mathbb{D}$ is said to be a <u>positive scattering function</u> if

(1) it is analytic in $\mathbb{D}$

(2) contractive in $\mathbb{D}$, i.e.

$$\tilde{S}(z)S(z) \le I_n \tag{1.2}$$

for every point $z \in \mathbb{D}$. It is said to be <u>lossless</u> if in addition to (1) and (2) we also have

$$(3)\ \tilde{S}(e^{i\theta})S(e^{i\theta}) = I_n \text{ on } \mathbb{T}. \tag{1.3}$$

It is well known that a lossless (resp. passive) linear, causal, time-invariant n-port can be described by an n x n lossless (resp. passive) scattering matrix which, by matrix multiplication, maps the z-transforms of the incident waves into the z-transforms of the reflected waves.

A lossless (resp. passive), linear, time-invariant, two port, as depicted in fig. 1, with incident waves a_1 and b_2 and reflected waves b_1 and a_2, can be described by a 2 x 2 lossless (resp. passive) scattering matrix $\Sigma(z)$ which relates the z-transforms of the incident waves to the z-transforms of the reflected waves by the rule:

$$\begin{bmatrix} A_2(z) \\ B_1(z) \end{bmatrix} = \Sigma(z) \begin{bmatrix} A_1(z) \\ B_2(z) \end{bmatrix} \tag{1.4}$$

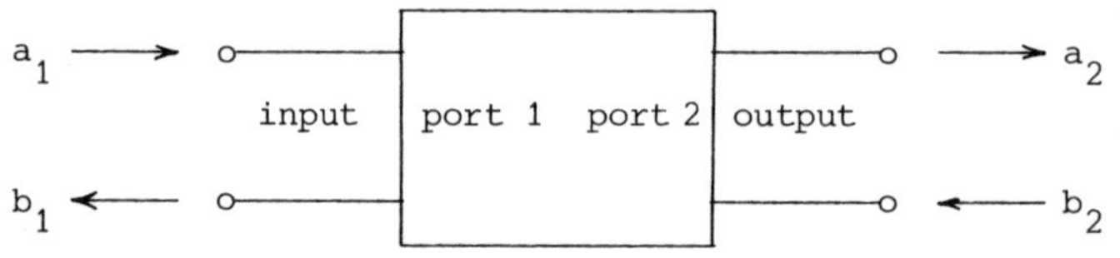

Fig. 1 Representation of a scattering system with two ports.

If a lossless two-port with lossless n x n scattering matrix Σ is loaded with a passive load (alias a 1 port) which is described by passive scattering function $S_L(z)$, as depicted in fig. 2, then the input scattering function $B_1(z) = S_1A_1(z)$ is again passive.

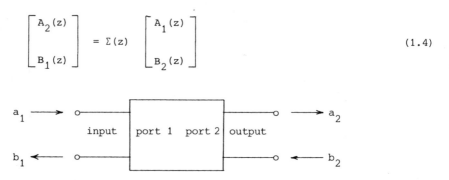

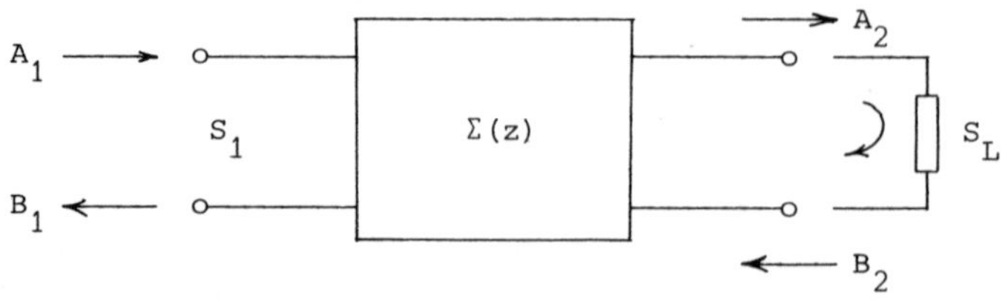

Fig. 2 A Lossless $\Sigma(z)$ with a passive load $S_2(z)$

Given A_1 *and* B_1*, such that the unknown scattering function* $S_1 = B_1A_1^{-1}$ *is passive, the LIS problem is to find a lossless* $n \times n$ Σ *and a passive* S_L *such that* S_1 *originates as the input scattering matrix in the configuration of fig. 2.*

It turns out that every solution of the LIS problem may be constructed recursively using a generalization of the Schur algorithm [1]. In order to discuss recursive solutions, it is convenient to introduce the $2n \times 2n$ chain scattering matrix $\Theta(z)$ which relates waves at the output port to waves at the input port by the rule:

$$\begin{bmatrix} A_2 \\ B_2 \end{bmatrix} = \Theta(z) \begin{bmatrix} A_1 \\ B_1 \end{bmatrix} \tag{1.5}$$

Θ is related to Σ by:

$$\Theta = [P\Sigma + P^{\perp}][P^{\perp}\Sigma + P]^{-1} \tag{1.6}$$

$$= [P - \Sigma P^{\perp}]^{-1}[\Sigma P - P^{\perp}] \tag{1.7}$$

where

$$P = \begin{bmatrix} I & \\ & 0 \end{bmatrix} \quad \text{and} \quad P = \begin{bmatrix} 0 & \\ & I \end{bmatrix}$$

In all but trivial cases, Θ will exist. Let

$$J = P - P^{\perp} \tag{1.8}$$

then it may be shown that if Σ is lossless and $\det \Sigma_{22} \neq 0$, Θ is defined and characterized by the following properties:

1. In D

$$\Theta(z)J\tilde{\Theta}(z) \leq I_{2n} \tag{1.9}$$

2. on T

$$\Theta(e^{i\theta})J\tilde{\Theta}(e^{i\theta}) = I_{2n} \tag{1.10}$$

The CSM's which we shall use shall always be rational. A pole of such a matrix will be called a transmission mode while a zero, a transmission zero. Poles and zeros are each other complements, if ω is a zero, $1/\bar{\omega}$ is a pole and vice versa. In the sequal we make use of the following normalized form of the J-lossless CSM $\Theta(z)$ introduce in [39] and [40]:

$$\Theta(z) = \frac{1}{2}\begin{bmatrix} R_*^{-1}(1+C_*) & R_*^{-1}(1-C_*) \\ F^{-1}(1-C) & F^{-1}(1+C) \end{bmatrix} \tag{1.11}$$

where F and R belong to $H^2(\mathbb{D})$ and C is a p.r. function (Caratheodory function) in $\mathbb{D}$, with moreover:

$$\frac{1}{2}(C + C_*) = FF_* = R_*R \tag{1.12}$$

and

$$V = R_*F^{-1} \tag{1.13}$$

is inner. In the terms of the CSM formalism, the LIS problem becomes:

A J-lossless matrix Θ and a passive scattering function S_L is a solution of the LIS problem for a given passive scattering function S_i if there exist a function A_L such that

$$\begin{bmatrix} A_L \\ S_L A_L \end{bmatrix} = \Theta \begin{bmatrix} 1 \\ S_i \end{bmatrix} \tag{1.14}$$

or equivalently if S_i admits a fractional representation:

$$S_i = (S_L\Theta_{12} - \Theta_{22})^{-1}(\Theta_{21} - S_L\Theta_{11}) \tag{1.15}$$

We make use of the following properties of p.r. functions. Every such function C(z) admits a Herglotz representation:

$$C(z) = \frac{1}{2\pi}\int \frac{e^{i\theta}+z}{e^{i\theta}-z}\,d\mu + id \tag{1.16}$$

where integration is (as always) over $\mathbb{T}$, μ is a finite non-negative measure on $\mathbb{T}$ and d is real constant (the imaginary part of C(0)). Furthermore the following is true:

$$\lim_{r\uparrow 1} (1-r)C(rb) = \frac{1}{\pi}\mu(\{b\}) \tag{1.17}$$

where $\mu(\{b\})$ is the measure of the single point $b \in \mathbb{T}$. If C is of p.r. class, then $S = (C-1)(C+1)^{-1}$ belongs to the "Schur class" S of contractive analytic functions in $\mathbb{D}$.

The 1 x 1 matrix $\begin{bmatrix} A(z) \\ B(z) \end{bmatrix}$ will be said to be <u>admissible</u> if:

(1) A and B are analytic in $\mathbb{D}$,

(2) $A \not\equiv 0$ and $BA^{-1} \in \mathcal{S}$.

$\begin{bmatrix} A \\ B \end{bmatrix}$ and $\begin{bmatrix} C \\ D \end{bmatrix}$ are said to be equivalent or $\begin{bmatrix} A \\ B \end{bmatrix}$ $\begin{bmatrix} C \\ D \end{bmatrix}$ if $BA^{-1} = DC^{-1}$

DEFINITION 1.1 *A point* $b \in \mathbb{T}$ *is said to be a PLL of degree* γ *for the pr function* C *given by (1.16) if either*

$$\mu\{(b)\} = 0 \text{ and } \int \frac{d\mu}{|e^{i\theta}-b|^2} < \infty \tag{1.18}$$

or

$$\mu\{(b)\} \neq 0 \text{ and } \int \frac{d\mu}{|e^{i\theta}-b|^{2(\gamma-1)}} < \infty \tag{1.19}$$

DEFINITION 1.2 *A rational scattering matrix* Σ *or CSM* Θ *of the form (1.11) is said to be fundamental if degree* Θ = *degree* C *and if* F *is outer.*

The following property may be shown: let Θ be a non-fundamental solution of the LIS problem, then there exist a fundamental solution $\hat{\Theta}$ and unitary functions U and V such that $\hat{\Theta}$ is a solution of the LIS problem and

$$\Theta = \begin{bmatrix} U & \\ & V \end{bmatrix} \hat{\Theta} \tag{1.20}$$

Moreover, $\hat{\Theta}$ is essentially unique.

2. FUNDAMENTAL SOLUTIONS AND THEIR PARAMETRIZATION

Let A and B be the given waves of the LIS problem $S = BA^{-1}$ the corresponding contractive function of type $\mathcal{S}$ $C = (1+S)(1-S)^{-1}$ the related pr function with measure μ from (1.16). Suppose that $b_j \in \mathbb{T}$, $j = 1 \ldots k$ are the PLL's of C of degree respect n_j. Then the LIS problem admits a fundamental solution (Θ, S_L) with a CSM having transmission zeros of degree at most n_j at the b_j's, and possibly elsewhere in $\mathbb{D}$. We shall show how such a fundamental solution can be constructed. For ease of discussion and without impairing the generality, we shall assume that the b_j's are not jump points of the measure under consideration.

Let

$$h_n(z) = \prod_{j=1}^{n} (1-\bar{\omega}_j z) \tag{2.1}$$

where the ω_j, j=1 .. n is an arbitrary set of points not necessarily distinct, belonging to $\mathbb{D} \cup \mathbb{T}$, and such that

$$\int \frac{d\mu}{|h_n(e^{i\theta})|^2} < \infty \tag{2.2}$$

Consider the space:

$$\mathcal{M}_n = \left\{ \frac{p(z)}{h_n(z)} \;\middle|\; p(z) \text{ is polynomial of degree} \leq n \right\} \tag{2.3}$$

as a subspace of $L^2(\mathbb{T}, d\mu)$, i.e. L^2 on the circle $\mathbb{T}$ with weight $d\mu$

and let $K_\omega(z)$ be the reproducing kernel for $\mathcal{M}_n$ in $L^2(\mathbb{T},d\mu)$ ($K_\omega(z) = \sum_{k=1}^{n} \overline{\psi_j(\omega)}\psi_j(z)$ where ψ_j is an orthonormal basis).

LEMMA 2.1 *For each* $\omega \in \mathbb{D}$, $[K_\omega(z)]^{-1}$ *belongs to the Hardy space* H^2 *and is outer.*

LEMMA 2.2 *If* f *and* g *belong to* $\mathcal{M}_n$, *then*

$$\frac{1}{2\pi}\int f(e^{i\theta})\overline{g(e^{i\theta})}\,d\mu = \frac{1}{2\pi}\int f(e^{i\theta})\overline{g(e^{i\theta})}\,d\mu_n \tag{2.3}$$

where

$$d\mu_n = \frac{K_0(0)}{|K_0(e^{i\theta})|^2}\,d\theta \tag{2.4}$$

LEMMA 2.3 *If* f *and* g *belong to* $\mathcal{M}_n$ *and* $d\mu_n$ *is as in (2.4), then*

$$f(z)g_*(z)\int \frac{e^{i\theta}+z}{e^{i\theta}-z}(d\mu - d\mu_n) =$$

$$= \int f(e^{i\theta})g_*(e^{i\theta})\,\frac{e^{i\theta}+z}{e^{i\theta}-z}(d\mu - d\mu_n) \tag{2.5}$$

for every $z \in \mathbb{D}$ *which is not a pole of* g_*.

(Lemma 2.3 may be interpreted as a generalized interpolation theorem for it says that C and C_n coincides in the points ω_j - points on $\mathbb{T}$ included).

Define now:

$$F_n(z) = \frac{\sqrt{K_0(0)}}{K_0(z)} \triangleq \frac{h_n(z)}{\psi_n(z)} \tag{2.6}$$

$$U_n(z) = \prod_{j=1}^{n} \frac{z-\omega_j}{1-\bar{\omega}_j z} \tag{2.7}$$

$$V_n(z) = \frac{z^n\psi_{n*}(z)}{\psi_n(z)} \tag{2.8}$$

$$R_{n*}(z) = V_n^{-1} F_n = \frac{h_n}{z^n \psi_{n*}(z)} \tag{2.9}$$

$$C_n(z) = \frac{1}{2\pi} \int \frac{e^{i\theta}+z}{e^{i\theta}-z} \, d\mu_n \tag{2.10}$$

and

$$\Theta_n(z) = \frac{1}{2} \begin{bmatrix} R_{n*}^{-1}(1+C_{n*}) & R_{n*}^{-1}(1-C_{n*}) \\ F_n^{-1}(1-C_n) & F_n^{-1}(1+C_n) \end{bmatrix} \tag{2.11}$$

then

$$\frac{1}{2}(C_n + C_{n*}) = F_n F_{n*} \tag{2.12}$$

and Θ_n is a J-lossless CSM. We have:

THEOREM 2.1 *Let* $\omega_1, \ldots, \omega_n$ *be an arbitrary set of points in* $\mathrm{D} \cup \mathrm{T}$ *and let*

$$C(z) = \frac{1}{2\pi} \int \frac{e^{i\theta}+z}{e^{i\theta}-z} \, d\mu \tag{2.13}$$

where μ *is a finite positive measure on* T *such that* $\dim L^2(d\mu) \geq n+1$ *and (2.2) is valid. Then the LIS problem for* $S = (C-1)(C+1)^{-1}$ *admits a fundamental solution of the form* (Θ_n, S_L) *where* Θ_n *is defined by (2.11) and* $S_L \in \mathcal{S}$.

From the collection of solutions described by Theorem 1, all fundamental solutions of the LIS problem may be derived. This is done with the use of the following theorems.

THEOREM 2.2 *Suppose that* Θ *is a rational J-lossless CSM with a transmission zero of degree* $m \geq 1$ *at the point* $b \in \mathbb{T}$ *and that* Θ *is loaded in a passive load* s_L. *Then the input pr. function* c *has a PLL of order at least* m *at* b.

THEOREM 2.3 *Let* $\hat{\Theta}$ *be a rational minimal outer normalized J-lossless matrix of degree* n *with transmission zeros* $\omega_1, \ldots, \omega_n$ *such that*

$$\hat{\Theta} \begin{bmatrix} c + 1 \\ c - 1 \end{bmatrix}$$

is admissible. Then

$$\Theta_n = \Theta' \hat{\Theta}$$

where Θ' *is a normalized J-lossless matrix with transmission zeros exclusively on* $\mathbb{T}$ *and where* $\deg \Theta_n = \deg \hat{\Theta}$.

In other words: all possible fundamental LIS solutions may be found by factoring out partially some of the transmission zeros on $\mathbb{T}$ to the left side. Hence: all fundamental solutions are right factors of the solutions constructed in Theorem 2.1. The proof of the last fact is somewhat involved and given in [42].

3. APPLICATIONS AND CONNECTIONS

Prediction Theory

Suppose that x_j, $j = 0, \pm 1, ..$ is a scalar, zero mean, finite variance second order stationary stochastic process with covariance:

$$r_j = E[x_k \overline{x_{k-j}}] \ , \ j = 0, \pm 1, \ldots \tag{3.1}$$

Then:

$$C(z) = r_0 + 2\sum_{j=1}^{\infty} r_j z^j \tag{3.2}$$

is a p.r. function in $\mathbb{D}$, and by the Herglotz theorem, there is a finite positive measure $d\mu$ on $\mathbb{T}$ such that

$$C(z) = \frac{1}{2\pi}\int \frac{e^{i\theta}+z}{e^{i\theta}-z}\, d\mu \tag{3.3}$$

for every $z \in \mathbb{D}$. The Hilbert space

$$\mathcal{X} = \text{closed span of } \{x_j,\ j=0,\pm 1,\ldots\} \tag{3.4}$$

is isomorphic to the space $L^2(d\mu/2\pi)$ under the Kolmogorov isomorphism:

$$x_k \to e^{-ik\theta},\ k = 0,\pm 1,\ldots \tag{3.5}$$

If the spectral measure is subject to (2.1) then μ_n belongs to the span of $\{e^{ik\theta},\ k \geq 0\}$, in $L^2(d\mu)$ whereas

$$\mathcal{M}_n' = \{f \in \mathcal{M}_n : f(0)\} = 0 \tag{3.6}$$

belongs to the span of $\{e^{ik\theta} : k \geq 1\}$ in $L^2(d\mu)$. (These spaces have an obvious meaning as the past and the strict past of the process. The function constant 1 is the Kolmogorov image of the "present" x_0). The projection of 1 onto μ_n' corresponds to the linear least squares estimate of x_0 given the space $\mathcal{M}_n'$ in the past of the process. With $K_0(z)$ as reproducing kernel for $\omega=0$ of $\mathcal{M}_n$ in $L^2(d\mu)$, we have that

$$(f,K_0(z))_{L^2(d\mu)} = f(0) \tag{3.7}$$

is zero for $f \in \mathcal{M}_n'$ and it follows that K_0 is orthogonal to $\mathcal{M}_n'$ in $L^2(d\mu)$. The orthogonal projection of 1 onto $\mathcal{M}_n'$ is given by:

$$1 - \frac{K_0(z)}{K_0(0)} = 1 - \frac{F_n^{-1\prime}(z)}{F_n^{-1}(0)} \tag{3.8}$$

the vector

$$\varepsilon_n(z) = \frac{F_n^{-1}(z)}{F_n^{-1}(0)} \tag{3.9}$$

corresponds to the forward innovations w.r.t. to $\mathcal{M}_n'$ [42], and the norm squared

$$\|\varepsilon\|^2 = K_0(0)^{-1} \tag{3.10}$$

is the square of the linear least squares prediction error.

The CSM Θ_n constructed in theorem 2.1 may be used as a prediction filter as shown in fig. 3.

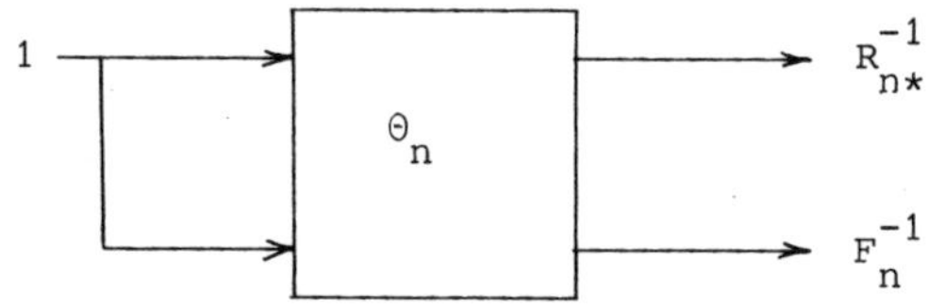

Fig. 3 The fundamental solution Θ_n used as a prediction filter.

In actual practice it will be realized as a cascade of elementary sections, see the literature on orthogonal filters. While F_n^{-1} may be interpreted as a prediction function, its inverse F_n produces a model for the original stochastic process. It is realized by the filter configuration shown in fig.4.

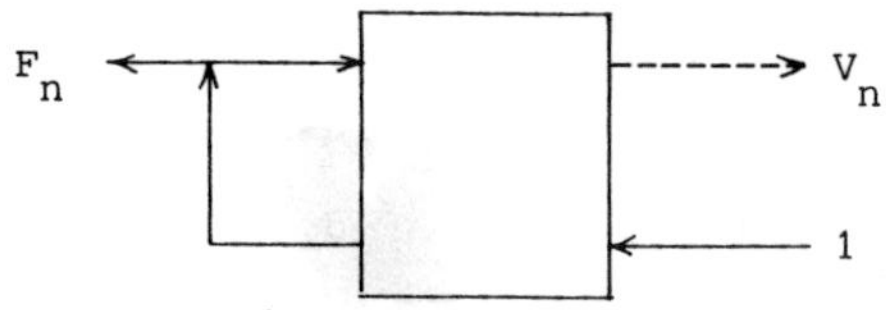

Fig. 4 The fundamental solution Σ_n used as a modeling filter.

The approximating properties of F_n are described by the following theorem.

Spectral Approximation

THEOREM 3.1 *If* $d\mu = |F(e^{i\theta})|^2 d\theta$ *where* F *is an outer Hardy function of class* H^2 *and* F_n *is a fundamental solution of the LIS problem, then*

$$\underset{f^{-1}\in\mathcal{M}_n}{\text{minimum}} \frac{1}{2\pi}\int \left|1 - \frac{F}{f}\right|^2 d\theta = 1 - \left|\frac{F(0)}{F_n(0)}\right|^2 \tag{3.11}$$

and the minimum is achieved by $kF_n(z)$ *where*

$$k = \frac{\overline{F_n(0)}}{\overline{F(0)}} \tag{3.12}$$

Fundamental LIS solutions have what one may call a "Maximum entropy-property" with respect to the space $\mathcal{M}_n$. This property is a generalization of the well-known maximum entropy property of Schur interpolants. The precise formulation is:

Maximum Entropy

THEOREM 3.2 *Let* Θ_n *be a fundamental LIS solution, let* C *be a p.r. function defined by*

$$\begin{bmatrix} 1 \\ S_L \end{bmatrix} \sim \Theta_n \begin{bmatrix} C+1 \\ C-1 \end{bmatrix} \tag{3.13}$$

(one would say: C *is the input impedance of the network loaded in* S_L*) and let* W *denote the density with respect of the Lebesgue measure of the Herglotz measure* d *for* C. *Then the entropy integral*

$$\frac{1}{2\pi}\int \log W(e^{i\theta})d\theta \le \frac{1}{2\pi}\int \log|F_n(e^{i\theta})|^2 d\theta \tag{3.14}$$

with equality only if $S_L = 0$ *i.e. if* $C = C_n$.

Extensions

A last connection to be mentioned is with extension theory. Suppose that

- Θ_n is loaded in S_L according to (3.13);
- all ω_j's are in $\mathcal{D}$
- U_n is defined as in (2.7)
- $S(z,S_L)$ is the input scattering function corresponding to S_L.
- Γ is the projection of $U_{n*}S$ on $(H^2)^\perp$
- for $F,G \in \mathfrak{M}_n$, $U_{n*}F = f_1 z^{-1} + f_2 z^{-2} + \ldots$ are in $(H^2)^\perp$ and we denote $f = (f_1, f_2, f_3, \ldots) \in \ell^2$
- H_Γ is the Hankel operator corresponding to

$\Gamma(z) = \gamma_1 z^{-1} + \gamma_2 z^{-2} + \ldots$ or

$$H_\Gamma = \begin{bmatrix} \gamma_1 & \gamma_2 & \gamma_3 & \cdots \\ \gamma_2 & \gamma_3 & \cdots & \\ \gamma_3 & \cdots & & \\ \cdots & & & \end{bmatrix} \tag{3.15}$$

We then have the following properties:

(i) $\Gamma(z)$ is independent of S_L.

(ii) For $F,G \in \mathfrak{M}_n$, the inner product $(F,G)_{L^2(d\mu)} = \int F(e^{i\theta})\overline{G(e^{i\theta})}\,d\mu$ is given by $\langle f,g\rangle = g^*(1-H_\Gamma^*)^{-1}(1-H_\Gamma^* H_\Gamma^*)(1-H_\Gamma)f$ (3.16)

(iii) All contractive extensions of $\Gamma(z) \in (H^2)^\perp$ to L^2 are given by loading Θ_n into an arbitrary contracting function S_L, obtaining S and computing $U_{n*}S$; This may also be expressed by saying that all extensions are fractional forms of the type:

$$U_{n*}\,\frac{-(1-C_n)+V_n(1+C_{n*})S_L}{(1+C_n)-V_n(1-C_{n*})S_L} \tag{3.17}$$

with S_L an arbitrary H^∞ contraction.

Given $\Gamma(z)$, one may construct $\mathcal{M}_n$, Θ_n and finally the fractional representation (3.17) by the following steps:

(i) Find a minimal inner $U(z)$ such that $U(z)\Gamma(z)$ is in H^2 (notice that $U(0) = 0$).

(ii) Define

$$\mathcal{M}_n = H^2 \ominus U(z)H^2$$

(iii) on $\mathcal{M}_n$ define the inner product induced by H_Γ according to (3.16).

(iv) Determine an orthonormal basis for $\mathcal{M}_n$, construct the reproducing Kernel and proceed according to Chapter 2.

REFERENCES

1. J. Schur, "Ueber Potenzreihen, die im Innern des Einheitskreises beschränkt sind", J. für dei Reine und Angewandte Mathematik, vol. 147, (Berlin), pp. 205-232, 1917.

2. E.A. Guillemin, "Synthesis of Passive Networks", J. Wiley, New York, 1957.

3. V. Belevitch, "Classical Network Theory", Holden-Day, San Francisco, 1968.

4. G. Pick, "Ueber die Beschränkungen Analytische Funktionen, welche durch vorgegebene Funktionwerte bewirkt werden", Math. Ann. 77, 7-23, (1916).

5. R. Nevanlinna, "Ueber beschrankte Analytische Funktionen", Ann. Acad.-Sci. Fenn., A32, 1-75 (1929).

6. C. Carathéodory and L. Fejer, "Ueber den Zusammenhang der Extremen von harmonischen Funktionen mit ihren Koeffizienten und über Picard-Landauschen Satz", Rend. Circ., Mat. Palermo 32 (1911), 218-239.

7. G. Szegö, "Orthogonal Polynomials", New York, 1939.

8. N.I. Akhiezer, "The Classical Moment Problem", Oliver and Boyd, Edinburgh, 1965.

9. L.Ya. Geronimus, "Orthogonal Polynomials", Consultants Bureau, New York, 1961.

10. N. Levinson, "The Wiener rms error criterion in filter design and prediction", J. Math. Phys., Vol. 25, pp. 261-278, Jan. 1947.

11. P. Dewilde, A.C. Vieira and T. Kailath, "On a Generalized Szegö-Levinson Realization Algorithm for Optimal Linear Predictors Based on a Network Synthesis Approach", IEEE Trans. on CAS, Vol. CAS-25, No. 9, Sept. 1978.

12. S. Darlington, "Synthesis of Reactance 4-poles which produce prescribed insertion loss characteristics", J. Math. Phys. 18, 257-355 (1939).

13. P.I. Richards, "A special class of functions with positive real part in a half-plane", Duke Math. J., vol. 14, no. 3, pp. 777-786, sept. 1947.

14. O. Brune, "Synthesis of a finite two-terminal network whose driving point impedance is a prescribed function of frequency", J. Math. Phys. 10 (3), 191-236, (Aug. 1931).

15. V. Belevitch, "Four dimensional transformations of 4-pole matrices with applications to the synthesis of reactance 4-poles", I.R.E. Trans. on Circuit Theory, Vol. CT-3, no. 2, pp. 105-111, June 1956.

16. J. Neirynck and Ph. van Bastelaer, "La Synthese des filtres par factorisation de la matrice de transfert", Revue MBLE, vol. X, No. 1, pp. 5-32.

17. P. Dewilde, "Cascade Scattering Matrix Synthesis", Techn. Rep. Stanford Info. Syst. Lab., June 1970.

18. P. Dewilde, "Input-Output Description of Roomy Systems", SIAM J. Opt. and Contr. 14, 712-736, (1976).

19. V.P. Potapov, "The Multiplicative Structure of J-contractive Matrix Functions", Amer. Math. Soc. Transl. 15, 131-243 (260).

20. D. Sarason, "Generalized Interpolants in H^∞", Trans. Amer. Math. Soc., 127, (1967), pp. 179-203.

21. B. Sz.-Nagy and C. Foias, "Harmonic Analysis of Operators on Hilbert Space", North Holland, Amsterdam, 1970.

22. B. Sz.-Nagy and C. Foias, "Sur les contractions de l'espace de Hilbert", Acta Sci. Math., 19-27.

23. V.M. Adamyan, D.Z. Arov and M.G. Krein, "Infinite Hankel Matrices and Generalized Caratheodory-Fejer and F. Riesz Problems", Funkcioncal Anal. i. Prilozen. 2 (1969), no. 1, 1-19= Funct. Anal. Appl. 2 (1968), 1-18.

24. V.M. Adamyan, D.Z. Arov and M.G. Krein, "Bounded Operators that Commute with a Contraction of Class C_{00} of Unit Rank of Nonunitarity" Funkcional Anal. i. Prilozen. 3 (1969), no. 3, 86-87 = Functional Anal. Appl. 3 (1969), 242-243.

25. V.M. Adamyan and D.Z. Arov, "On Unitary Couplings of Semi-unitary Operators", Amer. Math. Soc. Transl. (2), Vol. 95, 1970.

26. V.M. Adamyan, D.Z. Arov and M.G. Krein, "Analytic Properties of Schmidt Pairs for a Hankel Operator and the Generalized Schur-Takagi Problem", Math. USSR. Sbornik, Vol. 15 (1971), No. 1.

27. J.W. Helton, "Non-Euclidean Functional Analysis and Electronics", Bull. of the AMS, Vol. 7, No. 1, July 1982.

28. J.W. Helton, "Orbit structure of the Möbins transformation semigroup action on H^∞ (broadband matching)", Adv. in Math. Suppl. Studies, vol. 3, Academic Press, New York, 1978, pp. 129-197.

29. J.W. Helton, "Systems with infinite dimensional state space: the Hilbert Space Approach", Proc. IEEE 64 (1976), 145--160.

30. Ph. Delsarte, Y. Genin and Y. Kamp, "Orthogonal Polynomial Matrices on the Unit Circle", IEEE Trans. on CAS, Vol. 25, pp. 149-160, 1978.

31. Ph. Delsarte, Y. Genin and Y. Kamp, "Schur Parametrization of Positive-Definite Block-Toeplitz Systems", SIAM J. on Appl. Math. Vol. 36, pp. 33-46, 1979.

32. Ph. Delsarte, Y. Genin and Y. Kamp, "The Nevanlinna-Pick problem for Matrix-valued functions", SIAM J. on Appl. Math., Vol. 36, pp. 47-61, 1979.

33. I.C. Gohberg, "The Problem of Factorization of Operator Functions", Izvestija Akad. Nauk SSR, Series Math. 28 (1965), 1055-1082, (Russian).

34. H. Bart, I.C. Gohberg and M.A. Kaashoek, "Minimal Factorization of Matrix and Operator Functions", Birkhäuser, 1979.

35. T. Kailath, A. Vieira and M. Morf, "Inverse of Toeplitz Operators, Innovations and Orthogonal Polynomials", SIAM Rev., Vol. 20, no. 1, pp. 106-119, Jan. 1978.

36. T. Kailath, "Lectures on Wiener and Kalman Filtering", Springer-Verlag, CISM series, 1981.

37. L. Ljung, T. Kailath and B. Friedlander, "Scattering Theory and Linear Least Squares Estimation", Part. 1: continuous-time problems", Proc. IEEE, 64(1), 131-139.

38. H. Lev-Ari and T. Kailath, "Parametrization and Modeling of Non-stationary Processes", Techn. Rep. Stanford Univ., 1982.

39. Y. Genin and P. van Dooren, "On Σ-lossless Transfer Functions and Related Questions", Rept. R447, Philips Res. Labs., Brussels, Belgium, Nov. 1980.

40. P. Dewilde and H. Dym, "Schur Recursions, Error Formulas, and Convergence of Rational Estimators for Stationary Stochastic Sequences", IEEE Trans. on Info. Theory, Vol. IT-27, No. 4, July 1981.

41. P. Dewilde and H. Dym, "Lossless Chain Scattering Matrices and Optimum Linear Prediction: the Vector Case", Circuit Theory and Appl., Vol. 9, pp. 135-175 (1981).

42. P. Dewilde and H. Dym, "Lossless Inverse Scattering with Rational Networks: Theory and Applications", Techn. Rept., Delft Univ. of Technology, 1982.

43. T. Kailath, "Lectures on Wiener and Kalman Filtering", Lecture Notes CISM, Springer Verlag, 1981.

P.M. Dewilde
Delft University of Technology
Department of Electrical Engineering
Mekelweg 4, 2628 CD Delft
The Netherlands

Operator Theory:
Advances and Applications, Vol. 12

SUBISOMETRIC DILATIONS AND THE COMMUTANT LIFTING THEOREM

R. G. Douglas and C. Foias

A new class of dilations of contractions is introduced for which uniqueness and the commutant lifting theorem hold. The operator part added on is a uniform Jordan model.

0. The notion of a (strong) unitary dilation for a contraction was introduced by B. Sz.-Nagy in [5]. As application he obtained a direct proof of von Neumann's result that the unit disk is a spectral set for all contractions. As a consequence of the subsequent development by Sz.-Nagy and the second author as well as others, dilation theory has had important consequences in systems theory, and interpolation theory as well as in operator theory. For these purposes it is often convenient to consider isometric dilations. Two of the most important properties possessed by minimal isometric dilations is their uniqueness up to isomorphism and the commutant lifting property.

In this note we consider a larger class of dilations which we call subisometric because they are part of an isometric dilation. We obtain an alternate intrinsic characterization of this class and proceed to determine which of them possess the aforementioned uniqueness and commutant lifting properties. It turns out that both questions yield the same class. The key is that the "added operator" in the dilation be a uniform Jordan operator.

For $\mathcal{E}$ a Hilbert space let $H^2(\mathcal{E})$ denote the Hardy space on the unit circle of $\mathcal{E}$-valued functions. The shift operator $S \otimes I_{\mathcal{E}}$ in $H^2(\mathcal{E})$ is defined to be multiplication by z. For a (scalar) inner function m we set

$$\mathcal{H}(m) = H^2 \ominus mH^2$$

and define

$$S(m) = P_{\mathcal{H}(m)} S | \mathcal{H}(m)$$

where $P_{\mathcal{H}(m)}$ denotes the orthogonal projection of H^2 onto $\mathcal{H}(m)$. An operator $S(m) \otimes I_{\mathcal{E}}$ on $\mathcal{H}(m) \otimes \mathcal{E}$ is said to be a uniform Jordan operator. The subspaces $\mathcal{H}(m) \otimes \mathcal{E}$ have occurred before in this context in that they are precisely the hyperinvariant subspaces of $S \otimes I_{\mathcal{E}}$ on $H^2(\mathcal{E})$. Recall that a subspace $\mathfrak{M}$ of $\mathcal{H}$ is said to be hyperinvariant for the operator T on $\mathcal{H}$ if $\mathfrak{M}$ is invariant for every operator that commutes with T.

We expect this class of dilations to be important in the same areas as have been the isometric dilations. For that reason we believe it would be natural and interesting to consider extending the results of Arsene, Ceausescu and the second author [1], to this context, as well as to understand the connection between this concept and the remarkable new approach to interpolation theory of J. Ball and W. Helton [2].

The results in this note were obtained originally as an outgrowth of investigations begun in [4] by methods analogous to those presented there. A significant improvement in both the results and this proof resulted from a suggestion by A. E. Frazho to use the notion of hyperinvariant subspace in this study.

1. For S a contraction on a Hilbert space $\mathcal{G}$ let $D_S = (1-S^*S)^{\frac{1}{2}}$ and $\mathcal{D}_S = \overline{D_S\mathcal{G}}$. By a dilation of S we shall mean an operator T on a Hilbert space $\mathcal{H}$ containing $\mathcal{G}$ such that

(1.1) $$PT = SP$$

where P denotes the orthogonal projection of $\mathcal{H}$ onto $\mathcal{G}$. Obviously T has the following form

(1.2) $$T = \begin{bmatrix} S & 0 \\ X & T' \end{bmatrix} \text{ where } \mathcal{H} = \mathcal{G} \oplus \mathcal{H}'.$$

The dilation will be called contractive if T is a contraction which is the case if and only if T' is also a contraction and X is of the form

(1.3) $$X = D_{T'*}CD_S$$

where C is a contraction from $\mathfrak{D}_S$ to $\mathfrak{D}_{T'*}$ [6]. The dilation T is called <u>minimal</u> if

(1.4) $$\mathfrak{H} = \bigvee_{n=0}^{\infty} T^n\mathfrak{G}.$$

The most studied dilations are the minimal isometric ones. The dilation (1.2) is isometric if and only if T' and the contraction C in (1.3) are isometric. Moreover, T is minimal if and only if C is unitary and $T'^{*n} \to 0$ strongly for $n \to \infty$.

In this note we shall study minimal contractive dilations T for which $T'^{*n} \to 0$ strongly and C is a unitary operator. We shall call such a dilation a <u>minimal subisometric dilation</u>. The following result justifies this terminology.

THEOREM 1. *A contractive dilation* T *on* $\mathfrak{H}$ *of* S *on* $\mathfrak{G}$ *is minimal subisometric if and only if there exists a minimal isometric dilation* U *on* $\mathfrak{K}$ *of* S *such that*

(i) $\mathfrak{G} \subset \mathfrak{H} \subset \mathfrak{K}$ *and* U *is an isometric dilation of* T

(ii) $U' = U|\mathfrak{K}\ominus\mathfrak{G}$ *is a minimal dilation of* T'.

PROOF. Assume that T is a minimal subisometric dilation of S. Let U' on $\mathfrak{K}'$ be a minimal isometric dilation of T'. Since $T'^{*n} \to 0$ strongly, it follows from ([7], Chap. II) that U' is a unilateral shift such that $\ker U'^* = \mathfrak{L}'_* = \overline{(I-U'T'^*)\mathfrak{H}'}$. Therefore

(1.6) $$\mathfrak{K}' = \bigvee_{n=0}^{\infty} U'^n\mathfrak{L}'_*.$$

We set $\mathfrak{K} = \mathfrak{G} \oplus \mathfrak{K}'$ and $U|\mathfrak{K}' = U'$. Moreover on $\mathfrak{G}$ we define U to be

(1.7) $$Ug = Sg + \psi CD_S g \quad (g \in \mathfrak{G})$$

where ψ is the unitary isomorphism between $\mathfrak{D}_{T'*}$ and $\mathfrak{L}'_* = \ker U'^*$ defined by (see [7], Chap. II)

(1.8) $$D_{T'*}h' = (1-U'T'^*)h' \quad (h' \in \mathfrak{H}').$$

This makes $U|\mathfrak{G}$ isometric since

$$\|Uh\|^2 = \|Sh\|^2 + \|\psi CD_S h\|^2$$

$$= \|Sh\|^2 + \|D_T h\|^2 = \|h\|^2, \quad (h \in \mathfrak{G})$$

Moreover, since $U\mathfrak{G} \subset \mathfrak{G} \oplus \mathfrak{L}'_*$ and the latter is orthogonal to $U\mathcal{K}'$, it follows that U is isometric on $\mathcal{K}$. Further, because $U\mathcal{K}' \subset \mathcal{K}'$ we infer that U is a dilation of S. Finally, U is minimal because of (1.6) and the fact that

$$\mathfrak{G} \oplus \mathfrak{L}'_* = \mathfrak{G} + \overline{(\psi CD_T)} = \mathfrak{G} + \overline{(\psi CD_T\mathfrak{G})}$$

$$\subset \overline{\mathfrak{G} + (U-T)\mathfrak{G}} \subset \mathfrak{G} \vee U\mathfrak{G}.$$

Property (ii) follows directly from the definition of U. In order to verify (i) we notice that

$$U(\mathcal{K}\ominus\mathcal{H}) = U(\mathcal{K}'\ominus\mathcal{H}') \subset \mathcal{K}' \ominus \mathcal{H}' = \mathcal{K} \ominus \mathcal{H}$$

so that U is a dilation of $T'' = QU|\mathcal{H}$, where Q denotes the orthogonal projection of $\mathcal{K}$ onto $\mathcal{H}$. Since

(1.9) $$T''(g\oplus h') = QU(g\oplus h') = Q(Tg\oplus(\psi CD_S g + U'h'))$$
$$= Tg \oplus 0 + 0 \oplus P'(\psi CD_S g + U'h')(g \in \mathfrak{G}, h' \in \mathcal{H}')$$

where P' denotes the orthogonal projection of $\mathcal{K}'$ onto $\mathcal{H}'$. Obviously, we have $P'U'h' = T'h'$ for h' in $\mathcal{H}'$ so that $T = T'$ will follow from (1.9) once we verify

(1.10) $$P'(\psi CD_S g) = D_{T'*}CD_S g \quad (g \in \mathfrak{G}).$$

To do this we use the following fact which is valid for any isometric dilation U' on $\mathcal{K}'$ of a contraction T' on $\mathcal{H}'$, namely

(1.11) $$P'|\ker U'^* = D_{T'*}\psi^{-1}$$

where ψ was defined in (1.8). Indeed

$$P'(1-U'T'^*)h' = (1-T'T'^*)h' = D_{T'*}D_{T'*}h'$$
$$= D_{T'*}\psi^{-1}(1-U'T'^*)h'$$

for h in $\mathcal{H}'$.

Conversely, let T on $\mathcal{H}$ be a contractive dilation of S on $\mathfrak{G}$ such that there exists a minimal isometric dilation U on $\mathcal{K}$ of S satisfying the properties (i) and (ii). Since $U' = U|\mathcal{K}\ominus\mathcal{H}$ is always a unilateral shift ([7], Chap. II), we infer from (ii) that

$$T'^{*n} = U'^{*n}|\mathcal{H}' \to 0 \text{ strongly.}$$

So in order to verify that T is a minimal subisometric dilation of S, it remains to check that the contraction C in (1.3) is a unitary operator. For this purpose we observe that

$$D_{T'*}CD_S g = P'(U-S)g \quad (g \in \mathfrak{G}) \tag{1.12}$$

where P' denotes the orthogonal projection of $\mathcal{K}'$ onto $\mathcal{H}'$. But $\overline{((U-S)\mathfrak{G})}$ is the generating wandering subspace for U' (see [7], Chap. II), that is, the space $\mathcal{L}'_* = \ker U'^*$ with the above notation. On the other hand

$$\varphi D_S g = (U-S)g$$

defines a unitary operator from $\mathfrak{D}_S$ to $\overline{(U-S)\mathfrak{G}} = \ker U'^*$ so (1.12) becomes

$$D_{T'*}CD_S g = P'\varphi D_S g \ (g \in \mathfrak{G}).$$

This with (1.11) becomes

$$D_{T'*}C = P'\varphi = D_{T'*}\psi^{-1}\varphi \text{ or } D_{T'*}(C-\psi^{-1}\varphi) = 0.$$

Since $C-\psi^{-1}\varphi$ has its range in $\mathfrak{D}_{T'*}$, it follows that $C = \psi^{-1}\varphi$ is unitary and the proof is complete.

2. An important property of the minimal isometric dilation of a given contraction S on $\mathfrak{G}$ is the <u>uniqueness property</u> (up to isomorphism): Namely, if T_1 on $\mathcal{H}_1$ and T_2 on $\mathcal{H}_2$ are two minimal isometric dilations of S on $\mathfrak{G}$, then there exists a unitary operator W from $\mathcal{H}_1$ onto $\mathcal{H}_2$ such that

$$WT_1 = T_2W \quad \text{and} \quad W|\mathfrak{G} = I_{\mathfrak{G}}. \tag{2.1}$$

The following two results show that only a special class of minimal subisometric dilations share this property. A minimal subisometric dilation T on $\mathcal{H}$ of S on $\mathfrak{G}$ will be said to be a <u>minimal uniform m-Jordan dilation</u> if the operator T' in representation (1.2) of T is isomorphic to $S(m) \otimes I_{\mathcal{E}}$ for some inner function m and Hilbert space $\mathcal{E}$.

THEOREM 2. *If* T_1 *on* $\mathcal{H}_1$ *and* T_2 *on* $\mathcal{H}_2$ *are minimal uniform* m*-Jordan dilations of* S *on* $\mathfrak{G}$*, then there exists a unitary operator* W *from* $\mathcal{H}_1$ *to* $\mathcal{H}_2$ *such that* $WT_1 = T_2W$ *and* $W|\mathfrak{G} = I_{\mathfrak{G}}$.

PROOF. Let U_j on $\mathcal{K}_j$ be the minimal isometric dilation of T_j provided by Theorem 1. Since U_j' in representation (1.2) is the minimal dilation of T_j', we have by definition that

(2.2) $\mathbb{K}_j \ominus \mathbb{H}_j = \mathbb{K}_j' \ominus \mathbb{H}_j' = m(U_j')(\mathbb{K}_j \ominus \mathbb{G}) \quad (j=1,2).$

On the other hand let X be a unitary operator from $\mathbb{K}_1$ to $\mathbb{K}_2$ such that

(2.3) $$XU_1 = U_2X \text{ and } X|\mathbb{G} = I_{\mathbb{G}}.$$

By (2.3) we infer that $Y = X|\mathbb{K}_1 \ominus \mathbb{G}$ is a unitary operator from $\mathbb{K}_1 \ominus \mathbb{G}$ to $\mathbb{K}_2 \ominus \mathbb{G}$ having the property

(2.4) $$YU_1' = U_2'Y.$$

Therefore, by (2.2)

$$X(\mathbb{K}_1 \ominus \mathbb{H}_1) = Y(\mathbb{K}_1 \ominus \mathbb{H}_1) = Ym(U_1')(\mathbb{K}_1 \ominus \mathbb{G})$$

$$= m(U_2')Y(\mathbb{K}_1 \ominus \mathbb{G}) = m(U_2')(\mathbb{K}_2 \ominus \mathbb{G}) = \mathbb{K}_2 \ominus \mathbb{H}_2$$

so that X being unitary, we see that $W = X|\mathbb{H}_1$, is a unitary operator from $\mathbb{H}_1$ to $\mathbb{H}_2$ having the desired properties (2.1).

THEOREM 3. *If* T_1 *on* $\mathbb{H}$ *is a minimal subisometric dilation of* S *on* $\mathbb{G}$ *with the property that for any other minimal subisometric dilation* T_2 *of* S *such that* T_2' *is unitarily equivalent to* T_1' *it follows that* T_2 *is isomorphic to* T_1, *then* T_1' *is unitarily equivalent to some* $S(m) \otimes I_{\mathbb{E}}$.

PROOF. Let

$$T_1 = \begin{bmatrix} S & 0 \\ D_{T'*}C_1D_S & T_1' \end{bmatrix}$$

be the representation (1.2) of T_1. Let C_2 be any unitary operator from $\mathbb{D}_S$ to $\mathbb{D}_{T'*}$ and define the minimal subisometric dilation T_2 of S by

$$T_2 = \begin{bmatrix} S & 0 \\ D_{T'*}C_2D_S & T_1' \end{bmatrix}.$$

Let U_1 on $\mathbb{K}_1$ and U_2 on $\mathbb{K}_2$ be the minimal isometric dilations of T_1 and T_2 assured by Theorem 1. By the assumption on T_1, there exists a unitary operator W from $\mathbb{H}_1$

to $\mathfrak{H}_2 = \mathfrak{H}_1$ such that $WT_1 = T_2W$ and $W|\mathfrak{G} = I_{\mathfrak{G}}$. By the standard proof of the uniqueness property of the minimal isometric dilations, W extends to a unique unitary operator X from $\mathfrak{K}_1$ to $\mathfrak{K}_2$ such that $XU_1 = U_2X$. Setting $Y = X|\mathfrak{K}_1 \ominus \mathfrak{G}$ we get a unitary operator from $\mathfrak{K}_1 \ominus \mathfrak{G}$ to $\mathfrak{K}_2 \ominus \mathfrak{G}$ such that $YU_1' = U_2'Y$ and therefore uniquely determined by its restriction $Y_0 = Y|\ker U_1'^*$ which is a unitary operator from $\ker U_1'^*$ to $\ker U_2'^*$. Introducing the operators $\psi = \psi_1 = \psi_2$ defined in (1.8) for $T' = T_1' = T_2'$ respectively we have

$$\ker U_j'^* = \overline{(U_j-S)\mathfrak{G}} = \psi C_j \mathfrak{D}_S \qquad (j=1,2)$$

and

$$\begin{aligned} Y_0\psi C_1 D_S g &= Y_0(U_1-S)g = X(U_1-S)g \\ &= XU_1 g - Sg = U_2Xg - Sg \\ &= (U_2-S)g = \psi C_2 D_S g \qquad (g \in \mathfrak{G}) \end{aligned}$$

and therefore

$$Y_0 = \psi C_2 C_1^{-1}\psi^{-1}.$$

When C_2 ranges over all unitary operators from $\mathfrak{D}_S$ to $\mathfrak{D}_{T'*}$, Y_0 ranges over all unitary operators on $\ker U_1'^*$. Since $Y\mathfrak{H}_1' = W\mathfrak{H}_1' = \mathfrak{H}_1'$ we conclude that $\mathfrak{H}_1'$ is invariant for all unitary operators Y that commute with U_1'. The following lemma with $U_1' = V$, $\mathfrak{K}_1' = H^2(\mathfrak{E})$ and $\mathfrak{H}_1' = H^2(\mathfrak{E}) \ominus \mathfrak{M}$ completes the proof.

LEMMA. *Let* $\mathfrak{E}$ *be a Hilbert space and* V *denote the unilateral shift on* $H^2(\mathfrak{E})$. *If* $\mathfrak{M}$ *is an invariant subspace for* V *which is invariant for all unitary operators on* $H^2(\mathfrak{E})$ *commuting with* V, *then*

$$\mathfrak{M} = mH^2(\mathfrak{E})$$

for some (numerical) inner function m.

PROOF. One can quickly reduce to the case where $\mathfrak{E}$ is separable. By ([7], Chap. V) there exists an inner function $\{\mathfrak{F},\mathfrak{E},\Theta(z)\}$ such that

$$\mathfrak{M} = \Theta H^2(\mathfrak{F}).$$

By assumption for any unitary operator Y on $\mathfrak{E}$ we have $Y\mathfrak{M} = \mathfrak{M}$ so that by

(2.5) $Y\Theta u = \Theta v_u \quad (u \in H^2(\mathfrak{F}))$

we can define a unitary operator $u \to v_u$ from $H^2(\mathfrak{E})$ to $H^2(\mathfrak{F})$ which commutes with V. Therefore there exists a unitary operator U_Y on $\mathfrak{F}$ such that $U_Y u(z) = v_u(z)$ for u in $H^2(\mathfrak{F})$ and $|z| < 1$. From (2.5) it follows that

(2.6) $Y\Theta(z) = \Theta(z)U_Y$

But $\Theta(e^{it})$ is isometric on $\mathbb{T}$ a.e. and hence using the separability of $\mathfrak{F}$, we can find an e^{it_o} for which $U_Y = \Theta_0^* Y\Theta_0$, holds for all unitary Y on $\mathfrak{E}$ with $\Theta_0 = \Theta(e^{it_o})$. If Θ_0 is not unitary we can choose Y such that $\ker \Theta_0^* Y\Theta_0 \neq (0)$ which contradicts the unitarity of U_Y. Thus (2.6) becomes

(2.7) $Y\Theta(z) = \Theta(z)\Theta_0^* Y\Theta_0 \quad (|z|<1)$

and multiplication on the right by Θ_0^* yields

(2.8) $Y\Theta(z)\Theta_0^* = \Theta(z)\Theta_0^* Y \quad (|z|<1)$.

This implies that $\Theta(z)\Theta_0^*$ is a scalar operator $m(z)I$ and hence

$$\mathfrak{M} = mH^2(\mathfrak{E}).$$

3. A dilation T on $\mathfrak{H}$ of a contraction S on $\mathfrak{G}$ is said to have the <u>commutant lifting property</u> (CLP) if for every contraction A commuting with S there exists a contraction B commuting with T which dilates A, that is, $PB = AP$ where P is the orthogonal projection of $\mathfrak{H}$ onto $\mathfrak{G}$.

It is a remarkable and useful fact that if T is a minimal isometric dilation of S, then T has CLP ([7], Chap. II). The following results show that the minimal uniform Jordan dilations share CLP and indeed are the only dilations which do. We state the CLP for intertwining operators between two dilations.

THEOREM 4. *Let* T_i *on* $\mathfrak{H}_i$ *be a minimal uniform* m-*Jordan dilation of* S_i *on* $\mathfrak{G}_i$ $(i=1,2)$. *If* A *is a contraction from* $\mathfrak{G}_1$ *to* $\mathfrak{G}_2$ *satisfying* $S_2A = AS_1$, *then there exists a contraction* B *from* $\mathfrak{H}_1$ *to* $\mathfrak{H}_2$ *satisfying* $T_2B=BT_1$ *and* $P_2B = AP_1$, *where* P_i *is the orthogonal projection of* $\mathfrak{H}_i$ *onto* $\mathfrak{G}_i$.

PROOF. Let U_i on K_i be the minimal isometric dilation of T_i provided by Theorem 1 (i=1,2). If V_i is a unitary operator from $H(m) \otimes E_i$ to H_i' satisfying

$$T_i'V_i = V_i(S(m) \otimes I_{E_i}),$$

then there exists a unique unitary operator W_i from $H^2(E_i)$ to K_i' such that $W_i(S_i \otimes I_{E_i}) = U_iW_i$ which extends V_i (i=1,2). Obviously

$$H^2(E_i) \ominus (H(m) \otimes E_i) = m H^2 \otimes E_i$$

$$= m(S \otimes I_{E_i})H^2(E_i)$$

so that

$$K_i' \ominus H_i' = W_i[(H^2(E_i) \ominus (H(m) \otimes E_i)]$$

$$= W_i[m(S \otimes I_{E_i})(H^2(E_i))]$$

$$= m(U_i')W_iH^2(E) = m(U_i')K_i' \quad (i=1,2).$$

Therefore $\tilde{C}(K_1' \ominus H_1') \subset K_2' \ominus H_2'$ for any $\tilde{C}$ such that $\tilde{C}U_1' = U_2'\tilde{C}$. Now let A be a contraction from G_1 to G_2 satisfying $S_2A = AS_1$. Applying the lifting theorem for intertwining operators ([7], Chap. II), we obtain a contraction C from K_1 to K_2 satisfying $U_2C = CU_1$ and dilating A. Therefore $CK_i' \subset K_2'$, where $K_i' = K_i \ominus G_i$, and

$$\tilde{C} = C|K_1$$

defined from K_1' to K_2' and intertwining U_1' and U_2' satisfies

$$(3.1) \qquad \tilde{C}(K_1' \ominus H_1') \subset K_2' \ominus H_2'.$$

Hence

$$C(K_1 \ominus H_1) = \tilde{C}(K_1' \ominus H_1') \subset K_i' \ominus H_i' = K_2 \ominus H_2$$

and therefore the contraction $B = C|H_1$ defined from H_1 to H_2 is a dilation of A which intertwines T_1 and T_2.

THEOREM 5. *If a contraction* T_0 *satisfies* $T_0^{*n} \to 0$ *strongly and has the property that for any contraction* S, *any minimal subisometric dilation* T *of* S *such that* $T' = T_0$, *has the commutant lifting property, then* T_0 *is unitarily equivalent to some* $S(m) \otimes I_E$.

PROOF. Let T on $\mathfrak{H}$ be the adjoint of the minimal isometric dilation of T'^* (where $T' = T_0$). Let $\mathfrak{G} = \mathfrak{H} \ominus \mathfrak{H}'$ and $S = PT|\mathfrak{G}$, where P denotes the orthogonal projection of $\mathfrak{H}$ onto $\mathfrak{G}$. Obviously, since $T\mathfrak{H}' = T'\mathfrak{H}' \subset \mathfrak{H}'$, T is a contractive dilation of S. Since T^* is the minimal isometric dilation of T'^* it is easy to check (by duality) that the contraction C in representation (1.3) is a unitary operator. Then T is a minimal subisometric dilation of S. Let U on $\mathfrak{K}$ be the minimal isometric dilation of S provided by Theorem 1. Since $T'^{*n} \to 0$, it follows from ([7], Chap. II) that

$$\mathfrak{K} = \bigvee_{n=0}^{\infty} U^n \mathfrak{G}$$

and that U^* is a bilateral shift extending $T^*|\mathfrak{G}$. Therefore $U|\mathfrak{K} \ominus \mathfrak{G}$ is a unilateral shift the unitary extension of which is U. If Θ is any contraction on $\mathfrak{K}\ominus\mathfrak{G}$ commuting with $U|\mathfrak{K}\ominus\mathfrak{G}$, then there exists a unique extended extension $\tilde{\Theta}$ on $\mathfrak{K}$ commuting with U and $\tilde{\Theta}$ is necessarily a contraction. We set $A = P_{\mathfrak{G}}\tilde{\Theta}|\mathfrak{G}$, where $P_{\mathfrak{G}}$ is the orthogonal projection of $\mathfrak{K}$ onto $\mathfrak{G}$. Then $A^* = \Theta^*|\mathfrak{G}$ and Θ^* is the only bounded operator on $\mathfrak{K}$ commuting with U^* and extending A^*.

Since T is assumed to have the CLP, there exists a contraction B on $\mathfrak{H}$ commuting with T and dilating A. But U^* is also the unitary extension of T^*. Therefore there exists a unique operator $\tilde{B}$ on $\mathfrak{K}$ such that $\tilde{B}^*$ commutes with U^* and extends B^*. Obviously, $\tilde{B}$ is also a dilation of A so that $\tilde{B}^*|\mathfrak{G} = A^* = \tilde{\Theta}^*|\mathfrak{G}$. From the uniqueness of the extension, we have $\tilde{B} = \tilde{\Theta}$. On the other hand $\tilde{B}$ is a dilation of B so $\tilde{\Theta}$ is a dilation of B and therefore

$$\Theta(\mathfrak{K}\ominus\mathfrak{H}) = \tilde{\Theta}(\mathfrak{K}\ominus\mathfrak{H}) \subset \mathfrak{K} \ominus \mathfrak{H}.$$

We conclude that $\mathfrak{K} \ominus \mathfrak{H}$ is hyperinvariant for $U|\mathfrak{K}\ominus\mathfrak{G}$ and hence (see [3], [7])

$\mathfrak{K}\ominus\mathfrak{H} = m(U|\mathfrak{K}\ominus\mathfrak{G})\mathfrak{K}\ominus\mathfrak{G}$ for some inner function m. This implies ([7], Chap. VI) that T' is unitarily equivalent with $S(m) \otimes I_{\mathcal{E}}$, where $\mathcal{E} = \ker(U|\mathcal{K}\ominus\mathcal{G})^*$.

REFERENCES

1. Arsene, Gr., Ceausescu, Z. and Foias, C.: On intertwining dilations VIII, J. Operator Theory 4(1980), 55-91

2. Ball, J. A. and Helton, T. W.: A Beurling-Lax Theorem for the Lie group U(m,n) which contains most classical interpolation theory. J. Operator Theory 8(1983), 107-142.

3. Douglas, R. G. and Pearcy, C. M.: On a topology for invariant subspaces. J. Functional Anal. 2(1963), 323-341.

4. Douglas, R. G. and Foias, C.: A homological view in dilation theory, INCREST Preprint No. 15/1976.

5. Sz.-Nagy, B.: Sur les contractions de l'espace de Hilbert, Acta Sci. Math. (Szeged) 15(1953), 87-92.

6. Sz.-Nagy, B. and Foias, C.: Forme triangulaire d'une contraction et factorisation de la fonction caractéristique, Acta Sci. Math. (Szeged) 28(1967), 201-212.

7. _____: Harmonic Analysis of Operators on Hilbert Space, American Elsevier, New York, 1970.

R. G. Douglas
Department of Mathematics
State University of New York
Stony Brook, New York 11794

C. Foias
Department of Mathematics
Indiana University
Bloomington, Indiana 47405

Operator Theory:
Advances and Applications, Vol. 12

POSITIVE DEFINITE EXTENSIONS, CANONICAL EQUATIONS AND INVERSE PROBLEMS

Harry Dym and Andrei Iacob

The intimate connections between a form of the covariance extension problem and the spectral theory of canonical equations are developed and exploited in order to deduce a representation formula for the set of all solutions to the extension problem in an intuitively pleasing way. Enroute, an expository account of the forward and inverse spectral problem for canonical equations and the theory of Hilbert spaces of matrix valued entire functions is given and a number of related applications are discussed.

TABLE OF CONTENTS

1. INTRODUCTION

The main purpose of this paper is to clarify the fundamental connection between a certain positive definite extension problem (which occurs, for example, in the theory of covariance extensions) and the inverse spectral problem for a suitably normalized system of first order differential equations, and to use this connection to give a complete description of the solutions to the extension problem in an intuitively pleasing way.

The given data for the extension problem under study is an $n \times n$ matrix valued function $k(t) = k(-t)^*$ which is defined and summable on the interval $-\tau \leq t \leq \tau$ and is such that

$$\int_0^\tau [\varphi(t)]^* \{\varphi(t) - \int_0^\tau k(t-s)\varphi(s)ds\}dt > 0$$

for every nonzero $n \times 1$ vector-valued square summable function φ. The objective is to find a summable $n \times n$ matrix valued function $h(s)$ such that

$$h(s) = k(s) \quad \text{for} \quad |s| \leq \tau \tag{1.1}$$

and

$$I_n - \hat{h}(\lambda) > 0 \tag{1.2}$$

for every $\lambda \in \mathbb{R}$, in which $\hat{h}$ designates the Fourier transform of h. The existence of at least one solution to this extension problem was established by Krein [K5] and independently, though much later, by Dym-Gohberg [DG1] as a special case of an extensive study of extensions in a more general setting. In fact it turns out that for every $\tau < \infty$ there are infinitely many solutions to this problem. For a complete description of all the extensions to a more general scalar problem see [KL]; that paper also contains a brief history of the problem and a number of relevant references.

The fundamental fact which clarifies the description of all solutions to the stated extension problem is that the set of functions $I_n - \hat{h}(\lambda)$ with $h \in L^1_{n\times n}(\mathbb{R})$ which meet condition (1.2) is precisely equal to the set of spectral functions for a class of suitably normalized first order matrix differential equations, which will be referred to henceforth as canonical equations; see Section 2 for the precise definitions. Thus, in particular, for every solution h of the extension problem, $I_n - \hat{h}(\lambda)$ can be identified as the spectral function of exactly one canonical equation (2.1) with summable suitably normalized potential V. As we shall see, the given data for the "2T-extension problem" : $k(s)$ for $-2T \leq s \leq 2T$, determines the potential V on the interval $0 \leq s \leq T$ and vice versa. Moreover, the different solutions of the 2T-extension problem correspond precisely to the different ways of continuing the potential on the interval $T < s < \infty$ so that it remains summable and satisfies certain symmetry conditions. Because of this symmetry constraint, the freedom amounts to specifying a single $n \times n$ complex summable matrix function on the interval. This matrix function (which, because of its role as the "corner" of a resolvent kernel, and following the usage of Krein [K5], will be designated $\Gamma_{2t}(0,2t)$ in the sequel) is the continuous counterpart of the well known reflection coefficient which arises in the analogous discrete extension problem.

The theory of canonical equations leads in a natural way to a linear fractional representation for the set of all solutions to the 2T-extension problem, as is explained in detail in Section 7. This, in turn, upon making some auxiliary identifications, leads readily to the conclusion that h is a solution of the 2T-extension problem if and only if $I_n - \hat{h}$ can be expressed in the form

$$I_n - \hat{h} = [(I_n + \hat{c}_{2T})^*]^{-1}(I_n - \sigma\Sigma)^{-1}(I_n - \sigma\sigma^*) \times$$
$$\times (I_n - \Sigma^*\sigma^*)^{-1}[(I_n + \hat{c}_{2T})]^{-1} \tag{1.3}$$

where σ is an arbitrary strictly contractive $n \times n$ matrix valued function in the Wiener algebra W_+^0 (which will be defined below), and c_{2T} and Σ are determined from the data:

$$\Sigma = e^{2i\lambda T}(I_n + \hat{a}_{2T})^{-1}(I_n + \hat{c}_{2T})$$

and a_{2T} and c_{2T} are the (unique) solutions of the equations

$$a_{2T}(t) - \int_0^{2T} k(t-s)a_{2T}(s)ds = k(t)$$

for $0 \leqslant t \leqslant 2T$, and

$$c_{2T}(t) - \int_{-2T}^{0} k(t-s)c_{2T}(s)ds = k(t)$$

for $-2T \leqslant t \leqslant 0$, respectively. It is perhaps well to remark that, although formula (1.3) is not established in the text, it follows readily from (7.4), (6.22) and (6.23) and the identifications in Lemma 6.1 of [DG1].

The special choice $\alpha = 0$ in (1.3) corresponds to the band extension h_{2T} (in the language of [DG1] and [DG3]). The name derives from the fact that the inverse Fourier transform of $[I_n - \hat{h}_{2T}]^{-1} - I_n$ lives precisely inside the band $-2T \leqslant s \leqslant 2T$ where the data is specified. Subject to some technical constraints, the band extension maximizes the entropy integral introduced in [DG1], and is in fact characterized by this property.

The function

$$Z_T(\lambda) = I_n - 2 \int_0^{\infty} e^{i\lambda s} h_{2T}(s)ds$$

and its limit $Z_\infty(\lambda)$ play an important role in the study of square summable solutions of canonical equations, and the corresponding "Weyl circles", which are discussed in Section 8.

The subject of canonical equations was systematically developed over the past thirty years by M.G. Krein and his students and colleagues: [Ad], [Ar], [K2]-[K5], [KMaF1], [KMaF2], [MaF1], [MaF2], [MaP1], [MaP2]; see also [At] and [DaK]. Nevertheless, since much of this material appears in Doklady notes in abbreviated form and/or in journal articles which are either not easily accessible or have not been translated into English, we have attempted to give a detailed, self-contained, exposition of a number of aspects of this subject, which is suitable for the purposes of this paper and in fact suffice related applications. In our approach, we have made extensive use of the theory of Hilbert spaces of vector valued entire functions due to de Branges [dB2], [dB3]. In particular, we use it to develop the spectral theory and associated Fourier analysis for the canonical equation. All this material is also presented in a self-contained expository manner and we hope that this paper will also serve as an introduction to this very beautiful theory.

In point of fact, the first seven sections of this paper cover the extension problem formulated above. However, having developed the machinery, we were not able to resist the temptation to include several of the many beautiful applications which were within reach, especially since some of them could now be dealt with in a more unified and simpler way than seemed to be available in the literature. This is perhaps the proper place to mention that in spite of the length of this paper, it is still not complete: In Sections 6 and 9, which deal with the inverse spectral and scattering problems, we have followed the time-honored practice of skipping most of the technical details, which are difficult and lengthy. Instead, we have focused on the ideas which underlie the methods of Gelfand-Levitan, Krein, and Marchenko, with special attention paid to the role played by factorization.

Finally, before turning to notation, let us remark that a number of the results touched upon here have been extended to include extensions of more general Hermitian functions and other classes of canonical equations as well; see for example Chapter 6 and the Appendix of [GK2], [KL], and [KP].

We shall use the symbols $\mathbb{R}$ and $\mathbb{C}$ for the real and complex numbers respectively; $\mathbb{R}^+$ stands for the half line of nonnegative reals whereas $\mathbb{C}_+$ [resp. $\mathbb{C}_-$] denotes the open upper [resp. lower] half of the complex plane. If A is a set, then $\bar{A}$ stands for its closure. The symbol A^* denotes the conjugate transpose of A if A is a matrix (and hence just the ordinary complex conjugate if A is a number) and the adjoint of A with respect to the inner product in use if A is an operator. If $A(\omega)$ is a matrix valued function, then $A^{\#}(\omega)$ stands for $[A(\omega^*)]^*$. We shall use the symbols $\hat{f}$ and $f^{\wedge}$ [resp. $f^{\vee}$] for the usual Fourier [resp. inverse Fourier] transform:

$$\hat{f}(\lambda) = \int e^{i\lambda s} f(s)ds \quad [\text{resp. } f^{\vee}(t) = \frac{1}{2\pi} \int e^{i\xi t} f(\xi)d\xi] .$$

Moreover, we shall adopt the convention that the limits of all integrals (as above) will be $\pm\infty$ unless specifically indicated otherwise. The symbols $L^p_{n\times m}(\mathbb{R})$, $L^p_{n\times m}(\mathbb{R}^+)$, $H^p_{n\times m}$ and $K^p_{n\times m}$ stand for the spaces of $n \times m$ matrix valued functions with entries belonging to $L^p(\mathbb{R})$, $L^p(\mathbb{R}^+)$, H^p and K^p, respectively, where H^p [resp. K^p] denotes the Hardy space over $\mathbb{C}_+$ [resp. $\mathbb{C}_-$]; L^p_n, H^p_n and K^p_n are short for $L^p_{n\times 1}$, $H^p_{n\times 1}$

and $K^p_{n\times 1}$, respectively. The symbol $W_{m\times m}$ denotes the Wiener algebra of $m \times m$ matrix valued functions of the form $f(\lambda) = c_f + \hat{k}_f(\lambda)$, where c_f is a constant matrix and $k_f \in L^1_{m\times m}(\mathbb{R})$. It follows from the Riemann-Lebesgue lemma that $c_f = f(\infty)$ and hence that $k_f = [f - c_f]^\vee$. The spaces

$$W^0_{m\times m} = \{f \in W_{m\times m}: f(\infty) = 0\} ,$$

$$W^I_{m\times m} = \{f \in W_{m\times m}: f(\infty) = I_m\} ,$$

$$(W_{m\times m})_+ = \{f \in W_{m\times m}: f^\vee(s) = 0 \quad \text{for} \quad s < 0\}$$

and

$$(W_{m\times m})_- = \{f \in W_{m\times m}: f^\vee(s) = 0 \quad \text{for} \quad s > 0\} ,$$

will play an important role. In order to simplify the typography, however, we shall drop the subscripts $m \times m$ and rely on the context for the size. The notations

$$W^0_\pm = W^0 \cap W_\pm \quad \text{and} \quad W^I_\pm = W^I \cap W_\pm$$

will also prove useful. We shall also refer to the matrix versions of the Wiener theorems a number of times; a convenient formulation may be found in Section 2 of [DG1]. The symbol □ marks the end of a proof.

2. MATRIZANTS

In this section we shall establish a number of properties of the so-called matrizant $U(t,\lambda)$ of the canonical equation given below, and of a normalized version of it, $\Omega(t,\lambda)$, which will be introduced later. To be more precise, from here and now on, the term *canonical equation* will signify a differential equation of the form

$$J\frac{dY}{dt} = V(t)Y + \lambda Y \qquad (t \geq 0)\ , \tag{2.1}$$

where

$$J = \begin{bmatrix} 0 & I_n \\ -I_n & 0 \end{bmatrix} ,$$

the unknown $Y = Y(t,\lambda)$ is a $2n \times m$ matrix valued function with $1 \leq m \leq 2n$ and V, the *potential*, is a given $2n \times 2n$ matrix valued function on $\mathbb{R}^+$ with summable entries which is both selfadjoint and J-selfadjoint a.e.:

$$V = V^* \quad \text{and} \quad JV = (JV)^* \quad [= -VJ]\ .$$

These last two conditions force V to be of the $n \times n$ block form

$$V = \begin{bmatrix} V_1 & V_2 \\ V_2 & -V_1 \end{bmatrix}, \quad \text{with} \quad V_j = V_j^*\ ,\ j = 1,2\ . \tag{2.2}$$

We remark that equations of the form (2.1) with potentials which are selfadjoint only can be simply transformed into equations of the same form with potentials which are both selfadjoint and J-selfadjoint; see e.g. [Ad].

The $2n \times 2n$ matrix solution $U(t,\lambda)$ of (2.1) with $U(0,\lambda) = I_{2n}$ is termed the *matrizant*. It is convenient to consider first the *normalized matrizant*

$$\Omega(t,\lambda) = [U_0(t,\lambda)]^{-1}U(t,\lambda)\ , \tag{2.3}$$

where

$$U_0(t,\lambda) = e^{-\lambda tJ} = \begin{bmatrix} (\cos\lambda t)I_n & -(\sin\lambda t)I_n \\ (\sin\lambda t)I_n & (\cos\lambda t)I_n \end{bmatrix} \tag{2.4}$$

designates the matrizant of the canonical equation with $V = 0$.

It is readily checked that

$$\frac{d}{dt}\,\Omega(t,\lambda) = e^{2\lambda tJ}V(t)J\Omega(t,\lambda)$$

or, equivalently, since $\Omega(0,\lambda) = I_{2n}$, that

$$\Omega(t,\lambda) = I_{2n} + \int_0^t e^{2\lambda sJ}V(s)J\Omega(s,\lambda)\,ds\ . \tag{2.5}$$

By standard methods it may be shown that equation (2.5) admits a solution which is absolutely continuous in t and entire in λ; [At] and [DaK] are good sources of information for such matters. More precisely, upon setting

$$\Omega_0(t,\lambda) = I_{2n}\ ,$$

$$\Omega_{j+1}(t,\lambda) = \int_0^t e^{2\lambda sJ}V(s)J\Omega_j(s,\lambda)\,ds\ ,\ j=0,1,\dots\ ,$$

and, taking note of the bounds

$$|e^{2\lambda sJ}| \leqslant e^{2|b|s} \qquad (s \geqslant 0\ ,\quad \lambda = a+ib)\ ,$$

it is readily seen that

$$|\Omega_j(t,\lambda)| \leqslant \frac{1}{j!}\ [\int_0^t e^{2|b|s}\ |V(s)|ds]^j$$

and hence, that

$$\sum_{j=0}^{\infty} \Omega_j(t,\lambda)$$

converges to a solution of (2.5) with the stated properties. Sharper bounds may be obtained by taking advantage of the identity

$$e^{\alpha J}V(s) = V(s)e^{-\alpha J}\quad ,\quad \alpha \in \mathbb{C}\ ,$$

(which follows from $JVJ = V$), in order to express Ω_j in the form

$$\Omega_j(t,\lambda) = \int_0^t e^{2\lambda sJ}\omega_j(t,s)\,ds$$

for $j = 1,2,\dots$, where

$$\omega_1(t,s) = V(s)J \quad \text{for} \quad 0 \leqslant s \leqslant t\ ,$$

and

$$\omega_{j+1}(t,s) = \int_s^t V(u)J\omega_j(u,u-s)\,du \quad \text{for} \quad 0 \leqslant s \leqslant t\ . \tag{2.6}$$

This in turn leads to the representation

$$\Omega(t,\lambda) = I_{2n} + \int_0^t e^{2\lambda sJ}\omega(t,s)\,ds \ , \tag{2.7}$$

where

$$\omega(t,s) = \sum_{j=1}^{\infty} \omega_j(t,s) \tag{2.8}$$

is summable:

$$\begin{aligned}\int_0^t |\omega(t,s)|\,ds &\leq \sum_{j=1}^{\infty} \int_0^t |\omega_j(t,s)|\,ds \\ &\leq \sum_{j=1}^{\infty} \frac{1}{j!} \left[\int_0^t |V(s)|\,ds\right]^j \\ &= \nu(t)-1 \ ,\end{aligned}$$

where

$$\nu(t) = \exp\left\{\int_0^t |V(s)|\,ds\right\} .$$

This yields the bounds

$$\begin{aligned}|\Omega(t,\lambda)| &\leq 1 + e^{2|b|t} \int_0^t |\omega(t,s)|\,ds \\ &\leq \nu(t) \exp\{2|b|t\} \ .\end{aligned} \tag{2.9}$$

Therefore

$$\begin{aligned}U(t,\lambda) &= e^{-\lambda tJ}\Omega(t,\lambda) \\ &= e^{-\lambda tJ} + \int_0^t e^{\lambda(2s-t)J}\omega(t,s)\,ds\end{aligned}$$

is subject to the bound

$$\begin{aligned}|U(t,\lambda)| &\leq e^{|b|t} + \int_0^t e^{|b|(t-s)}e^{|b|s}\,|\omega(t,s)|\,ds \\ &\leq \nu(t) \exp\{|b|t\} \ .\end{aligned} \tag{2.10}$$

We now prepare a property list of the matrizant $U(t,\lambda)$ and of the normalized matrizant $\Omega(t,\lambda)$.

1°. *Every entry in the matrix* $U(t,\lambda)$ *is an entire function of exponential type no larger than* t *in the variable* λ .

PROOF. This is immediate from inequality (2.10). □

2°. *For every pair of complex numbers* λ *and* μ

$$J-U^{\#}(t,\lambda)JU(t,\mu^*) = (\lambda-\mu^*)\int_0^t U^{\#}(s,\lambda)U(s,\mu^*)ds \ . \qquad (2.11)$$

PROOF. It follows readily from the differential equation (2.1) that

$$\lambda \int_0^t U^{\#}(s,\lambda)U(s,\mu^*)ds = \int_0^t [JU'(s,\lambda)-V(s)U(s,\lambda)]^{\#}U(s,\mu^*)ds$$

and that

$$\int_0^t [JU'(s,\lambda)]^{\#}U(s,\mu^*)ds =$$

$$= [JU(s,\lambda)]^{\#}U(s,\mu^*)\Big|_0^t + \int_0^t U^{\#}(s,\lambda)JU'(s,\mu^*)ds$$

$$= J - U^{\#}(t,\lambda)JU(t,\mu^*) + \int_0^t U^{\#}(s,\lambda)[\mu^*U(s,\mu^*)+V(s)U(s,\mu^*)]ds.$$

The rest is plain. □

3°. *The relationships*

$$J-U^{\#}(t,\lambda)JU(t,\lambda) = 0 \qquad (2.12)$$

and

$$\frac{J-U(t,\omega)^*JU(t,\omega)}{\omega^*-\omega} \geq 0 \qquad (2.13)$$

hold for every $\lambda \in \mathbb{C}$ *and every* $\omega \notin \mathbb{R}$.

PROOF. The proof is immediate from (2.11). □

4°. *The matrizant* $U(t,\lambda)$ *is invertible for every point* $\lambda \in \mathbb{C}$ *and*

$$[U(t,\lambda)]^{-1} = -JU^{\#}(t,\lambda)J \ . \qquad (2.14)$$

Moreover, the blocks $U_{ij}(t,\lambda)$ *of* U *are invertible for* $\lambda \notin \mathbb{R}$ *and* $t > 0$.

PROOF. The asserted invertibility of U and formula (2.14) are immediate from (2.12).

To prove the asserted invertibility of the $n \times n$ blocks U_{ij} it suffices to show that the null space of each one is equal to zero for $t > 0$ and $\lambda \notin \mathbb{R}$. With this in mind let $\xi \in \mathbb{C}^n$ be any vector which belongs to the null space of either $U_{11}(t,\lambda^*)$ or $U_{21}(t,\lambda^*)$. Then, as follows readily upon setting $\lambda = \mu$ in (2.11) and multiplying through by $[\xi^* \; 0]$ on the left and $[\xi^* \; 0]^*$ on the right,

$$(\lambda - \lambda^*) \int_0^t \left| U(s,\lambda) \begin{bmatrix} \xi \\ 0 \end{bmatrix} \right|^2 ds = 0 .$$

Therefore the integral, and so too the integrand (for $0 < s < t$) vanish for $\lambda \notin \mathbb{R}$. Thus, by the already established invertibility of $U(s,\lambda)$, $\xi = 0$. This proves the invertibility of $U_{11}(t,\lambda)$ and $U_{21}(t,\lambda)$ for $\lambda \notin \mathbb{R}$ and $t > 0$.

Much the same argument shows that if ξ belongs to the null space of either $U_{12}(t,\lambda^*)$ or $U_{22}(t,\lambda^*)$, then

$$\left| U(s,\lambda^*) \begin{bmatrix} 0 \\ \xi \end{bmatrix} \right| = 0$$

for $0 < s < t$ and $\lambda \notin \mathbb{R}$ and this serves to prove the invertibility of the other two blocks. □

5°. *The matrizant and its inverse are subject to the bounds*

$$[\nu(t)\exp\{|b|t\}]^{-1} \leq |[U(t,\lambda)]^{\pm 1}|$$
$$\leq \nu(t)\exp\{|b|t\} \qquad (2.15)$$

for every complex number $\lambda = a+ib$.

PROOF. The upper bound follows from (2.10) and the fact that

$$|[U(t,\lambda)]^{-1}| = |JU^{\#}(t,\lambda)J| = |U^{\#}(t,\lambda)| = |U(t,\lambda^*)| .$$

The lower bound is immediate from the upper bound and the elementary inequality

$$1 = |U(t,\lambda)[U(t,\lambda)]^{-1}| \leq |U(t,\lambda)| \; |[U(t,\lambda)]^{-1}| . \; □$$

6°. *The matrizant* $U(t,\lambda)$ *admits a representation of the form*

$$U(t,\lambda) = U_0(t,\lambda) + \int_{-t}^{t} K_U(t,s)U_0(s,\lambda)ds \qquad (2.16)$$

where

$$K_U(t,s) = \frac{1}{2}[\omega_+(t,\tfrac{t-s}{2})+\omega_-(t,\tfrac{t+s}{2})] , \tag{2.17}$$

$\omega(t,s)$ *is the matrix defined by* (2.6) *and* (2.8), *and* ω_+ [*resp.* ω_-] *denotes the* J-*commuting* [*resp.* J-*anticommuting*] *part of* ω .

Before embarking on the proof it is convenient to introduce and discuss some notation which will be useful in the sequel: Let

$$P_+ = \frac{1}{2}\begin{bmatrix} I_n & -iI_n \\ iI_n & I_n \end{bmatrix} \quad \text{and} \quad P_- = \frac{1}{2}\begin{bmatrix} I_n & iI_n \\ -iI_n & I_n \end{bmatrix} \tag{2.18}$$

be the two orthogonal projections on $\mathbb{C}^{2n}$ which figure in the spectral decomposition of the self-adjoint matrix $-iJ$:

$$-iJ = P_+ - P_- . \tag{2.19}$$

Then $P_\pm^2 = P_\pm = P_\pm^*$, $P_+P_- = P_-P_+ = 0$ and $P_+ + P_- = I_{2n}$. Moreover, every $2n \times 2n$ matrix H admits a unique additive decomposition as the sum of a term

$$H_+ = \frac{1}{2}(H - JHJ) = P_+HP_+ + P_-HP_-$$

which commutes with J and a term

$$H_- = \frac{1}{2}(H + JHJ) = P_+HP_- + P_-HP_+$$

which anticommutes with J .

PROOF OF 6°. Identity (2.7) yields the representation formula

$$U(t,\lambda) = U_0(t,\lambda) + \int_0^t e^{\lambda(2s-t)J}\,\omega(t,s)\,ds$$

with ω defined by (2.6) and (2.8). Therefore, since

$$e^{\mu J} = e^{i\mu}P_+ + e^{-i\mu}P_- \quad \text{for any} \quad \mu \in \mathbb{C} ,$$

it follows readily that

$$\int_0^t e^{\lambda(2s-t)J}\,\omega(t,s)\,ds = \int_0^t e^{i\lambda(2s-t)}\,P_+\omega(t,s)\,ds + \int_0^t e^{-i\lambda(2s-t)}\,P_-\omega(t,s)\,ds$$

$$= \frac{1}{2}\int_{-t}^{t} P_+\omega(t,\tfrac{t+u}{2})e^{i\lambda u}du + \frac{1}{2}\int_{-t}^{t} P_-\omega(t,\tfrac{t+u}{2})e^{-i\lambda u}du$$

$$= \frac{1}{2}\int_{-t}^{t} P_+\omega(t,\tfrac{t+u}{2})(P_+e^{\lambda uJ} + P_-e^{-\lambda uJ})du \quad +$$

$$+ \frac{1}{2}\int_{-t}^{t} P_-\omega(t,\tfrac{t+u}{2})(P_+e^{-\lambda uJ} + P_-e^{\lambda uJ})du$$

$$= \frac{1}{2}\int_{-t}^{t} [P_+\omega(t,\tfrac{t-u}{2})P_+ + P_-\omega(t,\tfrac{t-u}{2})P_-]\ e^{-\lambda uJ}du \quad +$$

$$+ \frac{1}{2}\int_{-t}^{t} [P_+\omega(t,\tfrac{t+u}{2})P_- + P_-\omega(t,\tfrac{t+u}{2})P_+]\ e^{-\lambda uJ}du$$

$$= \frac{1}{2}\int_{-t}^{t} \omega_+(t,\tfrac{t-u}{2})e^{-\lambda uJ}du + \frac{1}{2}\int_{-t}^{t} \omega_-(t,\tfrac{t+u}{2})e^{-\lambda uJ}du \quad ,$$

as claimed. □

7°. *For each fixed choice of* $t \geq 0$, *the normalized matrizant* $\Omega(t,\lambda)$ *is an invertible element of the Wiener algebra* W^I .

PROOF. Formula (2.7) exhibits $\Omega(t,\lambda)$ as an element of W^I since $\omega(t,s)$ is a summable function of s and

$$e^{2\lambda sJ} = e^{i2\lambda s}P_+ + e^{-i2\lambda s}\ P_- .$$

Moreover, it follows from (2.12) that

$$\Omega^{\#}(t,\lambda)J\Omega(t,\lambda) = J \tag{2.20}$$

and hence that $\Omega(t,\lambda)$ is an invertible matrix for every point $\lambda \in \mathbb{C}$ and so, in particular, for every $\lambda \in \mathbb{R}$. The Riemann-Lebesgue lemma, applied to (2.7), further guarantees that $\Omega(t,\lambda)$ is invertible at $\lambda = \infty$. Thus, by the matrix version of the Wiener theorem, $[\Omega(t,\cdot)]^{-1} \in W^I$ also. □

We pause to emphasize that an important reason for considering the normalized matrizant $\Omega(t,\lambda)$ is that, as $t \uparrow \infty$, it tends uniformly in $\lambda \in \mathbb{R}$ to a continuous $2n \times 2n$ matrix valued function which we shall designate by $\Omega_\infty(\lambda)$. Indeed, this is evident from the bound

$$|\Omega(t,\lambda)-\Omega(r,\lambda)| \leq \left| \int_r^t e^{2\lambda sJ} V(s)J\Omega(s,\lambda)\,ds \right|$$

$$\leq \int_r^t |V(s)|ds\ \nu(\infty)\ ,$$

for $\lambda \in \mathbb{R}$, which is easily deduced from (2.5) and (2.9). The conclusions on convergence can be sharpened with the help of the following estimates.

LEMMA 2.1. *The matrices* $\omega_j(t,s)$ *and* $\omega(t,s)$ *defined by* (2.6) *and* (2.8) *(for* $0 \leq s \leq t$ *and* $0 \leq t \leq \infty$ *, and equal to zero elsewhere) are subject to the following bounds:*

(a) $\displaystyle\int_0^\infty |\omega_j(t,s)-\omega_j(\infty,s)|ds \leq \frac{1}{j!}\{(\int_0^\infty |V(s)|ds)^j - (\int_0^t |V(s)|ds)^j\}$

for $j = 1,2,\ldots$.

(b) $\displaystyle\int_0^\infty |\omega(\infty,s)|ds \leq \nu(\infty)-1$.

(c) $\displaystyle\int_0^\infty |\omega(t,s)-\omega(\infty,s)|ds \leq \nu(\infty) \int_t^\infty |V(s)|ds$.

(d) $\displaystyle\int_s^T |\omega(t,(t+s)/2)|dt \leq 2\{\nu(T)-\nu(s)\}$.

(e) $\displaystyle\int_0^T |\omega(t,(t-s)/2)|dt \leq 2\{\nu(T) - \nu(0)\}$.

PROOF. The first asserted bound is clearly valid for $j=1$. On the other hand, if $j+1 \geq 2$, then

$$\int_0^\infty |\omega_{j+1}(t,s) - \omega_{j+1}(\infty,s)|ds = \int_0^t \left| \int_t^\infty V(u)J\omega_j(u,u-s)du \right| ds +$$

$$+ \int_t^\infty \left| \int_s^\infty V(u)J\omega_j(u,u-s)du \right| ds$$

$$= ① + ② \ ,$$

in a self-evident notation. But now

$$\textcircled{1} \leq \int_t^\infty |V(u)| \{ \int_0^t |\omega_j(u,u-s)|ds \} du$$

$$\leq \int_t^\infty |V(u)| \{ \int_{u-t}^u |\omega_j(u,s)|ds \} du \quad ,$$

whereas

$$\textcircled{2} \leq \int_t^\infty \{ \int_s^\infty |V(u)| \ |\omega_j(u,u-s)|du \} ds$$

$$= \int_t^\infty |V(u)| \{ \int_t^u |\omega_j(u,u-s)|ds \} du$$

$$= \int_t^\infty |V(u)| \{ \int_0^{u-t} |\omega_j(u,s)|ds \} du \ .$$

Therefore,

$$\textcircled{1} + \textcircled{2} \leq \int_t^\infty |V(u)| \{ \int_0^u |\omega_j(u,s)|ds \} du$$

$$\leq \int_t^\infty |V(u)| \ \{ (\int_0^u |V(s)|ds)^j / j! \} du$$

which leads readily to (a).

Since (b) and (c) follow easily from (a), we turn next to (d). Observe first that

$$\int_0^T |\omega(t,(t+s)/2)|dt \leq \sum_{j=0}^\infty \int_s^T |\omega_{j+1}(t,(t+s)/2)|dt$$

and then, by (2.6), that

$$\int_s^t |\omega_{j+1}(t,(t+s)/2)|dt \leq \int_s^T \left\{ \int_{(t+s)/2}^t |V(u)||\omega_j(u,u-(t+s)/2)|du \right\} dt$$

$$\leq \int_s^T |V(u)| \left\{ \int_u^{2u-s} |\omega_j(u,u-(t+s)/2)|dt \right\} du$$

$$= 2 \int_s^T |V(u)| \left\{ \int_0^{(u-s)/2} |\omega_j(u,y)|dy \right\} du \ ,$$

for $j \geq 1$. But now, upon combining estimates, it follows readily that

$$\int_s^T |\omega(t,(t+s)/2)|dt \leq 2\int_s^T |V(u)|\left\{\int_0^u |\omega(u,y)|dy\right\}du \; +$$

$$+ \int_s^T |V((t+s)/2)|dt$$

$$\leq 2\int_s^T |V(u)| \; \nu(u)du$$

$$= 2\{\nu(T) - \nu(s)\} \; ,$$

as asserted. The proof of (e) is similar. □

We return to the list of properties enjoyed by the matrizants.

8°. *The normalized matrizant* $\Omega(t,\lambda)$ *converges in* W^I *(and hence also uniformly on* $\mathbb{R}$ *) as* $t \uparrow \infty$ *to a limit*

$$\Omega_\infty(\lambda) = I_{2n} + \int_0^\infty e^{2\lambda sJ}\,\omega(\infty,s)ds \qquad (2.21)$$

which is J-*unitary on* $\mathbb{R}$ *:*

$$\Omega_\infty(\lambda)^* J\Omega_\infty(\lambda) = J \; . \qquad (2.22)$$

PROOF. Formula (2.21) is immediate from Lemma 2.1 and the representation formula (2.7). Formula (2.22) may be obtained by passing to the limit in (2.20) for real λ . □

We remark that $\Omega_\infty(\lambda)^{-1}$ is called the *asymptotic equivalence matrix* or the A-*matrix* and is denoted by $A(\lambda)$ in [DaK]. The reason for this terminology is that, for fixed $\lambda \in \mathbb{R}$, two arbitrary solutions, $U(t,\lambda)\xi$ and $U_0(t,\lambda)\xi_0$, $(\xi,\xi_0 \in \mathbb{C}^{2n})$ of the canonical equations (2.1), and (2.1) with $V \equiv 0$, respectively, are asymptotic, i.e.,

$$\lim_{t\uparrow\infty} |U(t,\lambda)\xi - U_0(t,\lambda)\xi_0| = 0 \; ,$$

if and only if their initial conditions, ξ and ξ_0 , are related by $\xi = [\Omega_\infty(\lambda)]^{-1}\xi_0$. The proof is obvious.

9°. *The modified matrizants*

$$P_{\pm}\Omega(t,\lambda) = P_{\pm} + \int_0^t e^{\pm 2i\lambda s} P_{\pm}\omega(t,s)ds \tag{2.23}$$

converge in $W_{\pm}^I$ *(and hence uniformly in* $\bar{\mathbb{C}}_{\pm}$*) to*

$$P_{\pm}\Omega_{\infty}(\lambda) = P_{\pm} + \int_0^{\infty} e^{\pm 2i\lambda s} P_{\pm}\omega(\infty,s)ds \tag{2.24}$$

as $t \uparrow \infty$.

PROOF. This is immediate from the bound

$$\begin{aligned}|P_{\pm}\Omega_{\infty}(\lambda)-P_{\pm}\Omega(t,\lambda)| &\leq \int_0^{\infty} |P_{\pm}\omega(\infty,s)-P_{\pm}\omega(t,s)|ds \\ &\leq \int_0^{\infty} |\omega(\infty,s)-\omega(t,s)|ds ,\end{aligned}$$

which is valid for λ in the appropriate closed half plane, and Lemma 2.1. □

THEOREM 2.1. *For every* $t \geq 0$,

$$\Omega_{11}(t,\lambda) \mp i\Omega_{21}(t,\lambda) \quad \textit{and} \quad \Omega_{22}(t,\lambda) \pm i\Omega_{12}(t,\lambda)$$

are invertible elements of $W_{\pm}^I$, *as are*

$$(\Omega_{\infty})_{11}(\lambda) \mp i(\Omega_{\infty})_{21}(\lambda) \quad \textit{and} \quad (\Omega_{\infty})_{22}(\lambda) \pm i(\Omega_{\infty})_{12}(\lambda) .$$

Moreover,

$$[\Omega_{11}(t,\lambda) \mp i\Omega_{21}(t,\lambda)]^{\pm 1} \to [(\Omega_{\infty})_{11}(\lambda) \mp i(\Omega_{\infty})_{21}(t,\lambda)]^{\pm 1}$$

and

$$[\Omega_{22}(t,\lambda) \pm i\Omega_{12}(t,\lambda)]^{\pm} \to [(\Omega_{\infty})_{22}(\lambda) \pm i(\Omega_{\infty})_{12}(t,\lambda)]^{\pm 1}$$

uniformly on $\bar{\mathbb{C}}_{\pm}$ *as* $t \uparrow \infty$. □

PROOF. For the sake of brevity, let

$$\varepsilon_t(\lambda) = \Omega_{11}(t,\lambda) - i\Omega_{21}(t,\lambda) ,$$

$$\chi_t = \Omega_{22}(t,\lambda) + i\Omega_{12}(t,\lambda)$$

and

$$\delta_t(\lambda) = \det[\varepsilon_t(\lambda)]$$

for $0 \leq t \leq \infty$. Then it follows by inspection of the diagonal blocks in (2.23) and (2.24) that ε_t and χ_t belong to W_+^I for

$0 \leq t \leq \infty$. Moreover, property 9° implies that ε_t [resp. χ_t] converges to ε_∞ [resp. χ_∞] in W_+^I , and hence uniformly in $\bar{\mathbb{C}}_+$ as $t \uparrow \infty$.

Next we observe that

$$\varepsilon_t(\lambda) = e^{i\lambda t}[U_{11}(t,\lambda)-iU_{21}(t,\lambda)] ,$$

$$\chi_t(\lambda) = e^{i\lambda t}[U_{22}(t,\lambda)+iU_{12}(t,\lambda)]$$

and that, thanks to (2.12) and (2.13), $U(t,\lambda)$ is iJ-contractive at each point $\lambda \in \bar{\mathbb{C}}_+$. Therefore, as follows from Lemma 4.1 with $G = I$, $\varepsilon_t(\lambda)$ and $\chi_t(\lambda)$ are invertible matrices for every $\lambda \in \bar{\mathbb{C}}_+$. Thus, since also $\varepsilon_t(\infty) = \chi_t(\infty) = I_n$, it follows readily from the matrix version of Wiener's theorem for W_+ that ε_t and χ_t are invertible in W_+ for every choice of $0 \leq t < \infty$. The next step is to extend this to $t = \infty$. To this end we first apply Lemma 4.1 to $\Omega_\infty(\lambda)$, which is iJ-contractive (in fact, iJ-unitary) on $\mathbb{R}$ thanks to (2.12), to conclude that $\varepsilon_\infty(\lambda)$ is invertible for every $\lambda \in \mathbb{R}$. Thus, since $\varepsilon_\infty \in W_+^I$, there exists an $\alpha > 0$ and a $\rho > 0$ such that

$$|\delta_\infty(\lambda)| \geq \alpha$$

for all $\lambda \in \mathbb{R}$ and all $\lambda \in \bar{\mathbb{C}}_+$ with $|\lambda| \geq \rho$. The latter may be justified with the help of the Phragmén-Lindelöf theorem; see p.179 of [T] for a convenient statement. At the same time, the already established uniform convergence of ε_t to ε_∞ on $\bar{\mathbb{C}}_+$ guarantees the existence of a $T > 0$ such that

$$|\delta_t(\lambda) - \delta_\infty(\lambda)| < \alpha/2$$

for $\lambda \in \bar{\mathbb{C}}_+$ and $t \geq T$. In particular, this guarantees that

$$|\delta_\infty(\lambda)| > |\delta_T(\lambda) - \delta_\infty(\lambda)|$$

for all $\lambda \in \bar{\mathbb{C}}_+$ with $\operatorname{Im} \lambda \geq 0$, $|\lambda| \geq \rho$. Therefore, by Rouché's theorem, $\delta_\infty(\lambda)$ is root free inside the semicircle $\operatorname{Im} \lambda \geq 0$, $|\lambda| = \rho$, and it now follows easily from the already established bound on $|\delta_\infty(\lambda)|$ for $|\lambda| \geq \rho$ that $\delta_\infty(\lambda) \neq 0$ on $\bar{\mathbb{C}}_+$. Thus, ε_∞ is seen to be invertible in W_+ , by another application of Wiener's theorem, and

$$[\varepsilon_t(\lambda)]^{-1} - [\varepsilon_\infty(\lambda)]^{-1} = [\varepsilon_t(\lambda)]^{-1}\{\varepsilon_\infty(\lambda) - \varepsilon_t(\lambda)\}[\varepsilon_\infty(\lambda)]^{-1}$$

tends to zero uniformly on $\bar{\mathbb{C}}_+$ as $t \uparrow \infty$ by the already established convergence of ε_t to ε_∞ and some elementary estimates which take advantage of the fact that $|\delta_\infty(\lambda)| \geq \alpha' > 0$ on $\bar{\mathbb{C}}_+$. This completes the proof of the assertions involving ε_t . The remaining assertions may be established in much the same way. □

3. DE BRANGES SPACES

In this section the abstract theory of some classes of reproducing kernel Hilbert spaces of vector and matrix valued entire functions is developed and then connected to the theory of canonical equations.

We shall say that a pair of $n \times n$ matrix valued entire functions $\{E_+,E_-\}$ is a de Branges pair if

$$E_+E_+^\# = E_-E_-^\# \quad \text{on} \quad \mathbb{C} , \tag{3.1}$$

$$\det E_+ \neq 0 \quad \text{on} \quad \bar{\mathbb{C}}_+ , \tag{3.2}$$

$$\det E_- \neq 0 \quad \text{on} \quad \bar{\mathbb{C}}_- , \quad \text{and} \tag{3.3}$$

$$\Sigma = E_+^{-1}E_- \quad \text{is inner over} \quad \mathbb{C}_+ . \tag{3.4}$$

With every de Branges pair $\{E_+,E_-\}$ we associate the linear space $\mathcal{B}(E_+,E_-)$ of $n \times 1$ vector valued entire functions f such that

$$E_+^{-1}f \in H_n^2 \ominus \Sigma H_n^2$$

and shall refer to it as the de Branges space based on E_+ and E_- . If E is a scalar valued entire function such that

$$|E(\omega)| > |E(\omega^*)| \quad \text{for} \quad \omega \in \mathbb{C}_+$$

and E has no real roots, then $\{E,E^\#\}$ is a de Branges pair and $\mathcal{B}(E,E^\#)$ is just the usual de Branges space based on E ; see e.g. pages 50-55 of [dB1] for more information on the latter.

THEOREM 3.1. *Let* $\{E_+,E_-\}$ *be a de Branges pair of* $n \times n$ *matrix valued entire functions. Then* $\mathcal{B} = \mathcal{B}(E_+,E_-)$ *is a reproducing kernel Hilbert space of* $n \times 1$ *vector valued entire functions with respect to the inner product*

$$\langle f,g\rangle_{\mathcal{B}} = \int_{-\infty}^{\infty} [(E_+^{-1}g)(\lambda)]^*[(E_+^{-1}f)(\lambda)]d\lambda$$

and with reproducing kernel based on the matrix valued function

$$\Lambda_\omega(\lambda) = \frac{E_+(\lambda)E_+(\omega)^* - E_-(\lambda)E_-(\omega)^*}{-2\pi i(\lambda-\omega^*)} \tag{3.5}$$

i.e., for every $\omega \in \mathbb{C}$ *and every* $\xi \in \mathbb{C}^n$ *,* $\Lambda_\omega \xi \in \mathcal{B}$ *and*

$$\langle f,\Lambda_\omega \xi\rangle = \xi^* f(\omega) , \tag{3.6}$$

for every $f \in \mathcal{B}$.

PROOF. It is readily seen that $\Lambda_\omega(\lambda)\xi$ is an $n \times 1$ vector valued entire function of λ for every fixed choice of $\omega \in \mathbb{C}$ and $\xi \in \mathbb{C}^n$. Moreover, $E_+^{-1}\Lambda_\omega \xi \in L_n^2(\mathbb{R})$. The rest of the proof is broken into five steps.

STEP 1. $\Lambda_\omega \xi \in \mathcal{B}$ *for every fixed choice of* $\omega \in \mathbb{C}_+$ *and* $\xi \in \mathbb{C}^n$.

PROOF OF STEP 1. Clearly

$$[E_+(\lambda)]^{-1}\Lambda_\omega(\lambda)\xi = \left[\frac{E_+(\omega)^* - \Sigma(\lambda)E_-(\omega)^*}{-2\pi i(\lambda-\omega^*)}\right]\xi$$

$$= \frac{I - \Sigma(\lambda)\Sigma(\omega)^*}{-2\pi i(\lambda-\omega^*)} E_+(\omega)^*\xi$$

belongs to H_n^2 and is orthogonal to ΣH_n^2 for every $\omega \in \mathbb{C}_+$:

$$\eta^* \int \left[\frac{I - \Sigma(\omega)\Sigma(\lambda)^*}{2\pi i(\lambda-\omega)}\right] \Sigma(\lambda)g(\lambda)d\lambda$$

$$= \eta^* \int \left[\frac{\Sigma(\omega) - \Sigma(\omega)}{2\pi i(\lambda-\omega)}\right] g(\lambda)d\lambda = 0$$

for every $\eta \in \mathbb{C}^n$ and $g \in H_n^2$, by Cauchy's formula for H_n^2 . The same conclusions hold for $\omega \in \mathbb{R}$ but require just a little more care. The main points are that

$$\left[\frac{I - \Sigma(\lambda)\Sigma(\omega)^*}{\lambda-\omega}\right]\eta = \left[\frac{\Sigma(\omega) - \Sigma(\lambda)}{\lambda-\omega}\right] \Sigma(\omega)^*\eta$$

belongs to L_n^2 and, by Cauchy's formula, is orthogonal to $(\lambda-\beta)^{-1}\xi$ for every $\beta \in \mathbb{C}_-$ and $\xi \in \mathbb{C}^n$. Thus it belongs to

H_n^2 . Moreover, by dominated convergence,

$$\eta^* \int \left[\frac{I - \Sigma(\omega)\Sigma(\lambda)^*}{\lambda-\omega}\right] \Sigma(\lambda) g(\lambda) d\lambda$$

$$= \lim_{b\downarrow 0} \eta^* \int \left[\frac{\Sigma(\lambda) - \Sigma(\omega)}{\lambda-\omega-ib}\right] g(\lambda) d\lambda = 0$$

for $\omega \in \mathbb{R}$ and $g \in H_n^2$.

STEP 2. $E_+^{-1}f \in H_n^2 \ominus \Sigma H_n^2$ *if and only if*

$$E_-^{-1}f \in K_n^2 \ominus S^\# K_n^2 .$$

PROOF OF STEP 2. It is plain that $E_+^{-1}f$ belongs to

$$H_n^2 \ominus \Sigma H_n^2 = H_n^2 \cap \Sigma K_n^2$$

if and only if

$$f \in E_+ H_n^2 \cap E_- K_n^2$$

or equivalently if and only if $E_-^{-1}f$ belongs to

$$\Sigma^\# H_n^2 \cap K_n^2 = K_n^2 \ominus \Sigma^\# K_n^2 .$$

STEP 3. $\Lambda_\omega(\lambda)\xi \in \mathcal{B}$ *for every fixed choice of* $\omega \in \mathbb{C}_-$ *and* $\xi \in \mathbb{C}^n$.

PROOF OF STEP 3. In view of Step 2 it suffices to check that

$$E_-^{-1}\Lambda_\omega(\lambda)\xi = \frac{\Sigma^\#(\lambda)\Sigma^\#(\omega)^* - I}{-2\pi i(\lambda-\omega^*)} E_-(\omega)^*$$

belongs to $K_n^2 \ominus \Sigma^\# K_n^2$. But this is just a mirror image of the proof of Step 1.

STEP 4 *is to verify* (3.6).

PROOF OF STEP 4. It is readily checked, with the help of (3.1), that

$$\langle f, \Lambda_\omega \xi\rangle_{\mathcal{B}} = \xi^* E_+(\omega) \int \frac{E_+(\lambda)^{-1}f(\lambda)}{2\pi i(\lambda-\omega)} d\lambda - \xi^* E_-(\omega) \int \frac{E_-(\lambda)^{-1}f(\lambda)}{2\pi i(\lambda-\omega)} d\lambda .$$

But, by Step 2 and Cauchy's formula for H_n^2 and K_n^2 , the right hand side of the last formula is easily seen to be equal to

$\xi^* f(\omega)$, at least for $\omega \notin \mathbb{R}$. It remains therefore only to complete the evaluation for $\omega \in \mathbb{R}$. But this can be done by a straightforward limiting argument.

STEP 5. $\mathcal{B}$ *is a Hilbert space.*

PROOF OF STEP 5. Let $f_1, f_2, \ldots$ be a Cauchy sequence in $\mathcal{B}$. Then the inequality

$$|\xi^*\{f_k(\omega) - f_j(\omega)\}|^2 = |\langle f_k - f_j, \Lambda_\omega \xi \rangle_{\mathcal{B}}|^2$$
$$\leq \| f_k - f_j \|_{\mathcal{B}}^2 \, \| \Lambda_\omega \xi \|_{\mathcal{B}}^2$$
$$= \| f_k - f_j \|_{\mathcal{B}}^2 \, \xi^* \Lambda_\omega(\omega) \xi$$

implies that the given sequence converges uniformly on each compact subset of the complex plane. Therefore it converges to an $n \times 1$ vector entire function which clearly belongs to $\mathcal{B}$. Thus $\mathcal{B}$ is complete and is indeed a Hilbert space as asserted. □

For an alternate description of $\mathcal{B}(E_+, E_-)$ in terms of the pair

$$\bar{\bar{A}} = \frac{E_- + E_+}{2} \quad \text{and} \quad \bar{\bar{B}} = \frac{E_- - E_+}{2i} \tag{3.7}$$

see Theorem 1 of [dB3] and the references cited therein.

We shall refer to Λ_ω as the reproducing kernel for $\mathcal{B}$. For future applications it is useful to note that it can be expressed in the form

$$\Lambda_\omega(\lambda) = \frac{[E_+(\lambda) \;\; E_-(\lambda)] J_0 [E_+(\omega) \;\; E_-(\omega)]^*}{-2\pi i(\lambda - \omega^*)} \tag{3.8}$$

where

$$J_0 = \begin{bmatrix} I_n & 0 \\ 0 & -I_n \end{bmatrix} , \tag{3.9}$$

or equivalently, in terms of $\bar{\bar{A}}$ and $\bar{\bar{B}}$, as

$$\Lambda_\omega(\lambda) = \frac{[\bar{\bar{A}}(\lambda) \;\; \bar{\bar{B}}(\lambda)] \, J \, [\bar{\bar{A}}(\omega) \;\; \bar{\bar{B}}(\omega)]^*}{-\pi(\lambda - \omega^*)} \tag{3.10}$$

Shortly we shall associate a one parameter family of de Branges spaces, indexed by t , with the matrizant $U(t,\lambda)$ of the canonical equation (2.1). To put the connection in better perspective it seems worthwhile to develop the theory a little more fully than is needed for present purposes. With this in mind we now introduce the class $\mathcal{E}(\tilde{J})$ of $2n \times 2n$ entire matrix functions U which are subject to the following two conditions:

$$\tilde{J} - U(\lambda)\, \tilde{J}\, U^{\#}(\lambda) = 0 \quad \text{for all} \quad \lambda \in \mathbb{C}\ , \tag{3.11}$$

$$\frac{\tilde{J} - U(\omega)\, \tilde{J}\, U(\omega)^{*}}{-2\pi i(\omega-\omega^{*})} \geq 0 \quad \text{for all} \quad \omega \notin \mathbb{R}\ . \tag{3.12}$$

where $\tilde{J}$ is any signature matrix, i.e., any constant $2n \times 2n$ matrix which is both selfadjoint and unitary (such as iJ and J_0). It is readily checked that every member U of the class $\mathcal{E}(\tilde{J})$ is invertible on the whole complex plane and that $\mathcal{E}(\tilde{J})$ is closed under multiplication.

It is also useful to bear in mind that U and $U^{\#}$ [resp. U^*] can be interchanged in (3.11) [resp. (3.12)] without changing the space:

LEMMA 3.1. *A* $2n \times 2n$ *entire matrix function* U *belongs to* $\mathcal{E}(\tilde{J})$ *if and only if*

$$\tilde{J} - U^{\#}(\lambda)\, \tilde{J}\, U(\lambda) = 0 \quad \textit{for all} \quad \lambda \in \mathbb{C} \tag{3.13}$$

and

$$\frac{\tilde{J} - U(\omega)^{*}\, \tilde{J}\, U(\omega)}{-\ 2\pi i(\omega-\omega^{*})} \geq 0 \quad \textit{for all} \quad \omega \notin \mathbb{R}\ . \tag{3.14}$$

PROOF. The equivalence of (3.11) and (3.13) is easy. It amounts to the observation that $\tilde{J}U^{\#}\tilde{J}$ is a right inverse of U if and only if it is also a left inverse of U .

The equivalence of (3.12) with (3.14) is also well known, but lies a little deeper. One relatively quick proof utilizes the spectral decomposition of $\tilde{J}$ in terms of the orthogonal projectors of $\mathbb{C}^{2n}$ onto the eigenspaces corresponding to its two eigenvalues 1 and -1 . Let us, for the purposes of this argument, designate these projectors by $\tilde{P}$ and $\tilde{Q}$, respectively. Then

$$\tilde{J} = \tilde{P} - \tilde{Q}\ , \quad I = \tilde{P} + \tilde{Q}$$

and

$$\tilde{J} - U\tilde{J}U^* = (\tilde{P} + U\tilde{Q})(\tilde{P} + \tilde{Q}U^*) - (\tilde{Q} + U\tilde{P})(\tilde{Q} + \tilde{P}U^*)\ .$$

Now, if $\tilde{J} - U\tilde{J}U^* > 0$, then $\tilde{P} \pm U\tilde{Q}$ is invertible and consequently, upon setting

$$\Phi = (\tilde{P} + U\tilde{Q})^{-1}(U\tilde{P} + \tilde{Q})\ ,$$

it follows that

$$I - \Phi\Phi^* > 0 .$$

Therefore,

$$I - \Phi^*\Phi > 0$$

and substitution of the easily checked alternate formula

$$\Phi = (\widetilde{P}U - \widetilde{Q})(\widetilde{P} - \widetilde{Q}U)^{-1}$$

into the last inequality, leads rapidly to the conclusion that

$$(\widetilde{P} - U^*\widetilde{Q})(\widetilde{P} - \widetilde{Q}U) - (U^*\widetilde{P} - \widetilde{Q})(\widetilde{P}U - \widetilde{Q})$$

$$= \widetilde{J} - U^*\widetilde{J}U > 0 .$$

Since the argument can be reversed this effectively proves that (3.12) is equivalent to (3.14) for $\omega \in \mathbb{C}_+$ and hence too for $\omega \in \mathbb{C}_-$, because of the special form of U^{-1}. □

THEOREM 3.2. *Let*

$$U = \begin{bmatrix} A & C \\ B & D \end{bmatrix}$$

belong to $E(iJ)$. *Then each of the following pairs is a de Branges pair* $\{E_+,E_-\}$:

$$\{A+iC,A-iC\} ; \{B+iD,B-iD\} ; \{A^\#-iB^\#,A^\#+iB^\#\} ; \{C^\#-iD^\#,C^\#+iD^\#\} .$$

PROOF. Upon multiplying (3.12) with $\widetilde{J} = iJ$ by $[I_n \ 0]$ on the left and $[I_n \ 0]^*$ on the right it follows readily that

$$i(CA^* - AC^*) \geq 0 \quad \text{on} \quad \mathbb{C}_+$$

and

$$i(AC^* - CA^*) \geq 0 \quad \text{on} \quad \mathbb{C}_- .$$

But this in turn implies that

$$(A+iC)(A^*-iC^*) = AA^* + iCA^* - iAC^* + CC^*$$
$$\geq AA^* + CC^* \quad \text{on} \quad \bar{\mathbb{C}}_+$$

and similarly that

$$(A-iC)(A^* + iC^*) \geq AA^* + CC^* \quad \text{on} \quad \bar{\mathbb{C}}_- .$$

The next step is to check that $AA^* + CC^*$ is strictly positive on $\mathbb{C}$. If this were not the case, then there would exist a nonzero vector $\xi \in \mathbb{C}^n$ and a point $\omega \in \mathbb{C}$ such that

$$\xi^* A(\omega) A(\omega)^* \xi + \xi^* C(\omega) C(\omega)^* \xi = 0$$

and hence that

$$A(\omega)^* \xi = C(\omega)^* \xi = 0 \ .$$

But this in turn implies that

$$U(\omega)^* \begin{bmatrix} \xi \\ 0 \end{bmatrix} = \begin{bmatrix} A(\omega)^* \xi \\ C(\omega)^* \xi \end{bmatrix} = \begin{bmatrix} 0 \\ 0 \end{bmatrix}$$

which contradicts the invertibility of U . Therefore (3.2) and (3.3) are met with $E_+ = A+iC$ and $E_- = A-iC$. Next (3.1) follows readily from the auxiliary identity

$$AC^\# = CA^\# \quad ,$$

which is a consequence of (3.11), and (3.4) is an elementary estimate, thanks to the first inequality deduced above and the last identity which guarantees that $AC^* = CA^*$ on $\mathbb{R}$. This completes the proof that the first stated pair is a de Branges pair.

The proof that $\{B+iD, B-iD\}$ is a de Branges pair goes through in much the same way after multiplying (3.12) with $\widetilde{J} = iJ$ by $[0 \ I_n]$ on the left and $[0 \ I_n]^*$ on the right.

Finally the last two assertions may be deduced from the first two and the fact that $\mathcal{E}(iJ)$ is closed under the mapping of U into

$$\begin{bmatrix} 0 & I_n \\ I_n & 0 \end{bmatrix} U^\# \begin{bmatrix} 0 & I_n \\ I_n & 0 \end{bmatrix} = \begin{bmatrix} D^\# & C^\# \\ B^\# & A^\# \end{bmatrix} , \qquad (3.15)$$

as is readily checked with the help of Lemma 3.1. □

We remark that more de Branges pairs can be obtained by taking advantage of the fact that $\mathcal{E}(iJ)$ is closed under the mapping of U into MUN where M and N are any constant iJ unitary matrices:

$$MiJM^* = M^*iJM = iJ = NiJN^* = N^*iJN \ .$$

Matrices of the class $\mathcal{E}(iJ)$ figure in the theory of de Branges spaces in a second way. In order to explain this the unitary matrix

$$L = \frac{1}{\sqrt{2}} \begin{bmatrix} I_n & I_n \\ -iIn & iI_n \end{bmatrix} \tag{3.16}$$

which diagnoalizes iJ :

$$J_0 = L^*iJL \ , \tag{3.17}$$

will be useful. Moreover in the sequel we shall be interested in pairs of $n \times n$ matrix entire functions E and F which satisfy the following four conditions:

$$E^\# E = F^\# F \quad \text{on} \quad \mathbb{C} \tag{3.18}$$

$$\det E \neq 0 \quad \text{on} \quad \bar{\mathbb{C}}_+ \tag{3.19}$$

$$\det F \neq 0 \quad \text{on} \quad \bar{\mathbb{C}}_- \tag{3.20}$$

$$FE^{-1} \text{ is inner over } \mathbb{C}_+ \ . \tag{3.21}$$

It is readily checked that this is exactly the same as to say that $\{F^\# , E^\#\}$ is a de Branges pair.

THEOREM 3.3. *Let* $\{F^\#,E^\#\}$ *be a de Branges pair, let* $U \in E(iJ)$, *and let*

$$\begin{bmatrix} F_1 \\ E_1 \end{bmatrix} = L^*UL \begin{bmatrix} F \\ E \end{bmatrix} \ .$$

Then the pair $\{F_1^\#,E_1^\#\}$ *is again a de Branges pair.*

PROOF. To begin with it is readily checked that

$$\Theta = L^*UL$$

belongs to $E(J_0)$ and hence, as follows from (3.12) with $\tilde{J} = J_0$ upon decomposing Θ into $n \times n$ blocks Θ_{ij} , that

$$\Theta_{22}\,\Theta_{22}^* \geq I_n + \Theta_{21}\,\Theta_{21}^* \quad \text{on} \quad \bar{\mathbb{C}}_+ \ .$$

But this in turn implies that $\Theta_{22}(\omega)$ is invertible and

$$|\,[\Theta_{22}(\omega)]^{-1}\Theta_{21}(\omega)\,| < 1$$

at each point $\omega \in \bar{\mathbb{C}}_+$. Therefore

$$\begin{aligned} E_1 &= \Theta_{21}F + \Theta_{22}E \\ &= \Theta_{22}\{\Theta_{22}^{-1}\,\Theta_{21}FE^{-1} + I_n\}E \end{aligned}$$

is also invertible on $\bar{\mathbb{C}}_+$. This establishes (3.19) for E_1 .

We next observe that

$$F_1^\# F_1 - E_1^\# E_1 = [F_1^\# \; E_1^\#] \, J_0 \begin{bmatrix} F_1 \\ E_1 \end{bmatrix}$$

$$= [F^\# \; E^\#] \Theta^\# J_0 \Theta \begin{bmatrix} F \\ E \end{bmatrix}$$

$$= [F^\# \; E^\#] \, J_0 \begin{bmatrix} F \\ E \end{bmatrix}$$

$$= F^\# F - E^\# E = 0$$

on $\mathbb{C}$ and, by a similar argument,

$$F_1^* F_1 - E_1^* E_1 \leq F^* F - E^* E \leq 0$$

on $\bar{\mathbb{C}}_+$. This proves (3.18) and, thanks to the availability of (3.19), also (3.21).

The proof of (3.20) is similar to that of (3.19): You have only to take advantage of the inequality

$$\Theta_{11}\Theta_{11}^* - \Theta_{12}\Theta_{12}^* \geq I_n$$

on $\bar{\mathbb{C}}_-$ and then to express

$$F_1 = \Theta_{11} F + \Theta_{12} E$$

$$= \Theta_{11}\{I_n + \Theta_{11}^{-1}\Theta_{12} E F^{-1}\} F .$$

Since F_1 and E_1 are plainly entire matrix functions the proof is complete. □

COROLLARY. *If* $\{F^\#, E^\#\}$ *is a de Branges pair and if*

$$\begin{bmatrix} F_1 \\ E_1 \end{bmatrix} = \Theta \begin{bmatrix} F \\ E \end{bmatrix}$$

for some constant J_0 *unitary matrix* Θ *, then* $\mathcal{B}(F_1^\#, E_1^\#) = \mathcal{B}(F^\#, E^\#)$ *, norms and all.*

PROOF. Theorem 3.3 guarantees that $\{F_1^\#, E_1^\#\}$ is a de Branges pair. Moreover, it is plain from (3.8) that the reproducing kernel for $\mathcal{B}_1 = \mathcal{B}(F_1^{\#},E_1^{\#})$ is exactly the same as for $\mathcal{B} = \mathcal{B}(F^{\#},E^{\#})$. Thus

$$\|\sum_j \Lambda_{\omega_j}\xi_j\|_{\mathcal{B}}^2 = \sum_{k,j} \xi_k^* \Lambda_{\omega_j}(\omega_k)\xi_j = \|\sum \Lambda_{\omega_j}\xi_j\|_{\mathcal{B}_1}^2$$

and the assertion drops out from the fact that the set of all finite linear combinations of the $\Lambda_\omega\xi$, as ω runs over $\mathbb{C}$ and ξ ranges over $\mathbb{C}^n$, is dense in both spaces. □

The last corollary can be generalized:

THEOREM 3.4. *If* $\{F^{\#},E^{\#}\}$ *is a de Branges pair and if*

$$\begin{bmatrix} F_1 \\ E_1 \end{bmatrix} = \Theta \begin{bmatrix} F \\ E \end{bmatrix}$$

for some $\Theta \in E(J_0)$, *then*

(1) $\mathcal{B}_1 = \mathcal{B}(F_1^{\#},E_1^{\#}) \supset \mathcal{B}(F^{\#},E^{\#}) = \mathcal{B}$, *and*

(2) $\|g\|_{\mathcal{B}} \geq \|g\|_{\mathcal{B}_1}$ *for every* $g \in \mathcal{B}$ *, with equality if* g *belongs to the domain of multiplication by* λ .

For a proof of this theorem and additional information see Theorem 5 of [dB2] and Theorems 7 and 8 of [dB3]. An independent proof that the nested sequence of de Branges spaces which are of interest in the present study actually sit isometrically one inside the other is furnished in Section 5.

4. LINEAR FRACTIONAL TRANSFORMATIONS

The main purpose of this section is to introduce a linear fractional transformation based on the matrizant $U(t,\lambda)$ of the canonical equation (2.1). We pause first, however, to establish a few elementary facts which, among other things, serve to guarantee that the linear fractional transformation of interest is well defined.

LEMMA 4.1. *If* U *is a* $2n \times 2n$ *constant* iJ-*contractive matrix and* G *is a constant* $n \times n$ *matrix with* $G+G^* > 0$, *then each of the following matrices is invertible:*

$$U_{11} - iGU_{21} \quad , \quad U_{12} - iGU_{22}$$

$$U_{11} + iU_{12}G \quad , \quad U_{21} + iU_{22}G$$

PROOF. Let $\Gamma = U_{11} - iGU_{21}$ and $\Delta = iU_{12} + GU_{22}$. Then upon multiplying the inequality

$$UiJU^* \leq iJ$$

through by $[I_n \quad -iG]$ on the left hand and $[I_n \quad -iG]^*$ on the right, it is readily seen that

$$\Gamma\Delta^* + \Delta\Gamma^* \geq G + G^* . \tag{4.1}$$

Now, if Δ is not invertible, then there exists a vector $\xi \in \mathbb{C}^n$, $\xi \neq 0$, such that

$$\xi^*\Delta = 0$$

and hence

$$0 = \xi^*\Gamma\Delta^*\xi + \xi^*\Delta\Gamma^*\xi \geq \xi^*(G + G)\xi ,$$

which is impossible. This proves that Δ is invertible. By symmetry, Γ is also invertible.

The second pair of results may be deduced from the first by replacing U by U^* and G by G^* . □

LEMMA 4.2. *If* U *is a* $2n \times 2n$ *constant* iJ-*contractive matrix and if* G *is a constant* $n \times n$ *matrix with* $G + G^* > 0$, *then*

$$R_U[G] = (U_{11} - iGU_{21})^{-1}(GU_{22} + iU_{12}) \tag{4.2}$$

is well defined and

$$R_U[G] + R_U[G]^* > 0 .$$

PROOF. Lemma 4.1 guarantees that R_U is well defined and, by (4.1), that

$$(U_{11} - iGU_{21})\{R_U[G] + R_U[G]^*\}(U_{11}^* + iU_{21}^*G^*) \geq G + G^* . \ \square$$

LEMMA 4.3. *If* U_1 *and* U_2 *are both* $2n \times 2n$ *constant* iJ-*contractive matrices and if* G *is a constant* $n \times n$ *matrix with* $G+G^* > 0$, *then*

$$R_{U_2U_1}[G] = R_{U_1}[R_{U_2}[G]] .$$

PROOF. This is a straightforward calculation. □

It will be useful in the sequel to have an alternative

expression for the linear fractional transformation $R_U[G]$ defined in (4.2).

LEMMA 4.4. *If* U *is a constant invertible* iJ-*contractive matrix and if* G *is a constant* $n \times n$ *matrix with* $G+G^* > 0$, *then*

$$\Phi = \begin{bmatrix} U_{11} & -iU_{12} \\ iU_{21} & U_{22} \end{bmatrix}^{-1} \tag{4.3}$$

and

$$T_\Phi[G] = (\Phi_{11}G+\Phi_{12})(\Phi_{21}G+\Phi_{22})^{-1} \tag{4.4}$$

are well defined (i.e., the indicated inverses exist) and

$$R_U[G] = T_\Phi[G] . \tag{4.5}$$

PROOF. Let

$$M = \begin{bmatrix} I_n & 0 \\ 0 & -iI_n \end{bmatrix} .$$

Then

$$M^*UM = \begin{bmatrix} U_{11} & -iU_{12} \\ iU_{21} & U_{22} \end{bmatrix}$$

is clearly invertible. Thus, Φ is well defined, and the identity

$$0 = [I_n \quad -G]\begin{bmatrix} U_{11} & -iU_{12} \\ iU_{21} & U_{22} \end{bmatrix}\begin{bmatrix} \Phi_{11} & \Phi_{12} \\ \Phi_{21} & \Phi_{22} \end{bmatrix}\begin{bmatrix} G \\ I_n \end{bmatrix}$$

implies that

$$(U_{11}-iGU_{21})(\Phi_{11}G+\Phi_{12}) = (iU_{12}+GU_{22})(\Phi_{21}G+\Phi_{22}) .$$

Hence, to complete the proof, it suffices to show that

$$\Phi_{21}G + \Phi_{22}$$

is invertible. But now, as M is unitary and

$$M^*iJM = \begin{bmatrix} 0 & I_n \\ I_n & 0 \end{bmatrix} = J_1 ,$$

it follows readily from the presumed inequality

$$U^*iJU \leqslant iJ$$

that

$$\Phi^*J_1\Phi \geqslant J_1$$

and hence that

$$[G^* \; I_n]\Phi^* J_1 \Phi [G^* \; I_n]^* \geqslant [G^* \; I_n] J_1 [G^* \; I_n]^*$$

or, equivalently, that

$$(G^*\Phi_{11}^* + \Phi_{12}^*)(\Phi_{21}G + \Phi_{22}) + (G^*\Phi_{21}^* + \Phi_{22}^*)(\Phi_{11}G + \Phi_{12}) \geqslant G+G^* .$$

Therefore, by the argument following (4.1), $\Phi_{21}G + \Phi_{22}$ and $\Phi_{11}G + \Phi_{12}$ are both invertible. □

LEMMA 4.5. *If* U *is a* $2n \times 2n$ *constant invertible* iJ-*contractive matrix and* G *and* H *are constant* $n \times n$ *matrices with* $G+G^* > 0$ *and* $H+H^* > 0$ *, then*

$$R_U[G]-R_U[H] = (U_{11} - iGU_{21})^{-1}(G-H)(\Phi_{21}H + \Phi_{22})^{-1} \qquad (4.6)$$

where Φ *is defined in* (4.3).

PROOF. By Lemma 4.4,

$$R_U[G]-R_U[H] = R_U[G]-T_\Phi[H]$$

and the rest is a straightforward calculation based on (4.3). □

LEMMA 4.6. *Let* U *be a* $2n \times 2n$ *constant invertible* iJ-*contractive matrix with an invertible* $n \times n$ *block* U_{11} *and let* $Y = R_U[G]$ *, where* G *is a constant* $n \times n$-*matrix with* $G+G^* > 0$. *Then the matrices*

$$iU_{21}Y + U_{22} \quad \text{and} \quad U_{11}Y - iU_{12}$$

are invertible.

PROOF. Let $\xi \in \mathbb{C}^n$ belong to the null space of at least one of the two matrices of interest. Then, since

$$G(iU_{21}Y + U_{22}) = (U_{11}Y - iU_{12}) ,$$

thanks to (4.2), and G is invertible, it follows readily that ξ must belong to the null space of both:

$$(U_{11}Y - iU_{12})\xi = 0 \quad \text{and} \quad (iU_{21}Y + U_{22})\xi = 0 .$$

But now, as U_{11} is invertible, this in turn implies that

$$[U_{22} - U_{21}U_{11}^{-1}U_{12}]\xi = 0 ,$$

and hence that $\xi = 0$, since, as is both well known and evident from the representation

$$U = \begin{bmatrix} I_n & 0 \\ U_{21}U_{11}^{-1} & I_n \end{bmatrix} \begin{bmatrix} U_{11} & 0 \\ 0 & U_{22}-U_{21}U_{11}^{-1}U_{12} \end{bmatrix} \begin{bmatrix} I_n & U_{11}^{-1}U_{12} \\ 0 & I_n \end{bmatrix}$$

and the invertibility of U, the block $U_{22} - U_{21}U_{11}^{-1}U_{12}$ is invertible. □

Our next objective is to identify the set of matrices

$$\mathcal{D} = \mathcal{D}_U = \{R_U[G]: G+G^* > 0\} \tag{4.7}$$

as a matrix ball, under appropriate assumptions on U.

LEMMA 4.7. *Let* U *be a* $2n \times 2n$ *constant invertible* iJ-*contractive matrix with an invertible* $n \times n$ *block* U_{11}. *Then* $\mathcal{D}_U$ *is equal to the set of* $n \times n$ *constant matrices* Y *such that*

$$|\alpha^{\frac{1}{2}}(Y + \alpha^{-1}\beta)\delta^{\frac{1}{2}}| < 1 \quad ,$$

where

$$\alpha = i(U_{21}^*U_{11} - U_{11}^*U_{21}) \quad , \tag{4.8}$$

$$\beta = U_{21}^*U_{12} - U_{11}^*U_{22} \quad , \tag{4.9}$$

$$\delta = (\gamma + \beta^*\alpha^{-1}\beta)^{-1} \tag{4.10}$$

and

$$\gamma = i(U_{12}^*U_{22} - U_{22}U_{12}) \quad . \tag{4.11}$$

PROOF. If $Y = R_U[G]$, then it is readily seen that

$$G(iU_{21}Y + U_{22}) = U_{11}Y - iU_{12}$$

and hence, since $iU_{21}Y + U_{22}$ is invertible, by Lemma 4.5, that

$$G = (U_{11}Y - iU_{12})(iU_{21}Y + U_{22})^{-1} \quad .$$

But now, as $G+G^* > 0$, a straightforward computation yields

$$Y^*\alpha Y + \beta^*Y + Y^*\beta < \gamma$$

where α, β, and γ are given by (4.8), (4.9), and (4.11), respectively. Moreover, since

$$\begin{bmatrix} -\alpha & -i\beta \\ i\beta^* & \gamma \end{bmatrix} = U^*iJU < iJ \quad , \tag{4.12}$$

it follows easily that $\alpha > 0$, and hence that the inequality on Y can be reexpressed as

$$(Y^*\alpha^{\frac{1}{2}} + \beta^*\alpha^{-\frac{1}{2}})(\alpha^{\frac{1}{2}}Y + \alpha^{-\frac{1}{2}}\beta) < \gamma + \beta^*\alpha^{-1}\beta \quad ,$$

where $\alpha^{\frac{1}{2}}$ designates the positive square root of α. The right-hand side of the last inequality is thus seen to be positive and the inequality itself is readily checked to be equivalent to the statement that

$$|\alpha^{\frac{1}{2}}(Y + \alpha^{-1}\beta)\delta^{\frac{1}{2}}| < 1 \; ,$$

where δ is defined in (4.10) and $\delta^{\frac{1}{2}}$ denotes its positive square root. □

The auxiliary formula

$$\delta = [0 \quad I_n]\,[U^*iJU]^{-1}\,[0 \quad I_n]^* \tag{4.13}$$

which may be obtained by applying the representation formula used in the proof of Lemma 4.6 to the left hand side of (4.12), will be useful in Section 8.

It is convenient to introduce the class C of $n \times n$ matrix valued functions which belong to W_+^I and have strictly positive real part at each point $\mu \in \bar{\mathbb{C}}_+$.

THEOREM 4.1. *For each* $t \geq 0$, *the linear fractional transformation* $R_{U(t,\mu)}$ *based on the matrizant* $U(t,\mu)$ *of the canonical equation* (2.1) *is a one to one mapping of* C *into itself.*

PROOF. Let $G \in C$. Then, by Lemmas 4.1 and 4.2, which are applicable since $U(t,\mu)$ is iJ-contractive on $\bar{\mathbb{C}}_+$,

$$\mathrm{Re}\{R_{U(t,\mu)}[G(\mu)]\} > 0$$

at each point $\mu \in \bar{\mathbb{C}}_+$. Next, upon expressing

$$G = I + G_0$$

with $G_0 \in W_+^0$, it follows readily from (2.3) and (2.4) that

$$R_U[G] = (\Omega_{11} - i\Omega_{21} - ie^{i\lambda t}G_0U_{21})^{-1}(\Omega_{22} + i\Omega_{12} + e^{i\lambda t}G_0U_{22}) \; ,$$

and that

$$2e^{i\lambda t}U_{22} = e^{2i\lambda t}(\Omega_{22}-i\Omega_{12}) + (\Omega_{22}+i\Omega_{12})$$

and

$$2e^{i\lambda t}U_{21} = e^{2i\lambda t}(\Omega_{21}-i\Omega_{11}) + (\Omega_{21}+i\Omega_{11}) \ .$$

Moreover, upon writing out $P_{\pm}\Omega$ and $P_{\pm}$ explicitly (see (2.18) for the definitions), the auxiliary formulas

$$P_{\pm}\Omega(t,\lambda) = P_{\pm} + \int_0^t e^{\pm 2i\lambda s}P_{\pm}\omega(t,s)\,ds$$

exhibit $\Omega_{11}-i\Omega_{21}$ and $\Omega_{22}+i\Omega_{12}$ as elements of W_+^I and $\Omega_{11}+i\Omega_{21}$ and $\Omega_{22}-i\Omega_{12}$ as elements of W_-^I with the inverse transform parts restricted to the interval $[-2t,0]$. Thus, $e^{i\lambda t}G_0U_{22}$ and $e^{i\lambda t}G_0U_{21}$ are readily identified as elements of W_+^0, while $R_U[G]$ is seen to be a quotient of two elements in W_+^I, where the divisor has nonzero determinant on $\bar{\mathbb{C}}_+ \cup \{\infty\}$. Therefore, by the Wiener theorem, $R_U[G]$ is itself an element of W_+^I, as asserted. This completes the proof that R_U maps C into itself, and hence, since the one-to-oneness is plain from (4.6), the proof the theorem. □

LEMMA 4.8. *If* $U(t,\lambda)$ *is the matrizant of the canonical equation* (2.1) *and if* G *and* H *belong to* C, *then*

$$\{R_{U(t,\lambda)}[G(\lambda)] - R_{U(t,\lambda)}[H(\lambda)]\} \in e^{2i\lambda t}W_+^0 \ .$$

PROOF. The proof rests on an analysis of the terms in (4.6) under the assumptions of the lemma. To begin with, it follows much as in the proof of Theorem 4.1, that

$$\begin{aligned} U_{11}-iGU_{21} &= e^{-i\lambda t}(\Omega_{11}-i\Omega_{21}-ie^{i\lambda t}G_0U_{21}) \\ &= e^{-i\lambda t}G_1 \ , \end{aligned}$$

where $G_1^{\pm 1} \in W_+^I$. Moreover, since in the present circumstances

$$\begin{aligned} \Phi(t,\lambda) &= M^*J^*U^{\#}(t,\lambda)JM \\ &= \begin{bmatrix} U_{22}^{\#}(t,\lambda) & iU_{12}^{\#}(t,\lambda) \\ -iU_{21}^{\#}(t,\lambda) & U_{11}^{\#}(t,\lambda) \end{bmatrix} , \end{aligned} \qquad (4.14)$$

the term

$$\Phi_{21}H + \Phi_{22} = U_{11}^{\#} - iU_{21}^{\#}H$$

is subject to much the same sort of analysis. In particular, upon writing

$$H = I+H_0$$

with $H_0 \in W_+^0$, it follows that

$$U_{11}^{\#} - iU_{21}^{\#}H = (U_{11} + iU_{21} + iH_0^{\#}U_{21})^{\#}$$

$$= (e^{i\lambda t}\{\Omega_{11} + i\Omega_{21} + ie^{-i\lambda t}H_0^{\#}U_{21}\})^{\#}$$

belongs to $e^{-i\lambda t}W_+^I$. The rest is plain. □

We remark that this lemma implies that the linear fractional transformation R_U based on the matrizant $U = U(t,\lambda)$ of the canonical equation maps $\mathcal{C}$ properly into itself and not onto itself for $t > 0$. Indeed

$$R_{U(t,\lambda)}[G(\lambda)] - R_{U(t,\lambda)}[I_n]$$

belongs to $e^{2i\lambda t}W_+^0$ for every $G \in \mathcal{C}$.

We further remark that (4.14) and Lemma 4.4 guarantee that

$$R_U[G] = (U_{22}^{\#}G + iU_{12}^{\#})(-iU_{21}^{\#}G + U_{11}^{\#})^{-1}$$

for U a matrizant or, more generally, for any $U \in \mathcal{E}(iJ)$.

5. FOURIER ANALYSIS, SPECTRAL FUNCTIONS AND SPANS OF EXPONENTIALS

From now on we shall assume that

$$U(t,\lambda) = \begin{bmatrix} A(t,\lambda) & C(t,\lambda) \\ B(t,\lambda) & D(t,\lambda) \end{bmatrix} \tag{5.1}$$

is the matrizant of a canonical equation (2.1) (with, as always in this paper, a summable potential which is normalized as in (2.2)). The notations

$$E(t,\lambda) = A(t,\lambda) - iB(t,\lambda) \quad , \tag{5.2}$$

$$F(t,\lambda) = A(t,\lambda) + iB(t,\lambda) \quad , \tag{5.3}$$

$$\varepsilon_t(\lambda) = \Omega_{11}(t,\lambda) - i\Omega_{21}(t,\lambda) = e^{i\lambda t}E(t,\lambda) \qquad (5.4)$$

$$\varphi_t(\lambda) = \Omega_{11}(t,\lambda) + i\Omega_{21}(t,\lambda) = e^{-i\lambda t}F(t,\lambda) \qquad (5.5)$$

and

$$X(t,\lambda) = \begin{bmatrix} A(t,\lambda) \\ B(t,\lambda) \end{bmatrix} \qquad (5.6)$$

will prove useful. Since $U(t,\lambda)$ belongs to $\mathcal{E}(iJ)$ for each choice of $t \geq 0$, it follows from Theorem 3.2 that $\{F^{\#}(t,\lambda), E^{\#}(t,\lambda)\}$ is a de Branges pair for every choice of $t \geq 0$. We shall denote the corresponding de Branges space by $\mathcal{B}_t$ and the associated reproducing kernel by $\Lambda_\mu^t(\lambda)$:

$$\Lambda_\mu^t(\lambda) = \frac{F^{\#}(t,\lambda)F^{\#}(t,\mu)^* - E^{\#}(t,\lambda)E^{\#}(t,\mu)^*}{-2\pi i(\lambda-\mu^*)}$$

$$= \frac{X^{\#}(t,\lambda)JX^{\#}(t,\mu)^*}{-\pi(\lambda-\mu^*)} \qquad (5.7)$$

(see (3.8) and (3.10)). In view of (2.11), this can also be expressed as

$$\Lambda_\mu^t(\lambda) = \frac{1}{\pi}\int_0^t X^{\#}(s,\lambda)X(s,\mu^*)\,ds . \qquad (5.8)$$

We next define the transform

$$f^{\blacktriangle}(\lambda) = \int_0^\infty X^{\#}(s,\lambda)f(s)\,ds \qquad (5.9)$$

for functions $f \in L_{2n}^2(\mathbb{R}^+)$ with compact support, and set

$$\mathcal{K}_T = \{\int_0^T X^{\#}(s,\lambda)f(s)\,ds\colon f \in L_{2n}^2[0,T]\}$$

$$= \{f^{\blacktriangle}\colon f \in L_{2n}^2[0,T]\} .$$

A similar transform exists for the canonical equation (2.1) with zero potential. If

$$X_0(t,\lambda) = \begin{bmatrix} (\cos\lambda t)I_n \\ (\sin\lambda t)I_n \end{bmatrix}$$

denotes the first block-column of the matrizant $U_0(t,\lambda) = e^{-\lambda tJ}$, then the corresponding transform

$$f^{\Delta}(\lambda) = \int_0^\infty X_0^{\#}(s,\lambda)f(s)ds = \int_0^\infty [(\cos\lambda s)I_n \quad (\sin\lambda s)I_n]f(s)ds \ .$$

The latter is simply related to the ordinary Fourier transform

$$g^{\wedge}(\lambda) = \int_{-\infty}^{\infty} e^{i\lambda s}g(s)ds$$

on $L_n^2(\mathbb{R})$. If

$$f = \begin{bmatrix} f_1 \\ f_2 \end{bmatrix} \quad \text{and} \quad \sqrt{2}(\mathbb{N}f)(s) = \begin{cases} f_1(s)-if_2(s) & \text{for } s > 0 \\ f_1(-s)+if_2(-s) & \text{for } s < 0 \ , \end{cases}$$

then it is readily checked that $\mathbb{N}$ is an isometry from $L_{2n}^2(\mathbb{R}^+)$ onto $L_n^2(\mathbb{R})$ and

$$f^{\Delta} = \frac{1}{\sqrt{2}}(\mathbb{N}f)^{\wedge} \ . \tag{5.10}$$

Thus, by the Plancherel formula for the usual Fourier transform

$$\begin{aligned} 2\pi \|f\|^2_{L_{2n}^2(\mathbb{R}^+)} &= 2\pi\|\mathbb{N}f\|^2_{L_n^2(\mathbb{R})} = \|(\mathbb{N}f)^{\wedge}\|^2_{L_n^2(\mathbb{R})} \\ &= 2\|f^{\Delta}\|^2_{L_n^2(\mathbb{R})} \ . \end{aligned} \tag{5.11}$$

This effectively proves that the Δ-transform is (apart from a factor of $\sqrt{\pi}$) an isometric map of $L_{2n}^2(\mathbb{R}^+)$ onto $L_n^2(\mathbb{R})$. Our next objective is to prove an analogous statement for $\blacktriangle$. The proof is a little more elaborate, and we start with the $\blacktriangle$-transform of $L_{2n}^2[0,T]$.

THEOREM 5.1. *The* $\blacktriangle$*-transform defined in* (5.9) *maps* $L_{2n}^2[0,T]$ *isometrically (apart from a factor of* $\sqrt{\pi}$ *) onto* $\mathcal{B}_T$ *. That is,*

$$K_T = \mathcal{B}_T$$

and

$$\int_0^T g(s)^*g(s)ds = \frac{1}{\pi}\|g^{\blacktriangle}\|^2_{\mathcal{B}_T} \tag{5.12}$$

for all $g \in L_{2n}^2[0,T]$.

PROOF. The proof is broken into steps.

STEP 1 *is to establish* (5.12) *for functions of the form*

$$g(s) = \begin{cases} \frac{1}{\pi} \sum_{j=1}^{m} X^{\#}(s,\mu_j)^{*}\xi_j & \text{for } 0 \leq s \leq T , \\ 0 & \text{for } T < S , \end{cases}$$

where the μ_j *are arbitrary points in* $\mathbb{C}$ *and the* ξ_j *are arbitrary vectors in* $\mathbb{C}^n$.

PROOF OF STEP 1. We first observe that

$$\begin{aligned} g^{\blacktriangle}(\lambda) &= \frac{1}{\pi} \int_0^T X^{\#}(s,\lambda) \sum_{j=1}^{m} X^{\#}(s,\mu_j)^{*}\xi_j ds \\ &= \sum_{j=1}^{m} \Lambda_{\mu_j}^{T}(\lambda)\xi_j \end{aligned}$$

which clearly belongs to $\mathcal{B}_T$. Therefore, by (3.6) and (5.8)

$$\begin{aligned} \|g^{\blacktriangle}\|_{\mathcal{B}_T}^2 &= \langle \sum_{j=1}^{m} \Lambda_{\mu_j}^{T}\xi_j , \sum_{k=1}^{m} \Lambda_{\mu_k}^{T}\xi_k \rangle_{\mathcal{B}_T} \\ &= \sum_{j,k=1}^{m} \xi_k^{*} \Lambda_{\mu_j}^{T}(\mu_k)\xi_j \\ &= \sum_{j,k=1}^{m} \xi_k^{*} \frac{1}{\pi} \int_0^T X^{\#}(s,\mu_k)X^{\#}(s,\mu_j)^{*} ds\, \xi_j \\ &= \pi \int_0^T g(s)^{*}g(s) ds , \end{aligned}$$

as asserted.

STEP 2 *is to show that the set of all functions of the form considered in Step* 1 *is dense in* $L_{2n}^2[0,T]$.

PROOF OF STEP 2. If $f \in L_{2n}^2[0,T]$ is orthogonal to the indicated set of functions, then

$$\xi^{*} \int_0^T X^{\#}(t,\mu)f(t)dt = 0$$

for every $\xi \in \mathbb{C}^n$ and every point $\mu \in \mathbb{C}$. But now it follows from (2.16) (as is explained in more detail in the next section), that

$$X(t,\mu) = X_0(t,\mu) + \int_0^t K(t,s)X_0(s,\mu)ds ,$$

where $K(t,s)$ is given by (6.3), and hence that

$$\int_0^T X^{\#}(t,\mu)f(t)dt = g^{\blacktriangle}(\mu) = 0$$

for every $\mu \in \mathbb{C}$, where

$$g(t) = \begin{cases} f(t) + \int_t^T K(s,t)^* f(s)ds & \text{for } 0 \leq t \leq T, \\ 0 & \text{for } T < t. \end{cases}$$

Therefore, because of (5.11), $g(t) \equiv 0$, and hence, in view of Theorem 6.1, $f(t) \equiv 0$ also, as desired.

STEP 3 *is to complete the proof.*

PROOF OF STEP 3. Let $f \in L^2_{2n}[0,T]$. Then, in view of Steps 1 and 2, f can be approximated arbitrarily well in $L^2_{2n}[0,T]$ by a sequence of functions f_j for which (5.12) holds. In particular, this implies that

$$|\xi^*\{f^{\blacktriangle}(\lambda)-f_j^{\blacktriangle}(\lambda)\}|^2 = |\xi^* \int_0^T X^{\#}(s,\lambda)\{f(s)-f_j(s)\}ds|^2 \leq$$

$$\leq \xi^* \int_0^T X^{\#}(s,\lambda)X(s,\lambda)ds\xi \cdot ||f-f_j||^2_{L^2_{2n}[0,T]}$$

$$= \pi\xi^*\Lambda_\lambda^T(\lambda)\xi\ ||f-f_j||^2_{L^2_{2n}[0,T]}$$

for every $\lambda \in \mathbb{R}$ and $\xi \in \mathbb{C}^n$ and hence that $f_j^{\blacktriangle} \to f^{\blacktriangle}$ pointwise on $\mathbb{R}$ as $j \uparrow \infty$. Since $\{f_j^{\blacktriangle}\}$ is also Cauchy in $\mathcal{B}_T$, it follows that $f_j^{\blacktriangle} \to f^{\blacktriangle}$ in $\mathcal{B}_T$. In particular, this exhibits $f^{\blacktriangle}$ as an element of $\mathcal{B}_T$ and implies that $\blacktriangle$ maps $L^2_{2n}[0,T]$ isometrically (in the sense of (5.12) onto a closed subspace $\mathcal{K}_T$ of $\mathcal{B}_T$. Finally, since $\Lambda_\mu^T(\lambda)\xi$ belongs to $\mathcal{K}_T$ for every $\mu \in \mathbb{C}$ and $\xi \in \mathbb{C}^n$, thanks to (5.8), and linear combinations of these are dense in $\mathcal{B}_T$, it follows that $\blacktriangle$ is onto, i.e. $\mathcal{K}_T = \mathcal{B}_T$, and the proof is complete. □

The notation

$$\Delta_t(\lambda) = \{E^{\#}(t,\lambda)E(t,\lambda)\}^{-1} = \{\varepsilon_t^{\#}(\lambda)\varepsilon_t(\lambda)\}^{-1}$$
$$= \{F^{\#}(t,\lambda)F(t,\lambda)\}^{-1} = \{\varphi_t^{\#}(\lambda)\varphi_t(\lambda)\}^{-1} \qquad (5.13)$$

for $\lambda \in \mathbb{R}$, will prove convenient.

COROLLARY 5.1. *If* $f \in L^2_{2n}[0,T]$, *then*

$$\int_0^T f(s)^* f(s)\,ds = \frac{1}{\pi}\int_{-\infty}^{\infty} f^{\blacktriangle}(\lambda)^* \Delta_T(\lambda) f^{\blacktriangle}(\lambda)\,d\lambda \ . \qquad (5.14)$$

PROOF. This is immediate from (5.12), (5.13) and (3.18). □

COROLLARY 5.2. *If* $S < T$, *then* $\mathcal{B}_S$ *sits isometrically inside* $\mathcal{B}_T$ *: If* $g \in \mathcal{B}_S$, *then* $g \in \mathcal{B}_T$ *and*

$$\| g \|^2_{\mathcal{B}_S} = \int_{-\infty}^{\infty} g(\lambda)^* \Delta_S(\lambda) g(\lambda)\,d\lambda = \int_{-\infty}^{\infty} g(\lambda)^* \Delta_T(\lambda) g(\lambda)\,d\lambda = \| g \|^2_{\mathcal{B}_T} \ .$$

PROOF. This is immediate from the Theorem and Corollary 5.1. □

LEMMA 5.1. *The matrix valued functions* Δ_t *defined in* (5.13) *are invertible elements of* W^I *and, as* $t \uparrow \infty$, $[\Delta_t]^{\pm 1}$ *tends uniformly on* $\mathbb{R}$ *to a limit* $[\Delta_\infty]^{\pm 1}$ *of class* W^I . *Moreover, the inverse Fourier transform*

$$[\Delta_t^{-1} - I_n]^\vee(s) = 0 \ , \quad \textit{for} \quad |s| > 2t \ .$$

PROOF. By Theorem 2.1,

$$[\varepsilon_t(\lambda)]^{\pm 1} \to [\varepsilon_\infty(\lambda)]^{\pm 1} = [(\Omega_\infty)_{11}(\lambda) - i(\Omega_\infty)_{21}(\lambda)]^{\pm 1} \in W^I_+$$

uniformly on $\mathbb{R}$ as $t \uparrow \infty$. Therefore

$$[\varepsilon_t^\#(\lambda)]^{\pm 1} \to [\varepsilon_\infty^\#(\lambda)]^{\pm 1} \in W^I_-$$

uniformly on $\mathbb{R}$ as $t \uparrow \infty$ and

$$[\Delta_t(\lambda)]^{\mp 1} \to (\varepsilon_\infty^\#(\lambda)\varepsilon_\infty(\lambda))^{\pm 1} \in W^I$$

uniformly on $\mathbb{R}$ as $t \uparrow \infty$.

Finally, it follows from (2.23) that

$$\varepsilon_t(\lambda) = I_n + \int_0^t e^{2i\lambda s}\{\omega_{11}(t,s) - i\omega_{21}(t,s)\}ds$$

and hence that Δ_t^{-1} is of the asserted form. □

LEMMA 5.2. *Every* $f \in \mathcal{B}_T$ *also belongs to* $L^2_n(\mathbb{R}) = L^2_n(\mathbb{R}, d\lambda)$.

PROOF. If $g \in \mathcal{B}_T$, then

$$\int_{\infty}^{\infty} g(\lambda)^* \Delta_T(\lambda) g(\lambda) d\lambda < \infty$$

where, by Lemma 5.1, $\Delta_T \in W^I$. Therefore, by the Riemann-Lebesgue lemma, there exists an $R > 0$ such that

$$\Delta_T(\lambda) \geq \frac{1}{2} I_n \quad \text{for} \quad |\lambda| > R .$$

Thus

$$\int_{|\lambda|>R} g(\lambda)^* g(\lambda) d\lambda \leq 2 \int_{|\lambda|>R} g(\lambda)^* \Delta_T(\lambda) g(\lambda) d\lambda \quad .$$

The rest is plain, because $g(\lambda)^* g(\lambda)$ is continuous and so clearly summable over finite intervals. □

LEMMA 5.3. *$\mathcal{B}_S$ sits isometrically inside $L_n^2(\mathbb{R}, \Delta_\infty(\lambda) d\lambda)$: If $g \in \mathcal{B}_S$, then*

$$\int_{-\infty}^{\infty} g(\lambda)^* \Delta_\infty(\lambda) g(\lambda) d\lambda < \infty$$

and

$$\pi \| g \|_{\mathcal{B}_S}^2 = \int_{-\infty}^{\infty} g(\lambda)^* \Delta_S(\lambda) g(\lambda) d\lambda = \int_{-\infty}^{\infty} g(\lambda)^* \Delta_\infty(\lambda) g(\lambda) d\lambda .$$

PROOF. This follows from (5.14) upon letting $T \uparrow \infty$. The interchange of limit with integration is easily justified since Δ_T converges uniformly to Δ_∞ on $\mathbb{R}$, thanks to Lemma 5.1, and $g \in L_n^2(\mathbb{R})$, by Lemma 5.2. □

We will be also interested in an inverse transform for $\blacktriangle$. To guess its form, let $g \in \mathcal{B}_T$. Then, for any $\xi \in \mathbb{C}^n$ and $\mu \in \mathbb{C}$,

$$\begin{aligned} \xi^* g(\mu) &= \langle g, \Lambda_\mu^T(\cdot)\xi \rangle_{\mathcal{B}_T} = \xi^* \int_{-\infty}^{\infty} \Lambda_\mu^T(\lambda)^* \Delta_T(\lambda) g(\lambda) d\lambda \\ &= \xi^* \int_{-\infty}^{\infty} \{\frac{1}{\pi} \int_0^T X^\#(s,\lambda) X(s,\mu^*) ds\}^* \Delta_T(\lambda) g(\lambda) d\lambda \\ &= \xi^* \int_0^T X^\#(s,\mu) \{\frac{1}{\pi} \int_{-\infty}^{\infty} X(s,\lambda) \Delta_T(\lambda) g(\lambda) d\lambda\} ds . \end{aligned}$$

This suggests that for functions $g \in \mathcal{B}_T$ the inverse $\blacktriangle$-transform should be given by

$$g^{\blacktriangledown}(s) = \frac{1}{\pi} \int_{-\infty}^{\infty} X(s,\lambda)\Delta_T(\lambda)g(\lambda)d\lambda \quad ,$$

and that moreover, in view of Lemma 5.3, the last formula, with Δ_T replaced by Δ_∞ should hold simultaneously for all spaces $\mathcal{B}_T$. This is indeed the case.

THEOREM 5.2. *The transform* $f^{\blacktriangle}$ *which is defined in* (5.9) *for functions* $f \in L^2_{2n}(\mathbb{R}^+)$ *with compact support extends naturally to an isometric (apart from a factor of* $\sqrt{\pi}$ *) map of* $L^2_{2n}(\mathbb{R}^+)$ *onto* $L^2_n(\mathbb{R}, \Delta_\infty(\lambda)d\lambda)$:

$$\pi \| f \|^2_{L^2_{2n}(\mathbb{R}^+)} = \| f^{\blacktriangle} \|_{L^2_n(\mathbb{R},\Delta_\infty(\lambda)d\lambda)} \quad . \tag{5.16}$$

The inverse transform $g^{\blacktriangledown}$ *is given by the rule*

$$g^{\blacktriangledown}(\lambda) = \lim_{R\uparrow\infty} \frac{1}{\pi} \int_{-R}^{R} X(s,\lambda)\Delta_\infty(\lambda)g(\lambda)d\lambda \tag{5.17}$$

(where the limit is understood in $L^2_{2n}(\mathbb{R}^+)$*).*

PROOF. Given $f \in L^2_{2n}(\mathbb{R}^+)$, choose $f_k \in L^2_{2n}(\mathbb{R}^+)$ with compact support such that $\| f-f_k \|_{L^2_{2n}(\mathbb{R}^+)} \to 0$ as $k \to \infty$. Then, thanks to (5.12) and Lemma 5.3, the sequence $\{f_k^{\blacktriangle}\}$ is Cauchy in $L^2_n(\mathbb{R}, \Delta_\infty(\lambda)d\lambda)$. We define $f^{\blacktriangle}$ as the limit of this sequence, and check by standard arguments that the limit is independent of the choice of the sequence and that (5.16) holds.

Next, in order to see that the $\blacktriangle$-transform maps $L^2_{2n}(\mathbb{R}^+)$ onto $L^2_n(\mathbb{R}, \Delta_\infty(\lambda)d\lambda)$, use the formula

$$f^{\blacktriangle} = [(I + \mathbb{K}^*)f]^{\Delta} \quad ,$$

in which $\mathbb{K}$ is the operator based on the kernel K which intervened in the proof of Step 2 of Theorem 5.1, and the fact that $I + \mathbb{K}$ is a bounded invertible map of $L^2_{2n}(\mathbb{R}^+)$ onto itself (see Theorem 6.1). This yields the desired result because, as we have already noted, the Δ-transform maps $L^2_{2n}(\mathbb{R}^+)$ onto $L^2_n(\mathbb{R})$ which, by a simple adaptation of the proof of Lemma 5.2, is readily seen to contain the same set of functions as $L^2_n(\mathbb{R}, \Delta_\infty(\lambda)d\lambda)$.

It remains to verify the inversion formula. With the

help of identity (6.14) (the proof of which depends only on (5.16)) and formula (5.10), it is readily seen that

$$\begin{aligned}[(I+\mathbb{K})^{-1}f]^{\Delta} &= [\mathbb{N}^{*}(I-H)\mathbb{N}(I+\mathbb{K}^{*})f]^{\Delta} \\ &= 2^{-\frac{1}{2}}[(I-H)\mathbb{N}(I+\mathbb{K}^{*})f]^{\wedge} \\ &= 2^{-\frac{1}{2}}\Delta_{\infty}[\mathbb{N}(I+\mathbb{K}^{*})f]^{\wedge} \\ &= \Delta_{\infty}[(I+\mathbb{K}^{*})f]^{\Delta} \\ &= \Delta_{\infty}f^{\blacktriangle} ,\end{aligned}$$

for every $f \in L^2_{2n}(\mathbb{R}^+)$. On the other hand, formula (5.10) further implies that

$$g = 2^{\frac{1}{2}}\mathbb{N}^{*}[g^{\Delta}]^{\vee}$$

for every $g \in L^2_{2n}(\mathbb{R}^+)$, and hence, upon combining the last two formulas, it follows that

$$\begin{aligned}(I+\mathbb{K})^{-1}f &= 2^{\frac{1}{2}}\mathbb{N}^{*}[\Delta_{\infty}f^{\blacktriangle}]^{\vee} \\ &= \lim_{R\uparrow\infty} 2^{\frac{1}{2}}\mathbb{N}^{*}\frac{1}{2\pi}\int_{-R}^{R} e^{-i\lambda s}\Delta_{\infty}(\lambda)f^{\blacktriangle}(\lambda)d\lambda \\ &= \lim_{R\uparrow\infty}\frac{1}{\pi}\int_{-R}^{R} X_0(s,\lambda)\Delta_{\infty}(\lambda)f^{\blacktriangle}(\lambda)d\lambda .\end{aligned}$$

The desired formula drops out upon applying $I+\mathbb{K}$ to both sides and justifying the necessary interchanges, which is straightforward. □

It is readily checked that if $f \in L^2_m(\mathbb{R}^+)$ is absolutely continuous and has compact support, then

$$\left(J\frac{df}{dt} - Vf\right)^{\blacktriangle}(\lambda) = \lambda f^{\blacktriangle}(\lambda) - [0 \quad I_n]f(0) .$$

More precisely, if G denotes the restriction of $J\frac{df}{dt} - Vf$ to the set $\mathcal{D}(G)$ of $f \in L^2_m(\mathbb{R}^+)$ such that

(a) $\lambda f^{\Delta}(\lambda) \in L^2_n(\mathbb{R}, \Delta_{\infty}(\lambda)d\lambda)$ and

(b) $[0 \quad I_n]\, f(0) = 0$,

then it may be shown that G is selfadjoint and

$$(Gf)^{\blacktriangle}(\lambda) = \lambda f^{\blacktriangle}(\lambda) .$$

For additional information see [Ad], [MaF1] and [MaF2].

We shall refer to the matrix valued function $\Delta_\infty(\lambda)$ as the *spectral function* of the canonical equation (2.1), as is customary. For a different proof of the existence of a spectral function for (2.1) in terms of M.G. Krein's theory of directing functionals the papers of [L] and [MaF2] are suggested.

We now introduce the function

$$Z_t(\lambda) = R_{U(t,\lambda)}[I_n] \tag{5.18}$$

which will play an important role in the sequel. By Theorem 4.1, Z_t belongs to the class C introduced in Section 4, and is moreover readily seen to equal

$$R_{\Omega(t,\lambda)}[I_n] = [\Omega_{11}(t,\lambda)-i\Omega_{21}(t,\lambda)]^{-1}[\Omega_{22}(t,\lambda)+i\Omega_{12}(t,\lambda)], \tag{5.19}$$

which in turn converges uniformly on $\bar{\mathbb{C}}_+$ to the limit

$$Z_\infty(\lambda) = [(\Omega_\infty)_{11}(\lambda)-i(\Omega_\infty)_{21}(\lambda)]^{-1}[(\Omega_\infty)_{22}(\lambda)+i(\Omega_\infty)_{12}(\lambda)] , \tag{5.20}$$

thanks to Theorem 2.1.

THEOREM 5.3. *Let*

$$\Delta_\infty(\lambda) = I_n - \hat{h}(\lambda)$$

be the spectral function of a canonical equation with matrizant $U(t,\lambda)$. *Then the limit*

$$Z_\infty(\lambda) = \lim_{t\uparrow\infty} R_{U(t,\lambda)}[I_n]$$

discussed just above belongs to the class C *and can be expressed in the form*

$$Z_\infty(\lambda) = I_n - 2\int_0^\infty e^{i\lambda s}h(s)ds . \tag{5.21}$$

PROOF. It follows readily from the already established identification (5.20) and the J-unitarity of Ω_∞ on $\mathbb{R}$: (2.22), that

$$\begin{aligned}\frac{Z_\infty(\lambda)+Z_\infty(\lambda)^*}{2} &= \{(\Omega_\infty)_{11}(\lambda)-i(\Omega_\infty)_{21}(\lambda)\}^{-1}\{(\Omega_\infty)_{11}(\lambda)^*+i(\Omega_\infty)_{21}(\lambda)^*\}^{-1} \\ &= [\varepsilon_\infty(\lambda)]^{-1}[\varepsilon_\infty(\lambda)^*]^{-1} = \Delta_\infty(\lambda) ,\end{aligned} \tag{5.22}$$

for $\lambda \in \mathbb{R}$. Moreover, Theorem 2.1 guarantees that $Z_\infty \in W_+^I$ and hence can be expressed in the form

$$Z_\infty(\lambda) = I_n + \int_0^\infty e^{i\lambda s} p(s)\,ds$$

where $p \in L^1_{n\times n}(\mathbb{R}^+)$. Thus, by (5.22),

$$I_n + \int_0^\infty e^{i\lambda s} p(s)\,ds + \int_{-\infty}^0 e^{i\lambda s} p(-s)^* ds = 2\{I_n - \hat{h}(\lambda)\} ,$$

which in turn implies that

$$p(s) = -2h(s)$$

for $s > 0$, thus establishing (5.21).

Finally, since $h(-s) = h(s)^*$, it is readily checked that for complex $\omega = a+ib$ with $b \geq 0$

$$\begin{aligned} \operatorname{Re} Z_\infty(\omega) &= I_n - \int_{-\infty}^\infty e^{ias} h(s) e^{-b|s|}\,ds \\ &= \frac{b}{\pi} \int_{-\infty}^\infty \frac{I_n - \hat{h}(\lambda)}{(\lambda-a)^2+b^2}\,d\lambda \end{aligned} \tag{5.23}$$

and hence, since $I_n - \hat{h}(\lambda) > 0$ for every $\lambda \in \mathbb{R}$, that $\operatorname{Re} Z_\infty(\omega) > 0$ for every point $\omega \in \bar{\mathbb{C}}_+$. Thus $Z_\infty \in \mathcal{C}$ as asserted, and the proof is complete. □

We complete this section with another characterization of the space $\mathcal{B}_T = \mathcal{K}_T$ which is useful in the theory of estimation of stochastic processes.

THEOREM 5.4. *The space* $\mathcal{B}_T = \mathcal{K}_T$ *is also equal to the closed linear span in* $L^2_n(\mathbb{R}, \Delta_\infty(\lambda)\,d\lambda)$ *of the set of functions*

$$\left\{ \frac{e^{i\lambda t} - 1}{i\lambda}\,\xi : |t| \leq T \text{ and } \xi \in \mathbb{C}^n \right\} .$$

PROOF. Let χ be the indicator function of a subinterval $[c,d]$ of $[-T,T]$ and let $I + \mathbb{L} = (I + \mathbb{K})^{-1}$. Then

$$f = \sqrt{2}(I + \mathbb{L}^*)\mathbb{N}^* \chi\,\xi$$

belongs to $L^2_{2n}[0,T]$ and

$$f^\blacktriangle(\lambda) = [2^{-\frac{1}{2}} \mathbb{N}(I + \mathbb{K}^*) f]^\wedge(\lambda) = \frac{e^{id\lambda} - e^{ic\lambda}}{i\lambda}\,\xi .$$

This proves that the set of functions given in the statement of the theorem belongs to $\mathcal{K}_T = \mathcal{B}_T$.

Next, suppose that $g^{\blacktriangle} \in \mathcal{B}_T$ is orthogonal to the given set. Then, by Theorem 5.1 and the isometry noted in Lemma 5.3,

$$0 = \langle (I + \mathbb{L}^*)\, \mathbb{N}^* \chi \xi , g \rangle_{L^2_{2n}(\mathbb{R}^+)}$$

$$= \langle \mathbb{N}^* \chi \xi , (I + \mathbb{L})\, g \rangle_{L^2_{2n}(\mathbb{R}^+)}$$

for the indicator function χ of every subinterval of $[-T,T]$ and every $\xi \in \mathbb{C}^n$. This proves that

$$P_T(I + \mathbb{L})\, g = 0 ,$$

where P_T denotes the projection which is defined by the rule

$$(P_T f)(s) = \begin{cases} f(s) & \text{for } 0 \leq s \leq T , \\ 0 & \text{for } T < s , \end{cases} \tag{5.24}$$

and hence, since

$$P_T(I + \mathbb{K}) = P_T(I + \mathbb{K})\, P_T ,$$

that

$$P_T(I+\mathbb{K})\, P_T(I+\mathbb{L})\, g = P_T(I+\mathbb{K})(I+\mathbb{L})\, g = P_T g = g = 0 .$$

Thus the closed linear span of the given set fills out $\mathcal{B}_T$, as asserted. □

We remark that by Lemma 5.3 and the preceding Corollary 2, the theorem remains valid if Δ_∞ is replaced by Δ_S for any $S \geq T$.

6. THE INVERSE SPECTRAL PROBLEM

We are now in a position to establish formulas which relate the potential V of a given canonical equation (2.1) to its spectral function Δ_∞. These formulas form the basis of two methods for solving the inverse spectral problem, as will be explained a little later. Due to limitations in time and space we shall only sketch the main ideas and refer the reader to the literature for technical details. The paper [MaF2] contains a careful analysis of the spectral problem for the canonical equation.

To begin with let

$$P = \begin{bmatrix} I_n & 0 \\ 0 & 0 \end{bmatrix} \quad \text{and} \quad Q = \begin{bmatrix} 0 & 0 \\ 0 & I_n \end{bmatrix} . \tag{6.1}$$

It then follows readily from (2.16), with the help of the easily verified auxiliary identity

$$(P-Q)U_0(-s,\lambda) = U_0(s,\lambda)(P-Q)$$

that

$$U(t,\lambda)P = U_0(t,\lambda)P + \int_0^t K(t,s)U_0(s,\lambda)P\,ds , \tag{6.2}$$

where, by (2.17) and the ensuing discussion,

$$\begin{aligned} K(t,s) &= K_U(t,s) + K_U(t,-s)(P-Q) \\ &= \tfrac{1}{2}\{\omega(t,(t+s)/2)P+\omega(t,(t-s)/2)P + \\ &\quad +J\omega(t,(t+s)/2)JQ - J\omega(t,(t-s)/2)JQ\} \end{aligned} \tag{6.3}$$

for $0 \leq s \leq t$. The supplementary formula

$$X(t,\lambda) = X_0(t,\lambda) + \int_0^t K(t,s)X_0(s,\lambda)ds , \tag{6.4}$$

which was introduced in the last section, is immediate from (6.2).

THEOREM 6.1. *Let* $\mathbb{K}$ *denote the integral operator which is defined by the rule*

$$(\mathbb{K} f)(t) = \int_0^t K(t,s)f(s)ds , \quad \textit{for} \quad t \in \mathbb{R}^+ ,$$

where the kernel $K(t,s)$ *is specified in* (6.3) *Then* $I + \mathbb{K}$ *is a bounded invertible map of* $L^2_{2n}(\mathbb{R}^+)$ *onto itself.*

PROOF. The inequality

$$|K(t,s)| \leq |\omega(t,(t+s)/2)| + |\omega(t,(t-s)/2)|$$

is immediate from (6.3) and leads rapidly to the bounds

$$\int_0^t |K(t,s)|ds \leq 2\{\nu(t) - 1\} \tag{6.5}$$

and

$$\int_s^T |K(t,s)|\,dt \leq 4\{\nu(T) - 1\} . \tag{6.6}$$

The first of these is an easy consequence of the bound on ω which is given between (2.8) and (2.9); the second rests on parts

(d) and (f) of Lemma 2.1. The proof of the boundedness of $\mathbb{K}$ is now straightforward, with the help of (6.5) and (6.6):

$$\int_0^T \left| \int_0^t K(t,s)f(s)ds \right|^2 dt$$

$$\leqslant \int_0^T \{ \int_0^t |K(t,u)|du \int_0^t |K(t,s)||f(s)|^2 ds\} dt$$

$$\leqslant 2(\nu(T)-1) \int_0^T \{ \int_s^T |K(t,s)|dt\} |f(s)|^2 ds$$

$$\leqslant 8\{\nu(T)-1\}^2 \int_0^T |f(s)|^2 ds ,$$

for every $T \geqslant 0$, including $T = \infty$.

For the proof of the asserted invertibility we refer to [MaF2] where a formula for the inverse is provided. □

The inverse of $I + \mathbb{K}$ will be designated by $I + \mathbb{L}$. The kernels $K(t,s)$ of $\mathbb{K}$ and $L(t,s)$ of $\mathbb{L}$ live only on the triangle $0 \leqslant s \leqslant t < \infty$. In terms of the projections P_T defined by (5.24), this is the same as to say that

$$P_T \mathbb{K} = P_T \mathbb{K} P_T \quad \text{and} \quad P_T \mathbb{L} = P_T \mathbb{L} P_T$$

for every $T \geqslant 0$. From this it follows easily that

$$(I+P_T\mathbb{K} P_T)(I+P_T\mathbb{L} P_T)P_T = P_T(I+\mathbb{K})(I+\mathbb{L})P_T = P_T$$

and hence, since the same identity is valid if $\mathbb{K}$ and $\mathbb{L}$ are interchanged, that $I + P_T\mathbb{K} P_T$ is a bounded invertible map of $L^2_{2n}[0,T]$ onto itself with inverse $I + P_T \mathbb{L} P_T$.

We further remark that both $I + \mathbb{K}$ and $I + \mathbb{L}$ map locally absolutely continuous functions f with $Pf(0) = f(0)$ into themselves, and that, moreover,

$$(I + \mathbb{K}) G_0 = G(I + \mathbb{K})$$

and

$$(I + \mathbb{L}) G = G_0(I + \mathbb{L})$$

where $G = J \frac{d}{dt} - V$ and $G_0 = J \frac{d}{dt}$, restricted to suitable domains. Because of these intertwining relations $I + \mathbb{K}$ is often referred to as a *transformation operator*; see [MaF2] for additional information.

Our next objective is to express the potential in terms of the kernel $K(t,s)$.

THEOREM 6.2. *The potential* V *of the canonical equation* (2.1) *is determined from the kernel* $K(t,s)$ *in the representation* (6.2) *by the formula*

$$V(t) = [JK(t,t) - K(t,t)J] \quad . \tag{6.7}$$

PROOF. Let us suppose first that the potential V is smooth and that K is differentiable. Then, upon differentiating both sides of (6.2) with respect to t and invoking the differential equation (2.1) for U and (2.1) with $V \equiv 0$ for U_0 , it follows readily that

$$JK(t,t)U_0(t,\lambda)P + J\int_0^t \frac{\partial K(t,s)}{\partial t} U_0(s,\lambda)P\,ds$$

$$= V(t)U_0(t,\lambda)P + V(t)\int_0^t K(t,s)U_0(s,\lambda)P\,ds +$$

$$+ \lambda\int_0^t K(t,s)U_0(s,\lambda)P\,ds \quad .$$

But the last term on the right is also equal to

$$\int_0^t K(t,s)JU_0'(s,\lambda)P\,ds = K(t,t)JU_0(t,\lambda)P - K(t,0)JP - \int_0^t \frac{\partial K(t,s)}{\partial s} JU_0(s,\lambda)P\,ds \quad .$$

Thus

$$[JK(t,t)-K(t,t)J-V(t)]U_0(t,\lambda)P+K(t,0)JP =$$

$$= \int_0^t [V(t)K(t,s)-J\frac{\partial K(t,s)}{\partial t} - \frac{\partial K(t,s)}{\partial s}J]U_0(s,\lambda)P\,ds \quad . \tag{6.8}$$

Therefore, upon setting $\lambda = k2\pi/t$, [resp. $(2k+1)\pi/t$] , and letting $k \uparrow \infty$ through integer values, so that $U_0(t,\lambda) = I_{2n}$, [resp. $-I_{2n}$] , it follows from the Riemann Lebesgue lemma that

$$[JK(t,t) - K(t,t)J - V(t)]P \pm K(t,0)JP = 0$$

and hence that

$$K(t,0)JP = 0 \tag{6.9}$$

and

$$V(t)P = [K(t,t)J - JK(t,t)]P \ , \tag{6.10}$$

separately. But now, upon taking advantage of the fact that $JVJ = V$, $JP = QJ$ and $PJ = JQ$, it follows readily that (6.10) remains valid with P replaced by Q and hence that (6.7) holds, as asserted.

To complete the proof it is left to show that the formula (6.7) remains valid for summable potentials. This may be done by passing to the limit along an approximating sequence of smooth potentials. We shall not enter into all these details here. For a sample of such arguments the monographs [AM] and [M] may be consulted. □

We remark that (6.7) and (6.9) imply that the right hand side of (6.8) vanishes for all λ. This serves to exhibit K as the solution of the partial differential equation

$$J \frac{\partial K}{\partial t}(t,s) + \frac{\partial K}{\partial s}(t,s)J - V(t)K(t,s) = 0 \tag{6.11}$$

on the triangle $0 \leqslant s \leqslant t < \infty$, subject to the boundary conditions (6.7) and (6.9). All this is formal because the indicated derivatives do not exist unless the potential is presumed to be smooth and not merely summable. This difficulty can be circumvented by integrating (6.2) instead of differentiating and leads to the system of integral equations

$$K_+(t,u)P = -\frac{1}{2}JV\left(\frac{t-u}{2}\right)P - J\int_{(t-u)/2}^{t-u} V(s)K_+(s,t-s-u)P\,ds$$

$$- J\int_{t-u}^{t} V(s)K_-(s,s-t+u)P\,ds$$

$$K_-(t,u)P = -\frac{1}{2}JV\left(\frac{t+u}{2}\right)P - J\int_{(t+u)/2}^{t} V(s)K_+(s,t-s+u)P\,ds$$

for $0 \leqslant u \leqslant t$, in which $K_\pm$ denote the commuting and anticommuting parts of K with respect to J; for additional information see [MaF2].

We next express the spectral function in the form

$$\Delta_\infty(\lambda) = I - \hat{h}(\lambda) \tag{6.12}$$

with $h \in L^1_{n\times n}(\mathbb{R})$ and identify the operator $I+\mathbb{K}$ as a triangular factor of an integral operator based on h. Indeed, it follows readily from (6.4) (as has in fact already been noted in Section 5) that

$$f^{\blacktriangle}(\lambda) = [(I+\mathbb{K}^* f]^{\Delta}(\lambda) = 2^{-\frac{1}{2}}[\mathbb{N}(I+\mathbb{K}^*)f]^{\wedge}(\lambda)$$

for every $f \in L^2_{2n}(\mathbb{R}^+)$ with compact support. Therefore, by the Parseval formula (5.16) for the $\blacktriangle$-transform,

$$\int_0^\infty f(t)^* f(t)\,dt = \frac{1}{\pi}\int_{-\infty}^\infty f^{\blacktriangle}(\lambda)^*\Delta_\infty(\lambda) f^{\blacktriangle}(\lambda)\,d\lambda$$

$$= \frac{1}{2\pi}\int_{-\infty}^\infty [\mathbb{N}(I+\mathbb{K}^*)f]^{\wedge}(\lambda)^*(I_n - \hat{h}(\lambda))[\mathbb{N}(I+\mathbb{K}^*)f]^{\wedge}(\lambda)\,d\lambda$$

$$= \langle (I-H)[\mathbb{N}(I+\mathbb{K}^*)f], [\mathbb{N}(I+\mathbb{K}^*)f]\rangle_{L^2_n(\mathbb{R})} ,$$

where H is the convolution operator in $L^2_n(\mathbb{R})$ based on h:

$$(Hg)(t) = \int_{-\infty}^\infty h(t-s)g(s)\,ds . \tag{6.13}$$

But this in turn implies that

$$\langle f,f\rangle_{L^2_{2n}(\mathbb{R}^+)} = \langle (I+\mathbb{K})\mathbb{N}^*(I-H)\mathbb{N}(I+\mathbb{K}^*)f,f\rangle_{L^2_{2n}(\mathbb{R}^+)}$$

for every $f \in L^2_{2n}(\mathbb{R}^+)$ with compact support, and hence that

$$(I+\mathbb{K})\mathbb{N}^*(I-H)\mathbb{N}(I+\mathbb{K}^*) = I . \tag{6.14}$$

This exhibits $I+\mathbb{L}$ [resp. $I+\mathbb{L}^*$] as the lower [resp. upper] triangular factor of the positive operator

$$\mathbb{N}^*(I-H)\mathbb{N} = I - \mathbb{N}^* H \mathbb{N}$$

on $L^2_{2n}(\mathbb{R}^+)$, with respect to the chain of projectors $P_T : T \geq 0$, defined by (5.24).

The kernel $H(t,s)$ of $\mathbb{N}^* H \mathbb{N}$ can be readily identified as

$$H(t,s) = H_1(t-s) + H_2(t+s)$$

where

$$H_1(u) = \frac{1}{2}\begin{bmatrix} h(u)+h(-u) & -ih(u)+ih(-u) \\ ih(u)-ih(-u) & h(u)+h(-u) \end{bmatrix},$$

$$H_2(u) = \frac{1}{2}\begin{bmatrix} h(u)+h(-u) & ih(u)-ih(-u) \\ ih(u)-ih(-u) & -h(u)-h(-u) \end{bmatrix} ,$$

and H_1 [resp. H_2] commutes [resp. anticommutes] with J. Moreover, it follows from (6.14) that the kernel $K(t,s)$ of $\mathbb{K}$ may be obtained as the solution of the integral equation

$$K(t,s) - H(t,s) - \int_0^t K(t,u)H(u,s)\,du = 0 \quad , \tag{6.15}$$

on the triangle $0 \leqslant s \leqslant t < \infty$. In the literature on inverse problems (6.15) is known as the *Gelfand-Levitan equation*, though, as the present derivation indicates, it is even more fundamentally an equation for factorization; for additional discussion of this point and examples, the papers [DK] and [DI] are suggested.

Because of the special structure of the kernel H it turns out that the formulas for both K and V can be expressed in terms of the resolvent kernel $\Gamma_t(x,y)$ which is defined by the equation

$$\Gamma_t(x,y) - \int_0^t h(x-u)\Gamma_t(u,y)\,du = h(x-y) \quad , \tag{6.16}$$

for $0 \leqslant x,y \leqslant t$ and $0 \leqslant t \leqslant \infty$.

THEOREM 6.3. *The* $n \times n$ *blocks of the* $2n \times 2n$ *matrix kernel* $K(t,s)$ *of the operator* $\mathbb{K}$ *may be expressed in terms of the resolvent kernel defined by* (6.16) *as follows:*

$$K_{11}(t,s) = \{\,\Gamma_{2t}(2t,t+s) + \Gamma_{2t}(2t,t-s) + \Gamma_{2t}(0,t+s) + \Gamma_{2t}(0,t-s)\}/2$$

$$K_{12}(t,s) = \{\,\Gamma_{2t}(2t,t+s) - \Gamma_{2t}(2t,t-s) + \Gamma_{2t}(0,t+s) - \Gamma_{2t}(0,t-s)\}/2;$$

$$K_{21}(t,s) = \{-\Gamma_{2t}(2t,t+s) - \Gamma_{2t}(2t,t-s) + \Gamma_{2t}(0,t+s) + \Gamma_{2t}(0,t-s)\}/2;$$

$$K_{22}(t,s) = \{\,\Gamma_{2t}(2t,t+s) - \Gamma_{2t}(2t,t-s) - \Gamma_{2t}(0,t+s) + \Gamma_{2t}(0,t-s)\}/2,$$

on the triangle $0 \leqslant s \leqslant t < \infty$.

PROOF. Let $k(t,s)$ denote the $n \times n$ matrix valued kernel of the operator $\mathbb{N}\,\mathbb{K}\,\mathbb{N}^*$. It is readily checked that $\mathbb{N}\,\mathbb{K}\,\mathbb{N}^*$ is lower triangular with respect to the chain of projection operators

$$(\Pi_T f)(s) = \begin{cases} f(s) & \text{for } |s| \leq T \\ 0 & \text{for } |s| > T , \end{cases} \tag{6.17}$$

where $T \geq 0$ and $f \in L_n^2(\mathbb{R})$. This means that

$$\Pi_T \mathbb{N} \mathbb{K} \mathbb{N}^* = \Pi_T \mathbb{N} \mathbb{K} \mathbb{N}^* \Pi_T$$

for every $T \geq 0$ and hence that $k(t,s)$ lives inside the double cone shaped region of points (t,s) in $\mathbb{R}^2$ which satisfy the inequality $|s| \leq |t|$. Thus, the identity

$$\mathbb{N}(I + \mathbb{K}) \mathbb{N}^* (I-H) \mathbb{N} (I + \mathbb{K}^*) \mathbb{N}^* = I \tag{6.18}$$

implies that

$$k(t,s) - h(t-s) - \int_{-|t|}^{|t|} k(t,u) h(u-s) du = 0 \tag{6.19}$$

for $|s| \leq |t|$. The latter is, however, equivalent to the equation

$$k(t,y-|t|) - h(t+|t|-y) - \int_0^{2|t|} k(t,u-|t|) h(u-y) du = 0$$

for $0 \leq y \leq 2|t|$, which, upon comparison with (6.16), leads to the identification

$$k(t,s) = \Gamma_{2|t|}(t+|t|, s+|t|) , \tag{6.20}$$

for $|s| < |t|$. The asserted formulas for the blocks K_{ij} are now readily computed from (6.20) and the relationship between K and k which is inherited from the identity

$$\mathbb{K} = \mathbb{N}^* (\mathbb{N} \mathbb{K} \mathbb{N}^*) \mathbb{N} . \quad \square$$

THEOREM 6.4. *The potential* V *of a canonical equation with spectral function* $I_n - \hat{h}$ *is given by* (2.2), *where* V_1 *and* V_2 *(being selfadjoint) are determined by the formula*

$$\Gamma_{2t}(0,2t) = 2\{iV_1(t) - V_2(t)\} . \tag{6.21}$$

Moreover,

$$E(t,\lambda) = e^{-i\lambda t}\{I_n + \int_0^{2t} e^{i\lambda s} \Gamma_{2t}(2t,2t-s) ds\} \tag{6.22}$$

and

$$F(t,\lambda) = e^{i\lambda t}\{I_n + \int_0^{2t} e^{-i\lambda s} \Gamma_{2t}(0,s) ds\} . \tag{6.23}$$

PROOF. Formula (6.21) is immediate from (6.6) and Theorem 6.3, whereas (6.22) and (6.23) follow from (6.4) and Theorem 6.3. □

Formulas (6.6) and (6.21) can be used to solve the *inverse spectral problem* in which a function $h \in L^1_{n\times n}(\mathbb{R})$ with

$$I_n - \hat{h}(\lambda) > 0 \quad ,$$

for every $\lambda \in \mathbb{R}$, is given and the objective is to find a potential V such that the corresponding canonical equation (2.1), subject to the usual boundary conditions, has $I_n - \hat{h}(\lambda)$ as its spectral function.

Given such an h, the first method, which is the analog of the well-known Gelfand-Levitan method for solving the inverse spectral problem for the Schrödinger equation, is to solve (6.15) for K and then to read off V from (6.7). The second method, which is due to Krein, short cuts this procedure by taking advantage of Theorem 6.3: Given h, the algorithm is to compute $\Gamma_{2t}(0,2t)$ from (6.16) and then to obtain V from (6.21). For additional discussion relevant to the first method see [M], [Z1] and [Z2]†; [K2]-[K5], [MaF1] and [MaF2] pertain to the second method.

Given h as above, it remains to show that the $2n \times 2n$ matrix valued function V obtained by either of these algorithms is of the requisite form (2.2), which is immediate, and, in addition, that it belongs to the class $L^2_{2n}(\mathbb{R}^+)$. For the latter fact we rely on [KMa1] even though, to the best of our knowledge, a proof has not been published.

Next, let $I_n - \hat{g}$ denote the spectral function of the canonical equation (2.1) with potential V determined from the given h, as in the last paragraph. Then, in order to complete the circle, it still needs to be shown that $g = h$. Under the present assumptions, Theorem 6.4 guarantees that

$$\Gamma^h_{2t}(0,2t) = \Gamma^g_{2t}(0,2t) \quad ,$$

for $0 \leq t < \infty$, where for the sake of clarity, we have temporarily imposed superscripts on the resolvent kernels to indicate

† These should be [F1] and [F2], but were left out by mistake.

whether they are based on h or on g. Moreover, the solution $X(t,\lambda)$ of the canonical equation with this potential can also, in view of (6.22) and (6.23), be expressed in terms of $\Gamma^g_{2t}(2t,s)$ and $\Gamma^g_{2t}(0,s)$, for $0 \leq s \leq 2t$. Yet, a routine calculation based on the well-known Krein-Bellman identity

$$\frac{\partial \Gamma_t}{\partial t}(x,y) = \Gamma_t(x,t)\Gamma_t(t,y)$$

indicates that the function

$$\tilde{X}(t,\lambda) = \begin{bmatrix} \{\tilde{F}(t,\lambda) + \tilde{E}(t,\lambda)\}/2 \\ \{\tilde{F}(t,\lambda) - \tilde{E}(t,\lambda)\}/2i \end{bmatrix}$$

with $\tilde{E}$ and $\tilde{F}$ defined as in (6.22) and (6.23) but with Γ^h_{2t} in place of Γ^g_{2t} is also a solution of the given canonical equation, with $\tilde{X}(0,\lambda) = [I_n \quad 0]^*$. Thus, since there is only one such solution, it follows that

$$\tilde{X}(t,\lambda) = X(t,\lambda)$$

and hence that

$$\Gamma^h_{2t}(2t,s) = \Gamma^g_{2t}(2t,s) \quad \text{and} \quad \Gamma^h_{2t}(0,s) = \Gamma^g_{2t}(0,s) \quad ,$$

for $0 \leq s \leq 2t$ and $0 \leq t < \infty$. But this in turn implies, via the formulas in Theorem 6.3, that the kernels of triangular operators $\mathbb{K}$ which appear in the factorization formula (6.16) are the same for both the operator $I-H$ which is based on h and, in a self-evident notation the operator $I-G$ which is based on g. This proves that $g = h$, as desired, modulo some technical details which we cheerfully skip. The implications of this argument are summarized in the next theorem.

THEOREM 6.5. *Every* $n \times n$ *matrix valued function* $I_n - \hat{h}(\lambda)$ *of class* W^I *which is strictly positive for each point* $\lambda \in \mathbb{R}$ *is the spectral function of exactly one canonical equation* (2.1) *with summable potential* V, *normalized as in* (2.2). *Moreover,* V *is simply related to the resolvent kernel, which is defined through* (6.16), *by formula* (6.21).

We remark that the argument preceding the statement of Theorem 6.5 also indicates that knowledge of the "corners"

$\Gamma_t(0,t)$, $0 \leq t < \infty$, of the resolvent kernels suffice to fill in the rest:

$$\Gamma_t(x,y) \quad \text{for} \quad 0 \leq x,y \leq t \quad \text{and} \quad 0 \leq t < \infty .$$

Theorem 6.5 is evidently due to Krein. Variants thereof and much related information is outlined in a number of Doklady notes: [K2]-[K5]. There is a corresponding theorem which establishes a one-to-one correspondence between canonical equations with locally summable potentials and a precisely defined class of spectral functions which of course properly includes the class which arose here; see [MaFl] and [MaF2] for details.

Finally, we remark that formula (6.22), which can be reexpressed with the help of (6.20) as

$$E(t,\lambda) = e^{-i\lambda t} I_n + \int_{-t}^{t} k(t,u) e^{-i\lambda u} du \tag{6.24}$$

for $t \geq 0$, exhibits $E(t,\lambda)$ as an orthogonalized version of the exponentials. Indeed, if one seeks a function E of the form (6.24) such that, at least formally,

$$\int E(t,\lambda) \{I_n - \hat{h}(\lambda)\} e^{i\lambda s} \, d\lambda = 0$$

for $-t < s < t$, then (6.19) drops out. Thus the function $E(t,\lambda)$ may be viewed as the trigonometric counterpart (under the Kolmogorov isomorphism) of an innovations process. Moreover, much of the machinery which has been developed to this point purely in the context of spectral theory for canonical differential equations is applicable to the theory of linear least squares estimation for stationary $n \times 1$ vector valued stochastic processes of the form

$$y(t) = z(t) + \nu(t)$$

where

$$\begin{aligned} E[y(t)y(s)^*] &= E[\nu(t)\nu(s)] + E[z(t)z(s)^*] \\ &= I_n\delta(t-s) - h(t-s) . \end{aligned} \tag{6.25}$$

For additional information see [D], [LT], [KVM] and the references cited therein.

7. THE EXTENSION PROBLEM, BAND EXTENSIONS, AND ENTROPY

A central theme of this paper is that for every solution h of the τ-extension problem formulated in Section 1, $I_n - \hat{h}(\lambda)$ can (in view of Theorem 6.5) be identified as the spectral function $\Delta_\infty(\lambda)$ of a canonical equation. Moreover, as we shall now show, this connection can be put to good use to develop representation formulas for the sought-for extensions. The formula

$$Z_\infty(\lambda) = I_n - 2\int_0^\infty e^{i\lambda s}\, h(s)\,ds = \lim_{t\uparrow\infty} R_{U(t,\lambda)}[I_n]$$

of Theorem 5.3, which relates the extension h to the matrizant $U(t,\lambda)$ of the underlying canonical equation, is a convenient point of departure. To explain further, let $U_T(t,\lambda)$ denote the matrizant of the same canonical equation, but considered on the interval $T \leq t < \infty$ instead of $0 \leq t < \infty$:

$$J\frac{d}{dt}U_T(t,\lambda) = V(t)U_T(t,\lambda) + \lambda U_T(t,\lambda)$$

for $t > T$, and

$$U_T(T,\lambda) = I_{2n} .$$

Then

$$U(t,\lambda) = U_T(t,\lambda)U(T,\lambda) \quad \text{for} \quad t \geq T ,$$

and so, by Lemma 4.3,

$$Z_t(\lambda) = R_{U(t,\lambda)}[I_n] = R_{U(T,\lambda)}[R_{U_T(t,\lambda)}[I_n]] .$$

Now the proof of Theorem 5.3 can be readily adapted to prove that $R_{U_T(t,\lambda)}[I_n]$ tends uniformly on $\bar{\mathbb{C}}_+$ to a function $G \in \mathcal{C}$ as $t \uparrow \infty$, and hence that

$$Z_\infty(\lambda) = R_{U(T,\lambda)}[G] .$$

This formula exhibits the Z_∞ corresponding to each solution h of the extension problem as a linear fractional transformation of an element G of class $\mathcal{C}$. This is a full description of all the solutions to the 2T-extension problem:

THEOREM 7.1. *Let* $U(t,\lambda)$ *be the matrizant of the canonical equation with potential* $V(t)$ *which is specified by* (6.20), *for* $0 \leq t \leq T$ *(in terms of the given data*

$k \in L_n^1[-2T,2T]$). *Then the formula*

$$I_n - 2\int_0^\infty e^{i\lambda s}\, h(s)ds = R_{U(T,\lambda)}[G] \tag{7.1}$$

defines a one-to-one correspondence between the solutions h *of the* 2T-*extension problem with data* k *and the functions* $G \in \mathcal{C}$.

PROOF. Let h be any solution of the 2T-extension problem with data k . Then, as we have already explained, $I_n - \hat{h}$ is the spectral function of a canonical equation with potential $V(t)$ and matrizant $U(t,\lambda)$, and there exists an element $G_t \in \mathcal{C}$ such that

$$I_n - 2\int_0^\infty e^{i\lambda s}\, h(s)ds = R_{U(t,\lambda)}[G_t] .$$

Clearly, $U(t,\lambda)$ depends only upon knowledge of the potential $V(s)$ for $0 \leq s \leq t$, and so, in view of formula (6.20), only upon $h(s)$ for $0 \leq s \leq 2t$. In particular, $U(T,\lambda)$ is thus seen to be the same for every solution of the 2T-extension problem. Therefore, every solution of this extension problem can be expressed in the form (7.1) with some choice of $G \in \mathcal{C}$.

Now suppose conversely that $G \in \mathcal{C}$, and let h be any solution of the 2T-extension problem; the existence of at least one solution is guaranteed, for example, by Theorem 8.2 of [DG1]. Then, by the argument furnished just above, there exists an $H \in \mathcal{C}$ such that

$$I_n - 2\int_0^\infty e^{i\lambda s}\, h(s)ds = R_{U(T,\lambda)}[H] ,$$

while at the same time, by Theorem 4.1, there exists a $g \in L_{n\times n}^1(\mathbb{R}^+)$ such that

$$I_n - 2\int_0^\infty e^{i\lambda s}\, g(s)ds = R_{U(T,\lambda)}[G] .$$

Moreover, by Lemma 4.7, the difference

$$2\int_0^\infty e^{i\lambda s}\, \{g(s)-h(s)\}ds \in e^{2i\lambda T}\, W_+^0 .$$

But this in turn implies that

$$g(s) = h(s) \quad \text{for} \quad 0 \leq s \leq 2T$$

and therefore that g is also a solution of the 2T-extension problem. □

Special interest attaches to the solution of the 2T-extension problem corresponding to

$$Z_T(\lambda) = R_{U(T,\lambda)}[I_n] .$$

THEOREM 7.2. *Let* h_{2T} *denote the extension defined by*

$$I_n - 2\int_0^\infty e^{i\lambda s} h_{2T}(s)ds = Z_T(\lambda) . \tag{7.2}$$

Then

$$\{[I_n - \hat{h}(\lambda)]^{-1} - I_n\}^\vee(s) = 0$$

for $|s| > 2T$.

PROOF. It is readily checked, just as in the proof of (5.20), that

$$\frac{Z_T(\lambda)+Z_T(\lambda)^*}{2} = [\varepsilon_T(\lambda)]^{-1}[\varepsilon_T(\lambda)^*]^{-1}$$

for $\lambda \in \mathbb{R}$, and hence, as follows, by considering the upper left hand block of $P_+\Omega(T,\lambda)$ in formula (2.23), that

$$[I_n - \hat{h}_{2T}(\lambda)]^{-1} = \left[\frac{Z_T(\lambda)+Z_T(\lambda)^*}{2}\right]^{-1}, \text{ for } \lambda \in \mathbb{R} ,$$

is of the requisite form. □

The last theorem serves to identify h_{2T} as the special *band extension* of Dym-Gohberg for the positive definite case; see [DG1] and [DG3].

THEOREM 7.3. *The function* $I_n - \hat{h}_{2T}(\lambda)$ *based on the band extension* h_{2T} *of Theorem* 7.2 *is precisely the spectral function of the canonical equation with* $V(t)$ *specified by the data* k *for* $0 \leq t \leq T$, *and* $V(t) = 0$ *for* $t > T$.

PROOF. If $V(t) = 0$ for $t > T$, then

$$U(t,\lambda) = e^{-\lambda(t-T)J}U(T,\lambda)$$

for $t \geq T$. Thus

$$\Omega(t,\lambda) = U_0(t,\lambda)^{-1}U(t,\lambda) = e^{\lambda TJ}\,U(T,\lambda)$$

for $t \geq T$, which in turn implies that the upper left-hand block of

$$\Omega^{\#}(t,\lambda)\Omega(t,\lambda) = U^{\#}(t,\lambda)U(t,\lambda)$$

is equal to

$$E^{\#}(T,\lambda)E(T,\lambda)$$

for all $t \geqslant T$, and hence that

$$\Delta_{\infty}(\lambda) = \{E^{\#}(T,\lambda)E(T,\lambda)\}^{-1} = I_n - \hat{h}_{2T}(\lambda) \quad ,$$

as asserted. □

We turn next to entropy formulas and bounds. It is convenient, however, to first obtain a description of the set of all possible extensions in terms of the class S of $n \times n$ matrix valued functions σ in W_+^0 which are strictly contractive on $\mathbb{R}$:

$$\sigma(\lambda)^*\sigma(\lambda) < I_n$$

for every point $\lambda \in \mathbb{R}$ (and hence, also on $\bar{\mathbb{C}}_+$). For ease of reference we summarize the connection between S and C in the next Lemma, which we state without proof:

LEMMA 7.1. *The formula*

$$\sigma = (G-I_n)(G+I_n)^{-1} = I_n - 2(G+I_n)^{-1} \tag{7.3}$$

defines a one-to-one correspondence between the set of $G \in C$ *and the set of* $\sigma \in S$. *Moreover, if* $\sigma \in S$, *then* $I-\sigma$ *is an invertible element of* W_+^I.

THEOREM 7.4. *The formula*

$$I_n - \hat{h}(\lambda) = [E(T,\lambda)]^{-1}[I_n - \sigma(\lambda)\Sigma(T,\lambda)]^{-1}[I_n - \sigma(\lambda)\sigma(\lambda)^*] \times$$

$$\times \quad [I_n - \Sigma(T,\lambda)^*\sigma(\lambda)^*]^{-1}[E(T,\lambda)^*]^{-1} \tag{7.4}$$

where $\Sigma(T,\lambda) = F(T,\lambda)E(T,\lambda)^{-1}$, F *and* E *are given by* (6.21) *and* (6.22), *respectively, defines a one-to-one correspondence between the set of all solutions* h *of the* 2T-*extension problem, and the set of all* $\sigma \in S$.

PROOF. By Theorem 7.1, the set of all solutions h of the 2T extension problem is in one-to-one correspondence with the set of functions

$$I_n - \hat{h}(\lambda) = \mathrm{Re}\{R_{U(T,\lambda)}[G]\}$$

$$= \frac{1}{2}(U_{11} - iGU_{21})^{-1}(G+G^*)(U_{11}^* + iU_{21}^*G^*)^{-1}$$

where G runs over the class $\mathcal{C}$. The rest is plain from Lemma 7.1. □

Formula (7.4), or a variant thereof, is presumably known to Krein, although we do not have an explicit reference. A discrete version appears in Youla [Y]; another discrete variant is given in Theorem 9 of [DeD].

It is perhaps worth pointing out that, in terms of (7.4), the band extension corresponds to the choice $\sigma = 0$. Subject to some additional technical constraints the band extension can also be characterized as the extension which maximizes the entropy integral

$$\lim_{\varepsilon \downarrow 0} \frac{1}{\pi} \int_{-\infty}^{\infty} \frac{\log \det(\mathrm{Re}\{R_{U(T,\lambda)}[G]\})}{\varepsilon^2\lambda^2 + 1} \, d\lambda$$

over all $G \in \mathcal{C}$ [resp. $\sigma \in S$] : *the maximum is achieved by* $G = I$ *[resp.* $\sigma = 0$*] only.* For details, the interested reader is referred to [DG1]. Those that do so should note that there is a misprint in the definition of the entropy integral on page 202: the lower limit should be $-\infty$ and not 0 .

Arov and Krein [ArK] discuss a different set of entropy integrals. Since they do not present proofs, and their first result emerges as nice application of (7.4), we shall reformulate it to fit the present setting and sketch a proof in the next theorem. It is convenient, however, to first verify a preliminary inequality.

LEMMA 7.2. *Let* ξ *be an* $m \times n$ *contractive matrix:* $\xi^*\xi \leq I_n$, *and let* η *be an* $n \times m$ *strictly contractive matrix:* $\eta^*\eta < I_m$. *Then*

$$(I_m - \xi\eta)^{-1}(I_m - \xi\xi^*)(I_m - \eta^*\xi^*)^{-1} \leq (I_m - \eta^*\eta)^{-1}$$

with equality if and only if $\xi = \eta^*$.

PROOF. The matrix inequality

$$0 \leq (\xi-\eta^*)(I - \eta\eta^*)^{-1}(\xi^*-\eta) = \xi(I - \eta\eta^*)^{-1}\xi^* - (I - \eta^*\eta)^{-1}\eta^*\xi^* -$$
$$- \xi\eta(I - \eta^*\eta)^{-1} + (I - \eta^*\eta)^{-1}\eta^*\eta$$

implies that

$$(I-\xi\eta)(I-\eta^*\eta)^{-1}(I-\eta^*\xi^*) \geq (I-\eta^*\eta)^{-1}+\xi\eta(I-\eta^*\eta)^{-1}\eta^*\xi^* - \\ -(I-\eta^*\eta)^{-1}\eta^*\eta-\xi(I-\eta^*\eta)^{-1}\xi^* ,$$

which is readily seen to be equivalent to the stated inequality. The rest is plain. □

THEOREM 7.4. *For each point* $\omega = a+ib$ *in* $\mathbb{C}_+$ *and every* $G \in W_+$ *with* $\mathrm{Re}G > 0$ *on* $\bar{\mathbb{C}}_+$ *and at* ∞ ,

$$\frac{b}{\pi}\int \frac{\log\det\{\mathrm{Re}\, R_{U(T,\lambda)}[G]\}}{(\lambda-a)^2+b^2}\, d\lambda$$

$$\leq 2nbT - \log\det\{4\pi b\, \Lambda^T_{\omega^*}(\omega^*)\} , \tag{7.5}$$

with equality if and only if G *is constant with*

$$G(\lambda) = [I-\Sigma(T,\omega)^*]^{-1}[I+\Sigma(T,\omega)^*]$$

for every $\lambda \in \bar{\mathbb{C}}_+$.

PROOF. By formula (7.4), the left hand side of the asserted inequality is readily seen to be equal to

$$\int \log|\det \varepsilon_T(\lambda)|^{-2}\, d\mu_\omega + \int \log\det[I-\sigma(\lambda)^*\sigma(\lambda)]\, d\mu_\omega - \\ - \int \log|\det[I-\sigma(\lambda)\,\Sigma(T,\lambda)]|^2\, d\mu_\omega$$

$$= ① + ② - ③ ,$$

in a self-evident notation, where we have taken

$$d\mu_\omega(\lambda) = \frac{b}{\pi}\frac{1}{(\lambda-a)^2+b^2}\, d\lambda .$$

The evaluation

$$① = \log|\det[\varepsilon_T(\omega)]|^{-2}$$

is immediate since by Theorem 2.1, ε_T is outer. Next

$$③ = \log|\det[I-\sigma(\omega)\,\Sigma(T,\omega)]|^2$$

$$= \int \log|\det[I-\sigma(\lambda)\,\Sigma(T,\omega)]|^2\, d\mu_\omega ,$$

since

$$\det[I-\sigma(\lambda)\,\Sigma(T,\lambda)] \quad \text{and} \quad \det[I-\sigma(\lambda)\,\Sigma(T,\omega)]$$

are both outer functions in the variable λ for each fixed $\omega \in \mathbb{C}_+$ and every $\sigma \in S$. Thus, by Lemma 7.2,

(2) - (3) =

$$\int \log \det\{[I-\sigma(\lambda)\Sigma(T,\omega)]^{-1}[I-\sigma(\lambda)\sigma(\lambda)^*][I-\Sigma(T,\omega)^*\sigma(\lambda)^*]^{-1}\}d\mu_\omega$$

$$\leqslant -\int \log \det[I-\Sigma(T,\omega)^*\Sigma(T,\omega)]d\mu_\omega$$

$$= \det[I - \Sigma(T,\omega)^*\Sigma(T,\omega)] ,$$

with equality in the second line if and only if $\sigma(\lambda)$ is identically equal to $\Sigma(T,\omega)^*$. The proof of the only if part of the last assertion depends also upon the fact that if ξ and η are positive definite matrices with $\xi \leqslant \eta$, then equality prevails if and only if $\det \xi = \det \eta$.

The rest is plain upon combining bounds. □

We remark that the class of G's considered in the statement of Theorem 7.4 is larger than the class C because the requirement that $G(\infty) = I$ has been dropped. Indeed the choice of G which achieves the maximum does not belong to C and so does not correspond to a canonical equation (2.1).

Finally we remark that the computation of the term (1) in the proof of the last theorem yields the supplementary identity

$$\int \{\log \det[I - \hat{h}_{2T}(\lambda)]\}d\mu_\omega = \log |\det[\varepsilon_T(\omega)]|^{-2} \qquad (7.6)$$

for the band extension h_{2T}. Such formulas are useful for calculating the error in best least square approximation problems. Indeed for the process noted in (6.25) the causal least square estimate of $z(t)$, given $y(s)$ for $-t \leqslant s \leqslant t$, is

$$\tilde{z}(t) = -\int_{-t}^{t} k(t,u)y(u)du ,$$

and the corresponding mean square error is equal to

$$E\{[z(t) - \tilde{z}(t)]^*[z(t) - \tilde{z}(t)]\}$$

$$= \text{trace } E\{[z(t) - \tilde{z}(t)][z(t) - \tilde{z}(t)]^*\}$$

$$= -\text{trace } k(t,t) = -\text{trace } \Gamma_{2t}(2t,2t)$$

$$= \lim_{\delta\downarrow 0} \frac{1}{2\pi}\int \frac{\log \det[I - \hat{h}_{2t}(\lambda)]}{\delta^2\lambda^2 + 1} d\lambda ,$$

under suitable technical assumptions, but that is another story.

8. SQUARE INTEGRABLE SOLUTIONS AND MATRIX BALLS

In this section we characterize the square summable solutions of the canonical equation (2.1) with summable potential of the form (2.2). The results are not needed in the sequel. We have, however, included them because they give another application of the function $Z_\infty(\omega)$ and besides they are easily achieved from the analysis invested to this point. For another approach and a general investigation of the Weyl function for canonical equations see Hinton-Shaw [HS].

We start with a number of elementary identities involving the matrizant $U(t,\lambda)$ of the given equation and the corresponding F and Z functions, defined by (5.3) and (5.18).

LEMMA 8.1. *The identities*

$$U(t,\omega)\begin{bmatrix} iZ_t(\omega) \\ I_n \end{bmatrix} = \begin{bmatrix} iI_n \\ I_n \end{bmatrix} [F^\#(t,\omega)]^{-1} \tag{8.1}$$

and

$$[-iZ_t(\omega)^* \;\; I_n]U(t,\omega)^* iJU(t,\omega)[-iZ_t(\omega)^* \;\; I_n]^*$$
$$= 2\{F^\#(t,\omega)F^\#(t,\omega)^*\}^{-1} \tag{8.2}$$

hold for every point $\omega \in \bar{\mathbb{C}}_+$ *and every* $t > 0$.

PROOF. By definition (5.18) and formulas (4.4), (4.5) and (4.14),

$$\begin{aligned} Z_t(\omega) &= R_{U(t,\omega)}[I_n] \\ &= [U_{22}^\#(t,\omega) + iU_{12}^\#(t,\omega)][F^\#(t,\omega)]^{-1} \end{aligned}$$

for every $\omega \in \bar{\mathbb{C}}_+$ and every $t \geq 0$.

Thus, dropping the arguments for the moment,

$$U\begin{bmatrix} iZ \\ I_n \end{bmatrix} = U\begin{bmatrix} U_{22}^\# & -\;U_{12}^\# \\ -iU_{21}^\# & +\;U_{11}^\# \end{bmatrix}[F^\#]^{-1}$$

$$= UJU^\# J^* \begin{bmatrix} iI_n \\ I_n \end{bmatrix} [F^\#]^{-1}$$

which is equal to the right hand side of (8.1) as asserted, thanks to (2.12). The second identity is immediate from the first. □

LEMMA 8.2. *For each fixed choice of* $T > 0$, *the* $2n \times n$ *matrix valued function*

$$\Psi_T(t,\omega) = \begin{cases} U(t,\omega)\begin{bmatrix} iZ_T(\omega) \\ I_n \end{bmatrix} & \text{for} \quad 0 \leq t \leq T \\ 0 & \text{elsewhere} \end{cases}$$

is subject to the bound

$$\int_0^T \Psi_T(s,\omega)^* \Psi_T(s,\omega)\,ds \leq \frac{Z_T(\omega)+Z_T(\omega)^*}{2b} \tag{8.3}$$

for every point $\omega = a+ib$ *in* $\mathbb{C}_+$.

PROOF. You have to multiply (2.11) (with $t = T$, $\lambda = \omega^*$, and $\mu^* = \omega$) through by $[-iZ_T(\omega)^* \; I_n]$ on the left and its adjoint on the right in order to achieve the desired bound:

$$2b \int_0^T \Psi_T(s,\omega)^* \Psi_T(s,\omega)\,ds$$

$$= [-iZ_T(\omega)^* \; I_n]\{iJ-U(T,\omega)^* iJU(t,\omega)\} \begin{bmatrix} iZ_T(\omega) \\ I_n \end{bmatrix}$$

$$= Z_T(\omega) + Z_T(\omega)^* - 2\{F^\#(T,\omega)F^\#(T,\omega)^*\}^{-1}$$

$$\leq Z_T(\omega) + Z_T(\omega)^* . \quad \square$$

LEMMA 8.3. *The columns of the* $2n \times n$ *matrix valued function*

$$\Psi_\infty(t,\omega) = U(t,\omega)\begin{bmatrix} iZ_\infty(\omega) \\ I_n \end{bmatrix} \tag{8.4}$$

are square summable on $\mathbb{R}^+$.

PROOF. This is immediate from the inequality

$$\int_0^\infty \Psi_\infty(s,\omega)^* \Psi_\infty(s,\omega)\,ds \leq \frac{Z_\infty(\omega)+Z_\infty(\omega)^*}{2b} \tag{8.5}$$

which is obtained by first applying Fatou's lemma to (8.3) to get

$$\int_0^\infty \xi^* \Psi_\infty^*(s,\omega)\Psi_\infty(s,\omega)\xi\,ds \leq \liminf_{T \uparrow \infty} \int_0^\infty \xi^* \Psi_T(s,\omega)^* \Psi_T(s,\omega)\xi\,ds$$

$$\leq \liminf_{T \uparrow \infty} \xi^* \left\{ \frac{Z_T(\omega)+Z_T(\omega)^*}{2b} \right\} \xi$$

for $\omega = a+ib$ in $\mathbb{C}_+$ and every $\xi \in \mathbb{C}^n$, and then taking advantage of the fact that Z_T converges uniformly on $\bar{\mathbb{C}}_+$ to a limit Z_∞ of class C, as is explained in the discussion of (5.20). □

LEMMA 8.4. *The* $2n \times 1$ *vector valued functions*

$$U(t,\omega)\begin{bmatrix}\xi\\0\end{bmatrix} \quad \text{and} \quad U(t,\omega)\begin{bmatrix}0\\\xi\end{bmatrix}$$

are not square summable on $\mathbb{R}^+$ *for any choice of* $\omega \notin \mathbb{R}$ *and* $\xi \in \mathbb{C}^n$, $\xi \neq 0$.

PROOF. For any point $\omega = a+ib$,

$$\begin{aligned} U(t,\omega)^*U(t,\omega) &= \Omega(t,\omega)^* e^{(a-ib)tJ} e^{-(a+ib)tJ}\,\Omega(t,\omega) \\ &= \Omega(t,\omega)^* \{e^{2bt} P_+ + e^{-2bt} P_-\}\Omega(t,\omega) \\ &\geq \Omega(t,\omega)^* P_+ \Omega(t,\omega)\, e^{2bt}, \end{aligned}$$

where $P_\pm$ are the orthogonal projections of (2.18). But this in turn implies that

$$\int_0^t [\xi^*\ 0]U(t,\omega)^*U(t,\omega)[\xi^*\ 0]^* dt$$

$$\geq \int_0^t [\xi^*\ 0]\Omega(t,\omega)^* P_+\Omega(t,\omega)[\xi^*\ 0]^* e^{2bt}\, dt$$

$$= \frac{1}{2}\int_0^t |\varepsilon_t(\omega)\xi|^2 e^{2bt}\, dt,$$

which clearly diverges for $b > 0$ and nonzero $\xi \in \mathbb{C}^n$ since, by Theorem 2.1, $\varepsilon_t(\omega)$ converges, as $t \uparrow \infty$, to a limit $\varepsilon_\infty(\omega)$ which is invertible for $\omega \in \mathbb{C}_+$. This completes the proof of the first assertion for $\omega \in \mathbb{C}_+$. The second assertion is proved in much the same way. It depends upon the convergence of $\Omega_{12}(t,\omega)-i\Omega_{22}(t,\omega)$ to an invertible limit for $\omega \in \mathbb{C}_+$ as $t \uparrow \infty$, which is also supplied by Theorem 2.1. Both facts are easily established for $\omega \in \mathbb{C}_-$ also. □

THEOREM 8.1. *For each point* $\omega \in \mathbb{C}_+$ *the columns of the* $2n \times n$ *matrix valued function* $\Psi_\infty(t,\omega)$, *defined by* (8.4), *form a basis for the space of square summable solutions on* $\mathbb{R}^+$ *of the canonical equation* (2.1).

PROOF. Fix $\omega \in \mathbb{C}_+$. Then clearly every vector $y \in \mathbb{C}^{2n}$ admits a unique representation

$$y = \begin{bmatrix} iZ_\infty(\omega) \\ I_n \end{bmatrix} \xi + \begin{bmatrix} \eta \\ 0 \end{bmatrix}$$

with ξ and η in $\mathbb{C}^n$. Thus, the solution of the canonical equation with initial value y has the form

$$U(t,\omega)y = U(t,\omega) \begin{bmatrix} iZ_\infty(\omega) \\ I_n \end{bmatrix} \xi + X(t,\omega)\eta \ .$$

But now, by Lemma 8.3 the first term on the right always belongs to $L^2_{2n}(\mathbb{R}^+)$. Therefore, $U(t,\omega)$ is square summable on $\mathbb{R}^+$ if and only if $X(t,\omega)\eta$ is. However, in view of Lemma 8.4, this is only possible for $\eta = 0$.

COROLLARY. *If* $\omega \in \mathbb{C}_+$, *then* $U(t,\omega)\begin{bmatrix} \xi \\ i\xi \end{bmatrix}$ *is not square summable on* $\mathbb{R}^+$ *for any choice of* $\xi \in \mathbb{C}^n$, $\xi \neq 0$.

PROOF. By Theorem 8.1, a vector function of the given form is square summable if and only if there exists an $\eta \in \mathbb{C}^n$ such that

$$\begin{bmatrix} \xi \\ i\xi \end{bmatrix} = \begin{bmatrix} iZ_\infty(\omega) \\ I_n \end{bmatrix} \eta \ .$$

But this in turn is possible if and only if $i\xi = \eta$ and $[I_n + Z_\infty(\omega)]\xi = 0$, or, equivalently, since $I_n + Z_\infty(\omega)$ is invertible, if and only if $\xi = \eta = 0$. □

For the sake of completeness, we next evaluate the $\blacktriangle$-transform of the square summable solutions.

THEOREM 8.2. *For every point* $\omega = a+ib$ *in* $\mathbb{C}_+$,

$$\int_0^\infty X^\#(t,\lambda)\Psi_\infty(t,\omega)\,dt = \frac{I_n}{\lambda - \omega} \tag{8.6}$$

and

$$\int_0^\infty \Psi_\infty(t,\omega)^*\Psi_\infty(t,\omega)\,dt = \frac{Z_\infty(\omega) + Z_\infty(\omega)^*}{2b} \ . \tag{8.7}$$

PROOF. We first observe, with the help of (2.11) and (8.1), that

$$(\lambda-\omega) \int_0^\infty X^{\#}(t,\lambda)\Psi_T(t,\omega)\,dt =$$

$$= (\lambda-\omega)\ [I_n\ 0] \int_0^T U^{\#}(t,\lambda)U(t,\omega)\,dt \begin{bmatrix} iZ_T(\omega) \\ I_n \end{bmatrix}$$

$$= [I_n\ 0]\ \{J - U^{\#}(T,\lambda)JU(T,\omega)\} \begin{bmatrix} iZ_T(\omega) \\ I_n \end{bmatrix}$$

$$= I_n - F^{\#}(T,\lambda)\ [\ F^{\#}(T,\omega)\ ]^{-1} .$$

Next, since

$$F(t,\lambda) = e^{i\lambda t}\varphi_t(\lambda) = e^{i\lambda t}\{\Omega_{11}(t,\lambda)+\Omega_{21}(t,\lambda)\} ,$$

it follows readily from Theorem 2.1 that

$$\lim_{T\uparrow\infty} \int_0^\infty X^{\#}(t,\lambda)\ \Psi_T(t,\omega)\,dt = (\lambda-\omega)^{-1}\ I_n$$

and so, in order to complete the proof of (8.6), it suffices to show that

$$\int_0^\infty |\ [\Psi_\infty(t,\omega)-\Psi_T(t,\omega)]\xi|^2\ dt \to 0$$

as $T\uparrow\infty$ for every nonzero $\xi \in \mathbb{C}^n$. But now, as $\Psi_\infty(t,\omega)\xi$ is square summable, this reduces to showing that

$$\xi^*[Z_\infty(\omega)^*-Z_T(\omega)^*\ \ 0] \int_0^T U(t,\omega)^*U(t,\omega)\,dt \begin{bmatrix} Z_\infty(\omega)-Z_T(\omega) \\ 0 \end{bmatrix} \xi$$

tends to zero as $T\uparrow\infty$, which in turn follows from Theorem 8.5 and the identification of the term under consideration as

$$2b\pi\xi^*[Z_\infty(\omega)-Z_T(\omega)]^*\Lambda^T_{\omega^*}(\omega^*)\,[Z_\infty(\omega)-Z_T(\omega)]\xi .$$

Formula (8.7) is now immediate from (8.6), the Plancherel formula (5.16) and the connection between the spectral function and Z_∞ which was established in Theorem 5.3, especially formula (5.23).

By Lemma 4.7,

$$\mathcal{D}_t(\omega) = \{R_{U(T,\omega)}[G(\omega)]:\ G \in \mathcal{C}\} \tag{8.8}$$

is a matrix ball for every fixed $\omega \in \mathbb{C}_+$ and $t > 0$. Our next objective is to show that these balls are nested with respect to t and that, as $t \uparrow \infty$, they contract to the single matrix $Z_\infty(\omega)$. Because of this, the canonical equation (2.1) (with summable potential) is said to be in the limit point (as opposed to the limit circle) case. For a detailed analysis of similar problems, but in a more general setting, see Hinton-Shaw [HS] and Orlov [O].

LEMMA 8.5. *For every nonzero* $\xi \in \mathbb{C}^n$ *and every* $\omega \notin \mathbb{R}$,

$$\lim_{t\uparrow\infty} \xi^* \Lambda_\omega^t(\omega)\xi = \infty .$$

PROOF. By (5.8)

$$\xi^*\Lambda_\omega^t(\omega)\xi = [\xi^* \ 0] \frac{1}{\pi}\int_0^t U^\#(s,\omega)U(s,\omega^*)ds \begin{bmatrix}\xi\\0\end{bmatrix}$$

which diverges as $t \uparrow \infty$ for any point $\omega \notin \mathbb{R}$, by Lemma 8.4. □

THEOREM 8.3. *For any point* $\omega = a+ib \in \mathbb{C}_+$, *the set* $\mathcal{D}_t(\omega)$ *defined in* (8.8) *is equal to the matrix ball*

$$|[\alpha(t,\omega)]^{\frac{1}{2}}[Y+\alpha(t,\omega)^{-1}\beta(t,\omega)][\delta(t,\omega)]^{\frac{1}{2}}| < 1 \tag{8.9}$$

with

$$\alpha(t,\omega) = 2\pi b \ \Lambda_{\omega^*}^t(\omega^*) \tag{8.10}$$

$$\delta(t,\omega) = 2\pi b \ \Lambda_\omega^t(\omega) \tag{8.11}$$

and

$$\beta(t,\omega) = -\alpha(t,\omega)Z_t(\omega) + [F(t,\omega)]^*[F^\#(t,\omega)]^{-1} . \tag{8.12}$$

PROOF. In view of Lemma 4.7, it remains only to verify formulas (8.10)-(8.12). The first of these is immediate from (4.8) and (5.7). Next, with the help of (4.13) and (2.14), δ can be expressed in the form

$$\delta = i(U_{11}^\# U_{21} - U_{21}^\# U_{11}) .$$

The asserted formula (8.11) now drops out by another application of (5.7).

Finally, (8.12) may be obtained by writing out

$$\beta(t,\omega)-\alpha(t,\omega)Z_t(\omega) = \beta(t,\omega)-\alpha(t,\omega)T_{\Phi(t,\omega)}[I_n]$$

with the help of (4.4) and (4.14), and then invoking (2.12) (with $U^\#$ and U interchanged) to simplify the resulting expression. □

THEOREM 8.4. *For any point* $\omega = a+ib$ *in* $\mathbb{C}_+$ *the sets* $\mathcal{D}_t(\omega)$, $t \geq 0$, *are nested:*

$$\mathcal{D}_s(\omega) \supset \mathcal{D}_t(\omega) \quad \text{if} \quad s < t .$$

Moreover,

$$Z_t(\omega) \in \mathcal{D}_s(\omega) \quad \text{for every} \quad t \geq s$$

and

$$\bigcap_{t>0} \mathcal{D}_t(\omega) = Z_\infty(\omega) .$$

PROOF. If $s < t$, then, in terms of the notation introduced in Section 7,

$$U(t,\omega) = U_s(t,\omega)U(s,\omega)$$

and so, in view of Lemma 4.3 and Theorem 4.1,

$$R_{U(t,\omega)}[C] = R_{U(s,\omega)}[R_{U_s(t,\omega)}[C]] \subset R_{U(s,\omega)}[C] .$$

This proves that the $\mathcal{D}_t(\omega)$ are nested, and hence, since, by (5.18) and (8.8), $Z_t(\omega)$ clearly belongs to $\mathcal{D}_t(\omega)$,justifies the second assertion also. The final assertion follows from Lemma 8.5, Theorem 8.3, (5.18)-(5.20) (which explain the convergence of Z_t to Z_∞), and the fact that

$$[\alpha(t,\omega)]^{-1}[F(t,\omega)]^*[F^\#(t,\omega)]^{-1} \to 0 ,$$

as $t \uparrow \infty$. To prove the latter, it is convenient to reexpress the last two factors as

$$[E(t,\omega)]^*[\Sigma(t,\omega)]^*[F^\#(t,\omega)]^{-1} =$$

$$= e^{i\omega^* t}[\varepsilon_t(\omega)]^*[\Sigma(t,\omega)]^* e^{i\omega t}[\varphi_t^\#(\omega)]^{-1}$$

in order to deduce from Theorem 2.1 and the innerness of $\Sigma(t,\omega)$ that this part stays bounded as $t \uparrow \infty$ for $\omega \in \mathbb{C}_+$. The rest is then plain from Lemma 8.5. □

LEMMA 8.5. *For every point* $\omega \in \mathbb{C}_+$,

$$[Z_\infty(\omega)-Z_t(\omega)]^* \Lambda_{\omega^*}^t(\omega^*)[Z_\infty(\omega)-Z_t(\omega)] \to 0$$

as $t \uparrow \infty$.

PROOF. To begin with, both $Z_t(\omega)$ and $Z_\infty(\omega)$ belong to $\mathcal{D}_t(\omega)$, by Theorem 8.4. Therefore, by (8.9),

$$|[\alpha(t,\omega)]^{\frac{1}{2}}[Z_\infty(\omega)-Z_t(\omega)]| \leq$$

$$\leq |[\alpha(t,\omega)]^{\frac{1}{2}}[Z_\infty(\omega)-Z_t(\omega)][\delta(t,\omega)]^{\frac{1}{2}}|\cdot|[\delta(t,\omega)]^{-\frac{1}{2}}|$$

$$\leq 2|[\delta(t,\omega)]^{-\frac{1}{2}}| \ .$$

But now by (8.11) and Lemma 8.5, the right-hand side of the last inequality tends to zero as $t \uparrow \infty$, and this in turn is readily seen to be equivalent to the asserted statement. □

9. SCATTERING MATRIX AND THE INVERSE SCATTERING PROBLEM

In this section we shall show that there is a one-to-one correspondence between (normalized) canonical equations and the class of $n \times n$ unitary matrix functions in W^I with zero factorization indices. We shall refer to the latter as scattering matrices. The inverse scattering problem in this context is to recover the potential of the underlying canonical equation from the scattering matrix. One approach, as we shall explain below, is to compute the spectral function from the scattering matrix and then to use one of the two methods for solving the inverse spectral problem which were discussed in Section 6. We shall also outline a method which works directly with the scattering matrix which is due to Marchenko. This amounts to an upper-lower triangular factorization of an integral operator which is based on the scattering matrix. For a rigorous treatment of a similar problem, see Gasymov [Ga] and Gasymov-Levitan [GaL].

We begin with a description of the particular solution $X(t,\lambda)$ of the canonical equation (2.1) with a potential V which is, as always, summable and of the form (2.2).

LEMMA 9.1. *The solution* $X(t,\lambda)$ *of the canonical equation* (2.1) *can be expressed in the form*

$$X(t,\lambda) = [y_+(t,\lambda)S(\lambda)+y_-(t,\lambda)]\varepsilon_\infty(\lambda)+o(1) \tag{9.1}$$

for $t \uparrow \infty$ *and* $\lambda \in \mathbb{R}$, *where*

$$y_+(t,\lambda) = P_-U_0(t,\lambda)\begin{bmatrix} I_n \\ 0 \end{bmatrix} = \frac{1}{2}\begin{bmatrix} I_n \\ -iI_n \end{bmatrix} e^{i\lambda t} , \tag{9.2}$$

$$y_-(t,\lambda) = P_+U_0(t,\lambda)\begin{bmatrix} I_n \\ 0 \end{bmatrix} = \frac{1}{2}\begin{bmatrix} I_n \\ iI_n \end{bmatrix} e^{-i\lambda t} , \qquad (9.3)$$

and

$$S(\lambda) = \varphi_\infty(\lambda)[\varepsilon_\infty(\lambda)]^{-1} . \qquad (9.4)$$

PROOF. Clearly

$$X(t,\lambda) = U(t,\lambda)\begin{bmatrix} I_n \\ 0 \end{bmatrix} = e^{-\lambda tJ}\Omega(t,\lambda)\begin{bmatrix} I_n \\ 0 \end{bmatrix}$$

$$= e^{-\lambda tJ}\Omega_\infty(\lambda)\begin{bmatrix} I_n \\ 0 \end{bmatrix} + o(1)$$

uniformly on $\mathbb{R}$, as $t \uparrow \infty$, thanks to item 8° in Section 2 and the bound

$$|e^{-\lambda tJ}| \leq 1 \quad \text{for} \quad \lambda \in \mathbb{R} .$$

The rest drops out upon inserting

$$e^{-\lambda tJ} = e^{-i\lambda t}P_+ + e^{i\lambda t}P_-$$

and carrying out the indicated multiplications. □

The matrix S defined in (9.4) is termed the *scattering matrix* of the canonical equation (2.1).

LEMMA 9.2. *The scattering matrix* S *is a unitary matrix valued function of class* W^I *with zero factorization indices.*

PROOF. Since $[\varepsilon_\infty]^{\pm 1} \in W^I_+$ and $[\varphi_\infty]^{\pm 1} \in W^I_-$, by Theorem 2.1, formula (9.4) exhibits S as an invertible element of W^I with zero right factorization indices. Next, upon multiplying

$$\Omega_\infty(\lambda)^* J\Omega_\infty(\lambda) = \Omega_\infty(\lambda)^*(iP_+ - iP_-)\Omega_\infty(\lambda) = J , \quad \lambda \in \mathbb{R} ,$$

through by $[I_n \ 0]$ on the left and $[I_n \ 0]^*$ on the right, we observe that

$$\varepsilon_\infty(\lambda)^*\varepsilon_\infty(\lambda) = \varphi_\infty(\lambda)^*\varphi_\infty(\lambda) ,$$

and hence that

$$S(\lambda)^*S(\lambda) = I_n ,$$

for $\lambda \in \mathbb{R}$. This proves that S is unitary on $\mathbb{R}$ as claimed. Therefore the left factorization indices match the right factorization indices and so must vanish also.

THEOREM 9.1. *There is a one-to-one correspondence between the set of summable potentials of the form* (2.2) *and the set of unitary matrix functions in* W^I *with zero factorization indices.*

PROOF. In view of Theorem 6.5, it suffices to show that there is a one-to-one correspondence between the set of all scattering matrices S and the set of all spectral functions Δ_∞ . But this is clear from the formulas

$$S = \varphi_\infty \varepsilon_\infty^{-1} = [\varphi_\infty^*]^{-1} \varepsilon_\infty^*$$

and

$$\Delta_\infty = [\varepsilon_\infty^* \varepsilon_\infty]^{-1} = [\varphi_\infty^* \varphi_\infty]^{-1}$$

which hold on $\mathbb{R}$.

Indeed, every unitary $S \in W^I$ with zero factorization indices can be uniquely expressed in the form

$$S = S_- S_+ \quad (= [S_+^* S_-^*]^{-1})$$

with $(S_+)^{\pm 1} \in W_+^I$ and $(S_-)^{\pm 1} \in W_-^I$, and hence, upon setting

$$\Delta_\infty = S_+ S_+^* = [S_-^* S_-]^{-1} ,$$

it is readily checked that $\Delta_\infty \in W^I$ and $\Delta_\infty(\lambda) > 0$ for $\lambda \in \mathbb{R}$, i.e., Δ_∞ is a spectral function.

Conversely, if Δ_∞ is a spectral function, then it is well known (see e.g. [GK1]) that it admits a pair of unique factorizations of the form

$$\Delta_\infty = \delta_+ \delta_+^* = \delta_- \delta_-^*$$

with $(\delta_+)^{\pm 1} \in W_+^I$ and $(\delta_-)^{\pm 1} \in W_-^I$. Clearly

$$S = (\delta_-)^{-1} \delta_+$$

is a scattering matrix. □

The last theorem and its proof indicate that one method to solve the inverse scattering problem is to first construct the spectral function and then to solve the inverse spectral problem by the method of either Gelfand-Levitan or of Krein, as discussed in Section 6. We now turn to a third method, which is due to Marchenko.

THEOREM 9.2. *Every canonical equation* (2.1) *admits a*

pair of $2n \times n$ *matrix solutions*

$$Y_\pm(t,\lambda) = y_\pm(t,\lambda) + \int_t^\infty M(t,s)y_\pm(s,\lambda)ds \tag{9.5}$$

where $y_\pm$ *are defined in* (9.2)-(9.3) *and* $M(t,s)$ *is a* $2n \times 2n$ *matrix valued function which is subject to the bound*

$$\int_t^\infty |M(t,s)| \leqslant 2[\exp\{\int_t^\infty |V(s)|ds\} - 1] . \tag{9.6}$$

PROOF. Let us suppose first that the canonical equation admits a pair of solutions of the indicated form and, for the sake of definiteness, let us focus on one of them, and drop the subscripts for the purposes of this proof. Assuming further that M is differentiable, it follows readily upon substituting $Y(t,\lambda)$ into (2.1) that

$$[V(t)+JM(t,t)-M(t,t)J]y(t,\lambda) =$$

$$= \int_t^\infty [J\frac{\partial M(t,s)}{\partial t} + \frac{\partial M(t,s)}{\partial s} J - V(t)M(t,s)]y(s,\lambda)d\lambda ,$$

at least if $M(t,s) \to 0$ as $s \uparrow \infty$, and hence, much as in the proof of Theorem 6.2, that

$$J\frac{\partial M(t,s)}{\partial t} + \frac{\partial M(t,s)}{\partial s} J = V(t)M(t,s) \tag{9.7}$$

on the triangle $s > t \geqslant 0$, and

$$V(t) = M(t,t)J - JM(t,t) . \tag{9.8}$$

Next, upon splitting M into its J-commuting and J-anticommuting components

$$M_\pm = \frac{1}{2}(M \mp JMJ) ,$$

it is readily checked that $M_\pm$ must satisfy the system of equations

$$\frac{\partial M_+(t,s)}{\partial t} + \frac{\partial M_+(t,s)}{\partial s} + JV(t)M_-(t,s) = 0$$

$$\frac{\partial M_-(t,s)}{\partial t} - \frac{\partial M_-(t,s)}{\partial s} + JV(t)M_+(t,s) = 0 ,$$

on the triangle $s > t \geqslant 0$, subject to the boundary conditions

$$M_+(t,t) = 0 \quad \text{and} \quad M_-(t,t) = \frac{1}{2}JV(t) .$$

However, under the added assumption that $M_+(t,t+\tau) \to 0$ as $t \uparrow \infty$ for every $\tau \geq 0$, these are easily seen to be equivalent to the system of integral equations

$$M_+(t,s) = \int_t^\infty JV(u)M_-(u,u+s-t)\,du$$
$$M_-(t,s) = \frac{1}{2}JV\left(\frac{t+s}{2}\right) + \int_t^{(t+s)/2} JV(u)M_+(u,-u+s+t)\,du\,. \tag{9.9}$$

But now it is readily shown by standard iteration methods that for every summable V the system of integral equations (9.9) admits a unique solution pair $M_\pm(t,s)$ on the triangle $s > t \geq 0$, the sum of which: M, is subject to the bound (9.6) and that moreover the functions $Y_\pm$ specified by (9.5) are solutions of the corresponding canonical equation.

The technical details are left to the industrious. □

LEMMA 9.3. *The* $n \times n$ *matrix function*

$$Y_1^\#(t,\lambda)JY_2(t,\lambda)$$

is independent of t *for every pair* $Y_1(t,\lambda)$ *and* $Y_2(t,\lambda)$ *of* $2n \times n$ *matrix solutions of* (2.1).

PROOF. It is readily checked that the derivative with respect to t of the matrix of interest is equal to zero. □

Next, let us express the particular solution $X(t,\lambda)$ of the canonical equation (2.1) in terms of the two solutions, $Y_\pm(t,\lambda)$, given by (9.5). From the asymptotic behavior

$$Y_\pm(t,\lambda) = y_\pm(t,\lambda) + o(1)\,, \quad \text{as} \quad t \uparrow \infty\,,$$

for $\lambda \in \mathbb{R}$, it is clear that the columns of these two matrices yield a basis in the 2n-dimensional space of solutions of equation (2.1). Thus, we can write

$$X(t,\lambda) = Y_+(t,\lambda)C_+(\lambda) + Y_-(t,\lambda)C_-(\lambda)$$

with $n \times n$ matrices $C_\pm(\lambda)$ which do not depend on t. But now on the one hand, it is plain from (9.1) that

$$C_+(\lambda) = \varphi_\infty(\lambda) \quad \text{and} \quad C_-(\lambda) = \varepsilon_\infty(\lambda)\,.$$

Yet on the other hand, by Lemma 9.3,

$$Y_\pm^\#(t,\lambda)JX(t,\lambda) = Y_\pm^\#(t,\lambda)JY_+(t,\lambda)C_+(\lambda)+Y_\pm^\#(t,\lambda)JY_-(t,\lambda)C_-(\lambda) ,$$

is independent of t as are

$$Y_\pm^\#(t,\lambda)JX(t,\lambda) = Y_\pm^\#(0,\lambda)JX(0,\lambda) = -Y_{\pm 2}^\#(0,\lambda) ,$$

$$Y_\pm^\#(t,\lambda)JY_\pm(t,\lambda) = \lim_{t\uparrow\infty} Y_\pm^\#(t,\lambda)JY_\pm(t,\lambda) = \lim_{t\uparrow\infty} y_\pm^\#(t,\lambda)Jy_\pm(t,\lambda)$$
$$= \mp \tfrac{1}{2}iI_n$$

and

$$Y_+^\#(t,\lambda)JY_-(t,\lambda) = Y_-^\#(t,\lambda)JY_+(t,\lambda) = 0 ,$$

where, in the first of these formulas, $Y_{\pm 2}$ denotes the bottom $n \times n$ block of the $2n \times n$ matrix $Y_\pm$. Thus

$$Y_{+2}^\#(0,\lambda) = \tfrac{1}{2}iC_+(\lambda) \quad \text{and} \quad Y_{-2}^\#(0,\lambda) = -\tfrac{1}{2}iC_-(\lambda) .$$

Therefore, the scattering matrix

$$S(\lambda) = \varphi_\infty(\lambda)[\varepsilon_\infty(\lambda)]^{-1} = C_+(\lambda)[C_-(\lambda)]^{-1} = -Y_{+2}^\#(0,\lambda)[Y_{-2}^\#(0,\lambda)]^{-1}$$
$$= -Y_{-2}(0,\lambda)[Y_{+2}(0,\lambda)]^{-1} . \qquad (9.10)$$

This is the analogue for canonical equations of the familiar formula which expresses the asymptotic phase of the radial Schrödinger equations in terms of the Jost functions.

Our next objective is to relate the matrix kernel $M(t,s)$ figuring in the representation (9.5) of $Y_\pm(t,\lambda)$ to the scattering matrix $S(\lambda)$. To do this, observe first that, for every $g \in L_{2n}^2(\mathbb{R}^+)$ with compact support,

$$g^\blacktriangle(\lambda) = \int_0^\infty X^\#(t,\lambda)g(t)dt$$
$$= \int_0^\infty \{\psi(t,\lambda) + \int_t^\infty M(t,s)\psi(s,\lambda)ds\}^\# g(t)dt ,$$

where

$$\psi(t,\lambda) = y_+(t,\lambda)\varphi_\infty(\lambda)+y_-(t,\lambda)\varepsilon_\infty(\lambda) .$$

Hence,

$$g^\blacktriangle(\lambda) = \int_0^\infty \psi^\#(t,\lambda)\{g(t) + \int_0^t M(s,t)^* g(s)ds\}dt$$

$$= \int_0^\infty \psi^\#(t,\lambda)[(I + \mathbb{M}^*)g](t)dt ,$$

where $\mathbb{M}$ denotes the operator in $L^2_{2n}(\mathbb{R}^+)$ with kernel $M(t,s)$:

$$(\mathbb{M}g)(t) = \int_t^\infty M(t,s)g(s)ds . \tag{9.11}$$

Furthermore,

$$g^\blacktriangle(\lambda) = \varphi_\infty^\#(\lambda) \int_0^\infty y_+^\#(t,\lambda)[(I+\mathbb{M}^*)g](t)dt +$$

$$+ \varepsilon_\infty^\#(\lambda) \int_0^\infty y_-^\#(t,\lambda)[(I+\mathbb{M}^*)g](t)dt$$

$$= \frac{1}{\sqrt{2}}\varphi_\infty(\lambda)^* \int_{-\infty}^0 e^{i\lambda t}G(t)dt + \frac{1}{\sqrt{2}}\varepsilon_\infty(\lambda)^* \int_0^\infty e^{i\lambda t}G(t)dt$$

$$= \frac{1}{\sqrt{2}}\varphi_\infty(\lambda)^*(\underline{\underline{q}}\hat{G})(\lambda) + \frac{1}{\sqrt{2}}\varepsilon_\infty(\lambda)^*(\underline{\underline{p}}\hat{G})(\lambda) , \tag{9.12}$$

where

$$G(t) = [\mathbb{N}(I+\mathbb{M}^*)g](t)$$

and $\underline{\underline{p}}$ [resp. $\underline{\underline{q}}$] designates the orthogonal projection of $L^2_n(\mathbb{R},d\lambda)$ onto H^2_n [resp. K^2_n] .

Thus, by the Parseval formula (5.16),

$$2\pi \int_0^\infty g(t)^*g(t)dt = \langle\Delta_\infty[\varphi_\infty^*(\underline{\underline{q}}\hat{G})+\varepsilon_\infty^*(\underline{\underline{p}}\hat{G})],\varphi_\infty^*(\underline{\underline{q}}\hat{G})+\varepsilon_\infty^*(\underline{\underline{p}}\hat{G})]\rangle$$

$$= \langle S^*(\underline{\underline{q}}\hat{G})+(\underline{\underline{p}}\hat{G}),S^*(\underline{\underline{q}}\hat{G})+(\underline{\underline{p}}\hat{G})\rangle$$

$$= \langle\hat{G},\hat{G}\rangle + 2\,\mathrm{Re}\,\langle\underline{\underline{p}}S^*\underline{\underline{q}}\hat{G},\hat{G}\rangle ,$$

where $\langle\ ,\ \rangle$ denotes the standard inner product in $L^2_n(\mathbb{R})$. But now, upon expressing

$$S(\lambda) = I_n - \int e^{-i\lambda t}\sigma(t)dt \tag{9.13}$$

with $\sigma \in L^1_{n\times n}(\mathbb{R})$, and invoking the Plancherel formula for the conventional Fourier transform, it follows readily that the last two terms are equal to

$$2\pi\langle G,G\rangle - 4\pi\,\mathrm{Re}\int_0^\infty G(t)^*\left\{\int_0^\infty \sigma(t+s)^*G(-s)ds\right\}dt ,$$

and after a little manipulation, that this in turn leads to the conclusion

$$\langle g,g\rangle_{L^2_{2n}(\mathbb{R}^+)} = \langle (I+\mathbb{M})(I-H_S)(I+\mathbb{M}^*)g,g\rangle_{L^2_{2n}(\mathbb{R}^+)} , \tag{9.14}$$

where the inner products in (9.14) are taken in $L^2_{2n}(\mathbb{R}^+)$ as indicated, and H_S is the Hankel operator in $L^2_{2n}(\mathbb{R}^+)$ which is defined by the rule

$$(H_S g)(t) = \int_0^\infty \chi(t+u)g(u)\,du \tag{9.15}$$

with

$$\chi(u) = \frac{1}{2}\begin{bmatrix} \sigma(u)+\sigma(u)^* & i\sigma(u)^* - i\sigma(u) \\ i\sigma(u)^* - i\sigma(u) & -\sigma(u)-\sigma(u)^* \end{bmatrix} . \tag{9.16}$$

The fundamental identity

$$(I+\mathbb{M})\ (I-H_S)(I+\mathbb{M}^*) = I \tag{9.16}$$

which exhibits $I+\mathbb{M}$ as an upper triangular factor of $(I-H_S)^{-1}$ (with respect to the chain of projectors P_T , $T \geqslant 0$, defined in (5.24) is immediate from (9.14). Furthermore, (9.17) implies that the kernel M of $\mathbb{M}$ may be obtained as a solution of the *Marchenko equation*

$$M(t,s) - \chi(t+s) - \int_t^\infty M(t,u)\chi(u+s)\,du = 0 \tag{9.18}$$

on the triangle $0 \leqslant t \leqslant s < \infty$.

The Marchenko procedure for solving the inverse scattering problem is now clear, at least formally. Given a unitary $n \times n$ matrix valued $S \in W^I$ with zero factorization indices, the first step is to express it in terms of σ as in (9.13) and then to solve the Marchenko equation (9.18) for $M(t,s)$ where χ is taken from (9.16). The sought for potential $V(t)$ is then available from (9.8). This is just a sketch. To make things precise it remains to show that:

(1) The Marchenko equation (9.18) is actually solvable.

(2) The potential $V(t)$ specified by (9.8) has summable entries and that, as is in fact self-evident, it is of the form (2.2).

(3) The scattering matrix of the canonical equation with potential V specified by (9.8) is equal to the matrix S which was

given at the outset.

A complete rigorous proof of (1)-(3) is beyond the scope of this paper. Having come so far, however, we deem it worthwhile to add just a few more formal manipulations in order to obtain a variant of the Marchenko method which is due to Krein and Melik-Adamyan [KMaF1].

To begin with, observe that

$$P_+\chi P_+ = P_-\chi P_- = 0$$

and that this applied to (9.18) leads to the equation

$$P_+M(t,s)P_+ = \int_t^\infty [P_+M(t,u)P_-]\,[P_-\chi(u+s)P_+]du$$

or equivalently,

$$2\begin{bmatrix} m_1(t,s) & -im_1(t,s) \\ im_1(t,s) & m_1(t,s) \end{bmatrix} =$$

$$= \int_t^\infty \begin{bmatrix} m_2(t,u) & im_2(t,u) \\ im_2(t,u) & -m_2(t,u) \end{bmatrix} \begin{bmatrix} \sigma(u+s) & -i\sigma(u+s) \\ -i\sigma(u+s) & -\sigma(u+s) \end{bmatrix} du\ ,$$

where

$$m_1 = M_{11} + M_{22} + i(M_{12} - M_{21})$$

and

$$m_2 = M_{11} - M_{22} - i(M_{12} + M_{21})\ .$$

But this in turn implies that

$$m_1(t,s) - \int_t^\infty m_2(t,u)\sigma(u+s)du = 0$$

on the triangle $0 \leqslant t \leqslant s < \infty$, and by a similar analysis of the $P_+[\ \cdots\ ]P_-$ version of (9.18), that

$$m_2(t,s) - \int_t^\infty m_1(t,u)\sigma(u+s)^* du = 2\sigma(t+s)\ ,$$

on the same triangle. Therefore,

$$m_2(t,s)^* - \int_0^\infty \sigma(t+s+u)m_1(t,t+u)^* du = 2\sigma(t+s)$$

and (because $m_2(t,u)^* = 0$ for $u < t$)

$$m_1(t,s)^* - \int_0^\infty \sigma(s+u)^* m_2(t,u)^* du = 0$$

for $0 \leqslant t \leqslant s < \infty$. Finally, with the notation

$$\Phi_t(s) = \frac{1}{2} m_2(t,s)^*$$

and $\mathbb{\Gamma}_t$ for the Hankel operator in $L_n^2(\mathbb{R}^+)$ which is defined by the rule

$$(\mathbb{\Gamma}_t g)(s) = \int_0^\infty \sigma(t+s+u)g(u)du , \tag{9.19}$$

we see that

$$[(I - \mathbb{\Gamma}_t \mathbb{\Gamma}_t^*)\Phi_t](s) = \sigma(t+s) , \quad 0 \leqslant s < \infty . \tag{9.20}$$

A similar computation yields

$$[(I - \mathbb{\Gamma}_t^* \mathbb{\Gamma}_t)\Psi_t](s) = \sigma(t+s)^* , \quad 0 \leqslant s < \infty \tag{9.21}$$

where

$$\Psi_t(s) = \frac{1}{2} m_3(t,s)^* ,$$

and

$$m_3 = M_{11} - M_{22} + i(M_{12} + M_{21}) .$$

Next, from (2.2) and (9.8), it is readily seen that the blocks of the potential

$$\begin{aligned} V_1(t) &= -\{M_{12}(t,t)^* + M_{21}(t,t)^*\} \\ &= \{m_3(t,t)^* - m_2(t,t)^*\}/2i \\ &= i\{\Phi_t(t) - \Psi_t(t)\} \end{aligned} \tag{9.22}$$

and

$$\begin{aligned} V_2(t) &= M_{11}(t,t)^* - M_{22}(t,t)^* \\ &= \{m_3(t,t)^* + m_2(t,t)^*\}/2 \\ &= \Phi_t(t) + \Psi_t(t) . \end{aligned} \tag{9.23}$$

The method of Krein and Melik-Adamyan is thus to first solve the system of equations (9.19) and (9.20) for Φ_t and Ψ_t and then to obtain the potential via (9.22) and (9.23). By one of their theorems, $\|\mathbb{\Gamma}_0\| < 1$ for every scattering matrix S, where the norm of the Hankel operator $\mathbb{\Gamma}_0$ (and $\mathbb{\Gamma}_t$ below) is taken in $L_n^2(\mathbb{R}^+)$. This in turn implies that $\|\mathbb{\Gamma}_t\| < 1$ for all $t > 0$ and hence that the pair of equations (9.20) and (9.21) are uniquely solvable in $L_{n\times n}^1(\mathbb{R}^+)$. For a statement of this theorem and its relevance to factorization theory, see [KMaF1], and for further discussion [KaMF2]; a proof may be found in [DG2].

10. LAX-PHILLIPS SCATTERING

The purpose of this section is to illustrate the Lax-Phillips scattering theory via the canonical equation. In particular, we shall show that the scattering matrix for the Lax-Phillips theory is precisely the scattering matrix S which arose in the last section.

In the present setting, the appropriate starting point for the Lax-Phillips theory is the partial differential equation

$$-i\frac{\partial u}{\partial t} = J\frac{\partial u}{\partial r} - V(r)u \tag{10.1}$$

on the half plane $r \geqslant 0$, $t \in \mathbb{R}$, subject to the initial condition

$$u(r,0) = f(r) \quad \text{with} \quad f \in L^2_{2n}(\mathbb{R}^+) .$$

It is readily seen that the solution is given by

$$u(r,t) = \frac{1}{\pi}\int e^{i\lambda t}\, X(r,\lambda)\Delta_\infty(\lambda) f^{\blacktriangle}(\lambda)d\lambda .$$

Let

$$\mathcal{D}_+ = \{f \in L^2_{2n}(\mathbb{R}^+) : u(r,t) = 0 \quad \text{for} \quad t > r\}$$

and

$$\mathcal{D}_- = \{f \in L^2_{2n}(\mathbb{R}^+) : u(r,t) = 0 \quad \text{for} \quad t < r\} .$$

It turns out that these subspaces can be simply characterized in terms of the two spectral factorizations of Δ_∞ :

$$\Delta_\infty(\lambda) = \{\varepsilon_\infty(\lambda)^*\varepsilon_\infty(\lambda)\}^{-1} = \{\varphi_\infty(\lambda)^*\varphi_\infty(\lambda)\}^{-1} .$$

Before proceeding further it is well to recall that ε_r and ε_∞ are invertible elements of W^I_+, while φ_r and φ_∞ are invertible elements of W^I_-.

THEOREM 10.1. *The subspace*

$$\mathcal{D}_+ = \{f \in L^2_{2n}(\mathbb{R}^+) : (\varepsilon^*_\infty)^{-1}f^{\blacktriangle} \in H^2_n\}$$

while the subspace

$$\mathcal{D}_- = \{f \in L^2_{2n}(\mathbb{R}^+) : (\varphi^*_\infty)^{-1}f^{\blacktriangle} \in K^2_n\} .$$

PROOF. Let $f \in \mathcal{D}_+$. Then, by the definition of $\mathcal{D}_+$,

$$\frac{1}{\pi} \int e^{i\lambda t} X(r,\lambda)\Delta_\infty(\lambda) f^\blacktriangle(\lambda) d\lambda = 0 ,$$

for every $t > r$ and hence

$$e^{i\lambda r} X(r,\lambda)\Delta_\infty(\lambda) f^\blacktriangle(\lambda) \in H^2_{2n} ,$$

for every choice of $r > 0$.

Thus

$$e^{i\lambda r}[I_n \ -iI_n] X(r,\lambda)\Delta_\infty(\lambda) f^\blacktriangle(\lambda)$$

$$= \varepsilon_r(\lambda)[\varepsilon_\infty(\lambda)]^{-1}[\varepsilon_\infty(\lambda)^*]^{-1} f^\blacktriangle(\lambda)$$

belongs to H^2_n for every $r > 0$, as does $(\varepsilon^*_\infty)^{-1} f^\blacktriangle$, since $(\varepsilon_r)^{-1}$ and ε_∞ belong to W^I_+ .

Next, since ε_r and $(\varepsilon_\infty)^{-1}$ also belong to W^I_+ , the argument is easily reversed to complete the proof of the stated characterization of $\mathcal{D}_+$. The proof for $\mathcal{D}_-$ is similar. □

We next introduce the spaces

$$\mathcal{D}_\pm(t) = \{u(r,t): u(r,0) \in \mathcal{D}_\pm\} .$$

It is immediate from Theorems 10.1 and 5.2 that

$$\{\mathcal{D}_+(t)\}^\blacktriangle = e^{i\lambda t} \varepsilon^*_\infty H^2_n \tag{10.2}$$

and

$$\{\mathcal{D}_-(t)\}^\blacktriangle = e^{i\lambda t} \varphi^*_\infty K^2_n . \tag{10.3}$$

The fact that the $\blacktriangle$-transform is an isometry (up to a factor of $\sqrt{\pi}$) makes it plain that these spaces enjoy the following properties:

$$\mathcal{D}_+(t) \subset \mathcal{D}_+ \quad \text{and} \quad \mathcal{D}_-(-t) \subset \mathcal{D}_- \quad \text{for} \quad t > 0 \tag{10.4}$$

$$\bigcap_{t>0} \mathcal{D}_+(t) = \bigcap_{t>0} \mathcal{D}_-(-t) = 0 \tag{10.5}$$

$$\bigcup_t \mathcal{D}_\pm(t) = L^2_{2n}(\mathbb{R}^+) . \tag{10.6}$$

Moreover, it is clear from Theorem 5.2 and the above discussion that

$$f_+(\lambda) = [\sqrt{\pi}\, \varepsilon_\infty(\lambda)^*]^{-1} f^\blacktriangle(\lambda) \tag{10.7}$$

and

$$f_-(\lambda) = [\sqrt{\pi}\ \varphi_\infty(\lambda)^*]^{-1} f^\blacktriangle(\lambda) \tag{10.8}$$

are isometric mappings of $L^2_{2n}(\mathbb{R}^+)$ onto $L^2_n(\mathbb{R})$ such that the former [resp. the latter] maps $\mathcal{D}_+$ onto H^2_n [resp. $\mathcal{D}_-$ onto K^2_n] and such that in addition

$$[u(r,t)]_\pm = e^{i\lambda t}[u(r,0)]_\pm \quad .$$

The mappings (10.7) and (10.8) are the outgoing and incoming spectral representations of Lax-Phillips. The Lax-Phillips scattering matrix $S(\lambda)$, which is defined by the relation

$$f_-(\lambda) = S(\lambda)f_+(\lambda) \ ,$$

is thus seen to be precisely the same as the scattering matrix which was introduced in the last section:

$$S(\lambda) = [\varphi_\infty(\lambda)^*]^{-1}\ \varepsilon_\infty(\lambda)^* = \varphi_\infty(\lambda)[\varepsilon_\infty(\lambda)]^{-1} \ .$$

It is often convenient to work with subspaces $\mathcal{D}_\pm$ which are orthogonal in $L^2_{2n}(\mathbb{R}^+)$. This is decidedly not the case in the present setting unless $V = 0$, as the next theorem shows.

THEOREM 10.2. *Let* θ *denote the angle between the subspaces* $\mathcal{D}_+$ *and* $\mathcal{D}_-$ *in* $L^2_{2n}(\mathbb{R}^+)$. *Then*

$$\cos\theta = \|\Gamma_0\| < 1 \ ,$$

where $\|\Gamma_0\|$ *denotes the norm in* $L^2_n(\mathbb{R}^+)$ *of the Hankel operator* Γ_0 *based on* S *which is defined via* (9.13) *and* (9.19).

PROOF. In view of (5.16) and the identifications (10.2) and (10.3) it is readily checked that

$$\begin{aligned}\cos\theta &= \sup\ |\langle\Delta_\infty\varepsilon^*u\ ,\ \varphi_\infty^*v\rangle| \\ &= \sup\ |\langle u,S^*v\rangle| \\ &= \sup\ |\langle u,\underline{\underline{p}}S^*\underline{\underline{q}}v\rangle| \\ &= \|\Gamma_0^*\| \ ,\end{aligned}$$

where the sup is taken over all u in the unit ball of H^2_n and v in the unit ball of K^2_n , the inner product is taken in $L^2_n(\mathbb{R})$, and $\underline{\underline{p}}$ and $\underline{\underline{q}}$ denote the orthogonal projectors which are defined just below (9.12). The fact that $\|\Gamma_0\| < 1$ follows from the theorem of Krein and Melik-Adamyan which is discussed at the end of Section 9. □

Orthogonality may be obtained for potentials with compact support by working with the modified spaces

$$\mathcal{D}_+^\rho = \{f \in L_{2n}^2(\mathbb{R}^+): u(r,t) = 0 \quad \text{for} \quad t+\rho > r\}$$

$$\mathcal{D}_-^\rho = \{f \in L_{2n}^2(\mathbb{R}^+): u(r,t) = 0 \quad \text{for} \quad t < \rho-r\}$$

and

$$\mathcal{D}_\pm^\rho(t) = \{u(r,t): u(r,0) \in \mathcal{D}_\pm^\rho\} .$$

It is readily checked, by imitating the proof of Theorem 10.1, that

$$\mathcal{D}_+^\rho = \{f \in L_{2n}^2(\mathbb{R}^+): e^{-i\lambda\rho}(\varepsilon_\infty^*)^{-1} f^\blacktriangle \in H_n^2\}$$

$$\mathcal{D}_-^\rho = \{f \in L_{2n}^2(\mathbb{R}^+): e^{i\lambda\rho} (\varphi_\infty^*)^{-1} f^\blacktriangle \in K_n^2\} ,$$

and hence, by Theorem 5.2, that,

$$\{\mathcal{D}_+^\rho(t)\}^\blacktriangle = e^{i\lambda(t+\rho)} \varepsilon_\infty^* H_n^2 \tag{10.9}$$

and

$$\{\mathcal{D}_-^\rho(t)\}^\blacktriangle = e^{i\lambda(t-\rho)} \varphi_\infty^* K_n^2 . \tag{10.10}$$

It is moreover plain from the preceding analysis that (10.4)-(10.6) remain valid with $\mathcal{D}_\pm^\rho$ in place of $\mathcal{D}_\pm$, that the corresponding outgoing and incoming spectral representations are now

$$f_+^\rho(\lambda) = e^{-i\lambda\rho}(\sqrt{\pi}\ \varepsilon_\infty^*)^{-1} f^\blacktriangle$$

$$f_-^\rho(\lambda) = e^{i\lambda\rho} (\sqrt{\pi}\ \varphi_\infty^*)^{-1} f^\blacktriangle$$

and hence that the corresponding Lax-Phillips scattering matrix

$$S_\rho(\lambda) = e^{2i\lambda\rho} S(\lambda) .$$

Thus, if θ_ρ denotes the angle between $\mathcal{D}_+^\rho$ and $\mathcal{D}_-^\rho$, it is readily checked, just as in the proof of Theorem 10.2, that

$$\cos\theta_\rho = \|\underline{\underline{q}}\ S_\rho\ \underline{\underline{p}}\| . \tag{10.11}$$

But now, if $V(r) = 0$ for $r > \rho$, then $\Omega(r,\lambda) = \Omega(\rho,\lambda)$ for $r \geqslant \rho$ and so

$$\varphi_r(\lambda) = \varphi_\rho(\lambda) \quad \text{and} \quad \varepsilon_r(\lambda) = \varepsilon_\rho(\lambda)$$

for $r \geqslant \rho$. Thus

$$\begin{aligned} S_\rho(\lambda) &= e^{2i\lambda\rho}\varphi_\rho(\lambda)[\varepsilon_\rho(\lambda)]^{-1} \\ &= F(\rho,\lambda)[E(\rho,\lambda)]^{-1} \end{aligned} \tag{10.12}$$

is inner and, by (10.11),

$$\cos \theta_\rho = 0 ,$$

i.e., $\mathcal{D}_+^\rho$ and $\mathcal{D}_-^\rho$ are orthogonal.

We look next at the common part of the orthogonal complements $(\mathcal{D}_\pm^\rho)^\perp$ of the indicated sets in $L_{2n}^2(\mathbb{R}^+)$.

THEOREM 10.3. *An element* f *in* $L_{2n}^2(\mathbb{R}^+)$ *belongs to*

$$(\mathcal{D}_+^\rho)^\perp \cap (\mathcal{D}_-^\rho)^\perp$$

if and only if

$$e^{i\lambda\rho}(\varphi_\infty^*)^{-1}f^\blacktriangle \in H_n^2 \cap S_\rho K_n^2 .$$

PROOF. In view of Theorem 5.2 and (10.9) with $t=0$, $f \in (\mathcal{D}_+^\rho)^\perp$ if and only if

$$\int [e^{i\lambda\rho}\ \varepsilon_\infty(\lambda)^* g(\lambda)]^* \Delta_\infty(\lambda) f^\blacktriangle(\lambda)d\lambda = 0$$

for every choice of $g \in H_n^2$. But this is clearly the same as to say that

$$f^\blacktriangle \in e^{i\lambda\rho}\ \varepsilon_\infty^*\ K_n^2 .$$

A similar argument shows that $f \in (\mathcal{D}_-^\rho)^\perp$ if and only if

$$f^\blacktriangle \in e^{-i\lambda\rho}\ \varphi_\infty^* H_n^2 .$$

The rest is plain. □

We now summarize the implications of the preceding analysis for potentials with compact support and connect up with de Branges spaces.

THEOREM 10.4. *If the potential*

$$V(r) = 0 \quad \text{for} \quad r > \rho ,$$

then $\mathcal{D}_+^\rho$ *and* $\mathcal{D}_-^\rho$ *are orthogonal in* $L_{2n}^2(\mathbb{R}^+)$,

$$S_\rho(\lambda) = F(\rho,\lambda)[E(\rho,\lambda)]^{-1} = \Sigma(\rho,\lambda)$$

and

$$(\mathcal{D}_+^\rho)^\perp \cap (\mathcal{D}_-^\rho)^\perp = \mathcal{B}_\rho ,$$

where $\mathcal{B}_\rho$ *is the de Branges space based on the de Branges pair* $\{F^\#(\rho,\lambda), E^\#(\rho,\lambda)\}$.

PROOF. The first two assertions follow from (10.11) and (10.12), while the last assertion is an easy consequence of Theorem 10.3 and the fact that, for potentials which vanish outside the interval $[0,\rho]$,

$$e^{i\lambda\rho}\varphi_\infty(\lambda) = F(\rho,\lambda) \quad \text{and} \quad e^{-i\lambda\rho}\varepsilon_\infty(\lambda) = E(\rho,\lambda) .$$

□

Although there is much more than can be said about the Lax-Phillips method, we shall stop here, apart from a couple of closing observations.

The first of these is simply to remark that if $U(t)$ [resp. $U_0(t)$] denotes the operator which maps $f \in L_{2n}^2(\mathbb{R}^+)$ into the solution $u(r,t)$ of (10.1) [resp. $u_0(r,t)$ of (10.1) with $V = 0$], then the wave operators

$$W_\pm f = \lim_{t\to\pm\infty} U(t)U_0(-t)f$$

exist and the corresponding scattering operator

$$(W_+^* W_- f)(r) = \lim_{R\uparrow\infty} \frac{1}{\pi} \int_{-R}^{R} X_0(r,\lambda)S(\lambda)f^\Delta(\lambda)d\lambda ,$$

with the very same S as above; see [Ad], [MaP1] and [MaP2] for additional information.

Finally we remark that $u(r,t)$ is a solution of (10.1) if and only if

$$w(r,t) = L^* u(r,t) ,$$

with L as in (3.16), is a solution of

$$\frac{\partial w}{\partial t} = J_0 w + bw \tag{10.13}$$

where

$$b = -L^* iVL$$

is skew symmetric a.e. on $\mathbb{R}^+$. Equation (10.13) serves to connect the case under discussion with a much more general system of first order partial differential equations which have been analyzed by Lax-Phillips [LP].

11. EMBEDDING AND DUAL EQUATIONS

In Theorem 3.2 we have shown how to associate a de Branges pair (in fact several such) with every $U \in \mathcal{E}(iJ)$. In this section we shall prove a converse statement. For simplicity's sake we shall work in the setting of the Wiener algebra. For more general results [K1] and [KOv] should be consulted.

Let us assume for the rest of this section that $\{F^{\#},E^{\#}\}$ is a de Branges pair of matrix valued entire functions such that for some $T > 0$, $e^{i\lambda T}E$ is an invertible element of W_+^I and $e^{-i\lambda T}F$ is an invertible element of W_-^I . Then there exists an $h \in L^1_{n\times n}(\mathbb{R})$ such that

$$[E^{\#}(\lambda)E(\lambda)]^{-1} = I_n - \int_{-\infty}^{\infty} e^{i\lambda s}\, h(s)ds$$

for every point $\lambda \in \mathbb{R}$. Correspondingly, define

$$Z(\lambda) = \begin{cases} I_n - 2\int_0^{\infty} e^{i\lambda s}\, h(s)ds\ , & \text{for } \lambda \in \bar{\mathbb{C}}_+ \\ 2[E^{\#}(\lambda)E(\lambda)]^{-1} - I_n + 2\int_{-\infty}^{0} e^{i\lambda s}h(s)ds & \text{for } \lambda \in \mathbb{C}_- . \end{cases}$$

Clearly Z is analytic in $\mathbb{C}_+$ and meromorphic in $\mathbb{C}_-$.

LEMMA 11.1. *The* $n \times n$ *matrix valued functions* EZ *and* $FZ^{\#}$ *are analytic in the whole complex plane.*

PROOF. Clearly EZ is analytic in $\mathbb{C}_+$ and continuous on $\bar{\mathbb{C}}_+$. At the same time, in $\mathbb{C}_-$

$$E(\lambda)Z(\lambda) = 2[E^{\#}(\lambda)]^{-1} - E(\lambda)\{I_n - 2\int_{-}^{0} e^{i\lambda s}h(s)ds\}\ ,$$

and the right-hand side is clearly analytic in $\mathbb{C}_-$ and continuous on $\bar{\mathbb{C}}_-$. Therefore, since the boundary values obtained from above and below match:

$$2E(\lambda)\{I_n - \hat{h}(\lambda)\} - 2[E^{\#}(\lambda)]^{-1} = 0\ , \quad \lambda \in \mathbb{R}\ ,$$

EZ is analytic in $\mathbb{C}$, as asserted, as is $FZ^{\#}$, by a similar argument. □

THEOREM 11.1. *The* $2n \times 2n$ *matrix valued function with blocks*

$$U_{11} = \frac{1}{2}(F+E) \ , \quad U_{12} = \frac{1}{2i}(EZ - FZ^{\#}) \ ,$$

$$U_{21} = \frac{1}{2i}(F-E) \ , \quad U_{22} = \frac{1}{2}(EZ + FZ^{\#})$$

belongs to the class $E(iJ)$.

PROOF. To begin with, Lemma 11.1 guarantees that the entries of U are entire functions. Next, the verification of (3.13) is a routine calculation which rests on the identities (3.18) and

$$Z^{\#} + Z = 2(E^{\#}E)^{-1} \ . \tag{11.1}$$

The proof of (3.14) sits a little deeper. Let us first note that it suffices to prove that U is iJ-contractive in $\mathbb{C}_+$:

$$U^*iJU \leq iJ \ , \tag{11.2}$$

because this guarantees that $U^{-1} = JU^{\#}J$ is iJ-expansive in $\mathbb{C}_+$, and hence that U is iJ-expansive in $\mathbb{C}_-$. Now U satisfies (11.2) if and only if

$$\Theta = L^*UL = \frac{1}{2}\begin{bmatrix} F(I_n + Z^{\#}) & F(I_n - Z^{\#}) \\ E(I_n - Z) & E(I_n + Z) \end{bmatrix} \tag{11.3}$$

is J_0-contractive in $\mathbb{C}_+$, where L and J_0 are defined by (3.16) and (3.17), respectively. Next, let

$$\Sigma = (P\Theta + Q)(P + Q\Theta)^{-1} \ ,$$

where $P = L^*P_-L$ and $Q = L^*P_+L$ are given in (6.1), and observe that the indicated inverse exists in $\bar{\mathbb{C}}_+$ because the factor is block triangular with invertible diagonal blocks, in $\bar{\mathbb{C}}_+$. It is readily checked that

$$I_{2n} - \Sigma^*\Sigma = (P + \Theta^*Q)^{-1}(J_0 - \Theta^*J_0\Theta)(P + Q\Theta)^{-1}$$

and hence that Σ is unitary on $\mathbb{R}$ and, moreover, that Θ is J_0-contractive in $\mathbb{C}_+$ if and only if Σ is contractive in $\mathbb{C}_+$. But now as the diagonal blocks of

$$\Sigma = \begin{bmatrix} \Theta_{11} - \Theta_{12}\,\Theta_{22}^{-1}\,\Theta_{21} & \Theta_{12}\,\Theta_{22}^{-1} \\ -\Theta_{22}^{-1}\,\Theta_{21} & \Theta_{22}^{-1} \end{bmatrix}$$

belong to $e^{i\lambda T}W_+^I$ and the off diagonal blocks belong to W_+^0, (and so, in particular, to the Hardy space $H_{n\times n}^\infty$), Σ admits a Poisson representation

$$\Sigma(\omega) = \frac{b}{\pi}\int \frac{\Sigma(\omega)}{|\lambda-\omega|^2}\,d\lambda$$

for $\omega = a+ib \in \mathbb{C}_+$. The asserted contractiveness of Σ in $\mathbb{C}_+$ follows from its unitarity on $\mathbb{R}$ by elementary estimates. □

We remark that the matrices Θ and

$$\Theta_1 = \begin{bmatrix} 0 & I \\ I & 0 \end{bmatrix} J_0\Theta^{\#} J_0 \begin{bmatrix} 0 & I \\ I & 0 \end{bmatrix} = \begin{bmatrix} \Theta_{22}^{\#} & -\Theta_{12}^{\#} \\ -\Theta_{21}^{\#} & \Theta_{11}^{\#} \end{bmatrix}$$

both belong to the class $E(iJ_0)$. The latter is simply related to the linear fractional transformation

$$\begin{aligned} T_{\Theta_1}[0] &= -\Theta_{12}^{\#}[\Theta_{11}^{\#}]^{-1} \\ &= \{R_U[I_n] - I_n\}\{R_U[I_n] + I_n\}^{-1} . \end{aligned}$$

This last formula serves to exhibit the equivalence of $R_U[I_n]$ to the perhaps more familiar formalism of Darlington synthesis; see [Ar] for additional information on the latter.

Note that Theorem 11.1 can be restated as follows: Given any pair of $n \times n$ matrix valued entire functions A, B such that $\{A^{\#} - iB^{\#}, A^{\#} + iB^{\#}\}$ is a de Branges pair with $e^{i\lambda T}(A - iB)$ invertible in W_+^I and $e^{-i\lambda T}(A + iB)$ invertible in W_-^I, there is a matrix valued function $U \in E(iJ)$ with first block column

$$\begin{bmatrix} U_{11} \\ U_{21} \end{bmatrix} = \begin{bmatrix} A \\ B \end{bmatrix} .$$

In general, there are many such "embeddings". However, if one insists that the matrix valued function

$$Z_U = (U_{11} - iU_{21})^{-1}\,(U_{22} + iU_{12})$$

belong to $\mathcal{C}$ (i.e., $Z_U \in W^I$ and $Z_U + Z_U^* > 0$ in $\bar{\mathbb{C}}_+$), then the embedding becomes unique. Indeed, for a general $U \in \mathcal{E}(iJ)$, $Z_U + Z_U^* > 0$ in $\bar{\mathbb{C}}_+$ and hence admits the integral representation

$$Z_U(\omega) = \alpha + \beta\omega + \frac{1}{\pi i}\int \left[\frac{1}{\lambda-\omega} - \frac{\lambda}{1+\lambda^2}\right] d\sigma(\lambda) ,$$

for $\omega \in \mathbb{C}_+$, where α and β are constant $n \times n$ matrices such that

$$\alpha+\alpha^* = 0 \quad \text{and} \quad i\beta \geq 0$$

and $\sigma(\lambda)$ is a Hermitian matrix-valued distribution function which is subject to the constraint

$$\int \frac{d\sigma(\lambda)}{1+\lambda^2} < \infty .$$

Since

$$\frac{Z_U(\lambda) + Z_U^{\#}(\lambda)}{2} = [E^{\#}(\lambda)E(\lambda)]^{-1} , \quad E = A - iB ,$$

for every choice of $\lambda \in \mathbb{R}$, the Stieltjes' inversion formula implies that

$$\begin{aligned} d\sigma(\lambda) &= [E^{\#}(\lambda)E(\lambda)]^{-1}d\lambda \\ &= [I_n - \hat{h}(\lambda)]d\lambda . \end{aligned}$$

By a simple computation, $Z_U \in W^I$ forces $\beta = 0$ and

$$\alpha = \int_0^\infty e^{-t}[h(t)^* - h(t)]dt$$

and hence Z_U is uniquely determined by the matrix function E, i.e. by the first block-column of U. Moreover, one can readily reexpress U in the form

$$U = \begin{bmatrix} (F+E)/2 & (EZ_U - FZ_U^{\#})/2i \\ (F-E)/2i & (EZ_U + FZ_U^{\#})/2 \end{bmatrix}$$

which proves the uniqueness of the embedding in this special class of U's. In particular, since the matrizant $U(t,\cdot)$ for the canonical equation (2.1) belongs to $\mathcal{E}(iJ)$ and

$$Z_{U(t,\lambda)} = Z_t(\lambda)$$

belongs to W^I, it follows that

$$U(t,\lambda) = \begin{bmatrix} (F(t,\lambda)+E(t,\lambda))/2 & (E(t,\lambda)\ Z_t(\lambda)-F(t,\lambda)\ Z_t^{\#}(\lambda))/2i \\ (F(t,\lambda)-E(t,\lambda))/2i & (E(t,\lambda)\ Z_t(\lambda)+F(t,\lambda)\ Z_t^{\#}(\lambda))/2 \end{bmatrix} \tag{11.4}$$

for every $t \in \mathbb{R}^+$ and $\lambda \in \mathbb{C}$.

This concludes our discussion of embedding. For a different approach which makes use of the Hilbert space of $2n \times 1$ vector valued entire functions $\mathcal{H}(U)$ with reproducing kernel based on

$$\frac{J - U(\lambda) J U(\omega)^*}{-2\pi(\lambda-\omega^*)}$$

for $U \in \mathcal{E}(J)$, see [dB2] and [dB3]. Some discrete analogues may also be found in [DeD].

We turn next to the "dual" canonical equation which is again of the form (2.1) but with potential

$$V^\dagger = -NVN$$

in place of V, where

$$N = \begin{bmatrix} 0 & I_n \\ I_n & 0 \end{bmatrix}.$$

Since $JN = -NJ$, it is readily checked that $V^\dagger$ is also of the form (2.2) as well as being summable together with V. Moreover, the matrizant for the dual equation

$$U^\dagger(t,\lambda) = NU(t,-\lambda)N = \begin{bmatrix} D(t,-\lambda) & B(t,-\lambda) \\ C(t,-\lambda) & A(t,-\lambda) \end{bmatrix}.$$

The analysis developed in the preceding sections is applicable to the dual equation just as well as to the original and leads to de Branges spaces, transforms, entropy formulas and so forth in terms of $D(t,-\lambda)$ and $C(t,-\lambda)$ in place of A and B. There are interesting connections between the two sets of formulas, but for the present, we shall content ourselves with calculating the spectral function $\Delta_\infty^\dagger(\lambda)$ of the dual equation in terms of the spectral function Δ_∞ of (2.1). By the proof of Lemma 5.1

$$\Delta_\infty^\dagger = \{(\varepsilon_\infty^\dagger)^{\#}\varepsilon_\infty^\dagger\}^{-1} = \{(\varphi_\infty^\dagger)^{\#}\varphi_\infty^\dagger\}^{-1}$$

where

$$\varepsilon_\infty^\dagger = \lim_{t\uparrow\infty} \varepsilon_t^\dagger \,, \quad \varphi_\infty^\dagger = \lim_{t\uparrow\infty} \varphi_t^\dagger \,,$$

$$\varepsilon_t^\dagger(\lambda) = e^{i\lambda t}[D(t,-\lambda) - iC(t,-\lambda)]$$

and

$$\varphi_t^\dagger(\lambda) = e^{-i\lambda t}[D(t,-\lambda) + iC(t,-\lambda)] \,.$$

But now, by the parametrization (11.4) of the matrizant $U(t,\lambda)$ of (2.1),

$$\varepsilon_t^\dagger(\lambda) = e^{i\lambda t}F(t,-\lambda)\, Z_t^\#(-\lambda) \,,$$

whence, for $\lambda \in \mathbb{R}$,

$$\begin{aligned}\Delta_\infty^\dagger(\lambda) &= \lim_{t\uparrow\infty} [Z_t(-\lambda)F^\#(t,-\lambda)F(t,-\lambda)\, Z_t^\#(-\lambda)]^{-1} \\ &= [Z_\infty^\#(-\lambda)]^{-1} \lim_{t\uparrow\infty} [F^\#(t,-\lambda)F(t,-\lambda)]^{-1}[Z_\infty(-\lambda)]^{-1} \\ &= [Z_\infty^\#(-\lambda)]^{-1}\, \Delta_\infty(-\lambda)[Z_\infty(-\lambda)]^{-1} \,.\end{aligned}$$

Therefore, since

$$\Delta_\infty(-\lambda) = \frac{Z_\infty(-\lambda) + Z_\infty^\#(-\lambda)}{2}$$

for $\lambda \in \mathbb{R}$, we finally obtain that

$$\Delta_\infty^\dagger(\lambda) = \frac{[Z_\infty(-\lambda)]^{-1} + [Z_\infty^\#(-\lambda)]^{-1}}{2} \,.$$

This shows, in particular, that the Z_∞-function for the dual equation is the inverse of the $Z_\infty^\#$-function for the original canonical equation, composed with the symmetry $\lambda \longrightarrow -\lambda$. That is, if $\Delta_\infty^\dagger = I_n - \hat{h}^\dagger$, with $h^\dagger \in L^1_{n\times n}(\mathbb{R})$, then

$$\begin{aligned}Z_\infty^\dagger(\lambda) &= I_n - 2\int_0^\infty e^{i\lambda s}\, h^\dagger(s)ds = [Z_\infty^\#(-\lambda)]^{-1} \\ &= [I_n - 2\int_0^\infty e^{i\lambda s}\, h(-s)ds]^{-1} \,.\end{aligned}$$

12. MODEL THEORY AND CHARACTERISTIC OPERATOR FUNCTIONS

Canonical equations also play a central role in the theory of triangular models for abstract Volterra operators with finite rank imaginary part. The objective of the present section is to utilize the machinery developed to this point to clarify this connection. Towards this end we shall first show that the backward shift in the de Branges space $\mathcal{B}_T$ generated by the canonical equation is unitarily equivalent to a triangular integral operator Y in $L^2_{2n}[0,T]$ based on the matrizant $U(s,\lambda)$, $0 \leqslant s \leqslant T$. We will further show that the imaginary part of H has finite rank equal to 2n and identify its Brodskiǐ-Livšič characteristic operator function in terms of the matrizant.

THEOREM 12.1. *The de Branges spaces* $\mathcal{B}_t$ *which are generated by the canonical equation* (2.1) *are closed under the mapping*

$$\varphi \longrightarrow \frac{\varphi(\lambda) - \varphi(0)}{\lambda} .$$

PROOF. Theorem 5.1 guarantees that every $\varphi \in \mathcal{B}_t$ can be expressed in the form

$$\varphi(\lambda) = \int_0^t X^{\#}(s,\lambda) f(s) ds$$

for some choice of $f \in L^2_{2n}(\mathbb{R}^+)$, and hence that,

$$\frac{\varphi(\lambda) - \varphi(0)}{\lambda} = \int_0^t \frac{X^{\#}(s,\lambda) - X^{\#}(s,0)}{\lambda} f(s) ds .$$

But now, upon setting $\mu = 0$ in (2.11) and multiplying through by $[I_n \; 0]$ on the left, it follows readily that

$$[0 \; I_n] - X^{\#}(t,\lambda) JU(t,0) = \lambda \int_0^t X^{\#}(s,\lambda) U(s,0) ds$$

and so too that

$$[0 \; I_n] - X^{\#}(t,0) JU(t,0) = 0 .$$

Therefore

$$X^{\#}(t,\lambda) - X^{\#}(t,0) = \lambda \int_0^t X^{\#}(s,\lambda) U(s,0) [U(t,0)]^{-1} J ds$$

which in turn implies that

$$\frac{\varphi(\lambda)-\varphi(0)}{\lambda} = \int_0^t \left\{ \int_0^s X^{\#}(y,\lambda)U(y,0)[U(s,0)]^{-1}Jdy \right\} f(s)ds$$

$$= \int_0^t X^{\#}(y,\lambda)\left\{U(y,0)\int_y^t [U(s,0)]^{-1}Jf(s)ds\right\}dy \ .$$

This completes the proof since the term in the curly brackets is readily seen to belong to $L^2_{2n}(\mathbb{R}^+)$ and so, by another application of Theorem 5.1, its transform belongs to $\mathcal{B}_t$, as asserted. □

Formula (12.1) exhibits the important fact that "left translation" in $\mathcal{B}_T$ is unitarily equivalent to a Volterra transformation in $L^2_{2n}[0,T]$. Thus, if Y denotes the Volterra operator in $L^2_{2n}[0,T]$ which is defined by the rule

$$(Yg)(t) = U(t,0)J^* \int_0^t U(s,0)^* g(s)ds \ , \tag{12.2}$$

then its adjoint (with respect to the standard inner product in $L^2_{2n}[0,T]$) is

$$(Y^*g)(t) = U(t,0)\ J \int_t^T U(s,0)^* g(s)ds \ , \tag{12.3}$$

and formula (12.1) is seen to be equivalent to the statement that

$$\frac{f^{\blacktriangle}(\lambda)-f^{\blacktriangle}(0)}{\lambda} = (Y^*f)^{\blacktriangle}(\lambda) \tag{12.4}$$

for any $f \in L^2_{2n}[0.T]$.

It is readily checked further that

$$\left\{\left(\frac{Y-Y^*}{2i}\right)f\right\}(t) = \frac{1}{2}\ U(t,0)iJ \int_0^T U(s,0)^*\ f(s)ds$$

and thus if M denotes the linear transformation which sends $\eta \in \mathbb{C}^{2n}$ into

$$(M\eta)(s) = \frac{1}{\sqrt{2}}\ U(s,0)\ \eta \ , \tag{12.5}$$

that

$$\frac{Y-Y^*}{2i}\ f = MiJM^*f \ ,$$

for every $f \in L^2_{2n}[0,T]$, and hence that $\text{Im}Y$ is a finite-rank operator.

Our next objective is to show that the matrizant of the

canonical equation under consideration is simply related to the Brodskiĭ-Livšic characteristic operator function

$$W_\Theta(\lambda) = I_{2n} - 2i\ iJM^*(Y-\lambda I)^{-1}M$$

$$= I_{2n} + 2JM^*(Y-\lambda I)^{-1}M ,$$

of the operator node

$$\Theta = \begin{pmatrix} Y & M & iJ \\ L^2_{2n}[0,T] & & \mathbb{C}^{2n} \end{pmatrix} ,$$

see Brodskiĭ [B] for more information on the latter.

THEOREM 12.2. *Let* Y *denote the Volterra operator* (12.2) *in* $L^2_{2n}[0,T]$. *Then its characteristic opeator function*

$$W_\Theta(\lambda) = [U(T,0)]^{-1}\ U(T,1/\lambda) . \tag{12.7}$$

PROOF. The main effort is to obtain the formula

$$\{(Y-\lambda I)^{-1}f\}(t) = -\lambda^{-1}f(t)+\lambda^{-2}U(t,0)v(t,\lambda)\int_0^t [v(s,\lambda)]^{-1}JU(s,)^*f(s)ds \tag{12.8}$$

in which

$$v(t,\lambda) = [U(t,0)]^{-1}U(t,1/\lambda) , \tag{12.9}$$

which can also be characterized as the solution of the equation

$$J\frac{d}{dt}v(t,\lambda) = \lambda^{-1}\ U(t,0)^*U(t,0)v(t,\lambda) \tag{12.10}$$

for $t \geqslant 0$, with boundary condition

$$v(0,\lambda) = I_{2n} .$$

To this end, let $(Y-\lambda I)^{-1}f = g$. Then

$$f(t) = (Y-\lambda I)g$$

$$= U(t,0)J^*\int_0^t U(s,0)^*g(s,\ )ds - \lambda g(t,\lambda) .$$

Next, following Brodskiĭ-Livšic [BL], let

$$y(t,\lambda) = J^*\int_0^t U(s,0)^*g(s,\lambda)ds$$

so that

$$Jy'(t,\lambda) = U(t,0)^*g(t,\lambda) = \lambda^{-1}U(t,0)^*\{U(t,0)y(t,\lambda)-f(t) .$$

This exhibits v as a solution of the homogeneous differential equation for y, and hence, by the method of variation of constants, we can search for a solution of the form

$$y(t,\lambda) = v(t,\lambda)z(t,\lambda)$$

with $z(0,\lambda) = 0$. Upon substituting into the general differential equation for y this leads rapidly to the conclusion that

$$z(t,\lambda) = \lambda^{-1} \int_0^t [v(s,\lambda)]^{-1} J\, U(s,0)^* f(s)\,ds$$

and hence that

$$\begin{aligned}\{(Y-\lambda I)^{-1}f\}(t) &= g(t,\lambda) \\ &= \lambda^{-1}\{U(t,0)v(t,\lambda)z(t,\lambda)-f(t)\}\end{aligned}$$

is indeed given by formula (12.8).

Now it follows readily from (12.5) that

$$M^*f = \frac{1}{\sqrt{2}} \int_0^T U(s,0)^* f(s)\,ds$$

and hence, that for any $\xi \in \mathbb{C}^{2n}$,

$$\lambda M^*(Y-\lambda I)^{-1}M\xi = -\frac{1}{2}\int_0^T U(t,0)^*U(t,0)\,\xi\,dt + \frac{\lambda^{-1}}{2}\int_0^T U(t,0)^*U(t,0) \times$$
$$\times \{v(t,\lambda)\int_0^t [v(s,\lambda)]^{-1}JU(s,0)^*U(s,0)\xi ds\}dt$$

which, upon changing the order of integration in the second integral on the right and utilizing (12.10), reduces to

$$\begin{aligned}\lambda M^*(Y-\lambda I)^{-1}M\xi &= \frac{1}{2}J\,v(T,\lambda)\int_0^T [v(s,\lambda)]^{-1} J\,U(s,0)^*\,U(s,0)\,\xi\,ds \\ &= \frac{1}{2}J\,v(T,\lambda)\int_0^T \lambda\frac{d}{ds}[v(s,\lambda)]^{-1}\,ds\,\xi \\ &= \frac{\lambda}{2}J\{I - v(T,\lambda)\}\,\xi\ .\end{aligned}$$

The rest is immediate from (12.6) and (12.9). □

Finally we remark that every simple Volterra operator with imaginary part of finite rank is unitarily equivalent to a triangular integral operator of the form considered above, but based on a wider class of canonical equations than was considered here. This is a major topic in itself which is certainly beyond the scope of this paper and is in any event well treated in the literature; see [B], [BL] and [GK2] for good introductions. For a different approach in which the general theme is to model the Volterra transformation by the difference quotient in a reproducing kernel Hilbert space of entire functions based on the characteristic function of the given operator, see [dB3], and for a number of related issues, [dBR] and [dB4].

REFERENCES

[Ad] Adamjan, V.M.: Canonical differential operators in Hilbert space, Dokl. Akad. Nauk SSSR 178 (1968), 9-12. [English Trans.: Soviet Math. Dokl. 9 (1968), 1-5].

[AM] Agranovich, Z.S. and V.A. Marchenko: The Inverse Problem of Scattering Theory, Gordon & Breach, New York, 1963.

[Ar] Arov, D.Z.: The realization of a canonical system with dissipative boundary conditions at one end of a segment in terms of the coefficient of dynamic flexibility, Sib. Math. Zh., 16 (1975), 440-463. [English Trans.: Siberian Mat. J., 16 (1975), 335-352].

[ArK] Arov, D.Z. and M.G. Krein: Problem of search of the minimum of entropy in indeterminate extension problems, Funkts. Anal. Prilozhen., 15 (1981), 61-64. [English Trans.: Functional Anal. Appl., 15 (1981), 123-126.

[At] Atkinson, F.V.: Discrete and Continuous Boundary Problems, Academic Press, New York, 1964.

[dB1] de Branges, L.: Hilbert Spaces of Entire Functions, Prentice Hall, Englewood Cliffs, N.J., 1968.

[dB2] de Branges, L.: The expansion theorem for Hilbert spaces of entire functions, in: Entire Functions and Related Parts of Analysis, Proc. Symp. Pure Math., Vol. 11, Amer. Math. Soc., Providence, R.I., 1968, 79-148.

[dB3] de Branges, L.: The comparison theorem for Hilbert spaces of entire functions, Integral Equations and Operator Theory (1983), 603-646.

[dB4] de Branges, L.: Square Summable Power Series, Addison Wesley, in preparation.

[dBR] de Branges, L. and J. Rovnyak: Canonical models in quantum scattering theory, in: Perturbation Theory and its Application in Quantum Mechanics, John Wiley, New York, 1966, 295-392.

[B] Brodskiĭ, M.S.: Triangular and Jordan Representations of Linear Operators, Trans. Math. Monographs, Vol.32, Amer. Math. Soc., Providence, R.I., 1971.

[BL] Brodskiĭ, M.S. and M.S. Livšič: Spectral analysis of non-selfadjoint operators and intermediate systems, Uspekhi Mat. Nauk 13 (1958), 3-85. [English Trans.: Amer. Math. Soc. Translations (2), 13 (1960),265-346].

[DaK] Daleckiĭ Ju.L. and M.G. Krein: Stability of Solutions of Differential Equations in Banach Space, Trans. Math. Monographs, Vol.43, Amer. Math. Soc., Providence, R.I., 1974.

[DeD] Dewilde, P. and H. Dym: Lossless chain scattering matrices and optimum linear prediction: The vector case, Circuit Theory and Appl., 9 (1981), 135-175.

[D] Dym, H.: Applications of factorization theory to the inverse spectral problem, in: Proc. Third Int. Symp. on Mathematical Theory of Networks and Systems, Delft, 1979, Western Periodicals, N. Hollywood, California, 1979, 188-193.

[DG1] Dym, H. and I. Gohberg: On an extension problem, generalized Fourier analysis, and an entropy formula, Integral Equations Operator Theory 3 (1980),143-215.

[DG2] Dym, H. and I. Gohberg: Extensions of triangular operators and matrix functions, Indiana Univ. Math. J., 31 (1982), 579-606.

[DG3] Dym, H. and I. Gohberg: Extensions of kernels of Fredholm operators, J. Analyse Math., in press.

[DI] Dym, H. and A. Iacob: Applications of factorization and Toeplitz operators to inverse problems, in: Toeplitz Centennial, Operator Theory Advances and Applications, Vol. 4, Birkhäuser, Basel, 1982, 233-260.

[DK] Dym, H. and N. Kravitsky: On recovering the mass distribution of a string from its spectral function, Adv. Math. Suppl. 3 (1978), 45-90.

[Ga] Gasymov, M.G.: The inverse scattering problem for a system of Dirac equations of order 2n, Trudy Mosk. Mat. Obsc., 19 (1968), 41-112. [English Trans.: Trans. Moscow Math. Soc., 19 (1968), 41-119].

[GaL] Gasymov, M.G. and Levitan, B.M.: Determination of the Dirac system from the scattering phase, Dokl. Acad. Nauk SSSR 167 (1966), 1219-1222. [English Trans.: Soviet Math. Dokl. 7 (1966), 543-547].

[GK1] Gohberg, I.C. and M. G. Krein: Systems of integral equations on a half line with kernel depending on the difference of arguments, Uspekhi Mat. Nauk 13 (1958), 3-72. [English Trans.: Amer. Math. Soc. Transl.(2) 14 (1960), 217-287].

[GK2] Gohberg, I.C. and M.G. Krein: Theory and Applications of Volterra operators in Hilbert space, Transl. Math. Monographs, Vol. 24, Amer. Math. Soc., Providence, R.I., 1970.

[HS] Hinton, D. and K. Shaw: Titchmarsh-Weyl theory for Hamiltonian systems, in: Spectral Theory of Differential Operators, North Holland, Amsterdam, 1981, 219-231.

[KMV] Kailath, T., A. Vieira and M. Morf: Inverses of Toeplitz operators, innovations and orthogonal polynomials, SIAM Review 20 (1978), 106-119.

[KP] Kovalishna, I.V. and V.P. Potapov, Integral Representation of Hermitian Positive Functions, Private Transl. by T. Ando, Sapporo, Japan, 1982.

[K1] Krein, M.G.: On an indeterminate case of the Sturm-Liouville boundary value problem in the interval $(0,\infty)$, Izv. Akad. Nauk SSSR, Ser. Mat., 16 (1952), 293-324.

[K2] Krein, M.G.: On integral equations that generate differential equations of second order, Dokl. Akad. Nauk SSSR 97 (1954), 21-24.

[K3] Krein, M.G.: Continuous analogs of theorems on polynomials orthogonal on the unit circle, Dokl. Akad. Nauk SSSR 105 (1955), 637-640.

[K4] Krein, M.G.: On the determination of a potential of a particle from its S-function, Dokl. Akad. Nauk SSSR 105 (1955), 433-436.

[K5] Krein, M.G.: A contribution to the theory of accelerants and S-matrices of canonical differential systems, Dokl. Akad. Nauk SSSR 111 (1956), 1167-1170.

[KL] Krein, M.G. and H.K. Langer: Continuous analogues of orthogonal polynomials with respect to an indefinite weight on the unit circle, and extension problems associated with them, Dokl. Akad. Nauk SSSR 258 (1981), 537-541. [English Trans.: Soviet Math. Dokl. 23 (1981), 553-557].

[KMaF1] Krein, M.G. and F.E. Melik-Adamyan: A contribution to the theory of S-matrices of canonical differential equations with summable potential, Dokl. Akad. Nauk Armyan, SSSR 46 (1968), 150-155. [English Trans.: in: M.G. Krein, Topics in Differential and Integral Equations and Operator Theory, Operator Theory Advances and Applications, Vol. 7, Birkhäuser, Basel, 1983, 295-302].

[KMaF2] Krein, M.G. and F.E. Melik-Adamyan: Some applications of the theorem on factorization of a unitary matrix, Funkts. Anal. Prilozhen., 4 (1970), 73-75. [English Trans.: Functional Anal. Appl. 4 (1970), 327-329.

[KOv] Krein, M.G. and I.E. Ovcharenko: On the theory of inverse problems for canonical differential equations, Dokl. Akad. Nauk Ukr. SSR, Ser. A (1982), 14-18.

[L] Langer, H.: Über die Methode der richtenden Funktionale von M.G. Krein, Acta Math. Acad. Scient. Hungaricae 21 (1970), 207-224.

[LP] Lax, P.D. and R.S. Phillips: Scattering theory, Rocky Mountain J., 1 (1971) 173-223.

[LT] Levy, B.C. and J.N. Tsitsiklis, Vibrating strings and the recursive linear estimation of stationary stochastic processes, Preprint (1981).

[M] Marchenko, V.A.: Sturm-Liouville Operators [in Russian] Naukova Dumka, Kiev, 1977.

[MaF1] Melik-Adamyan, F.E.: On the theory of matrix accelerants and spectral matrix functions for canonical differential systems, Dokl. Akad. Nauk Armyan SSR 45 (1967), 145-151.

[MaF2] Melik-Adamyan, F.E.: On canonical differential operators in Hilbert space, Izv. Akad. Nauk Armyan SSR, Ser. Mat. 12 (1977), 10-31.

[MaP1] Melik-Adamyan, P.E.: On the properties of the S-matrix of canonical differential equations on the full axis, Dokl. Akad. Nauk Armyan. SSR 58 (1974), 199-205.

[MaP2] Melik-Adamyan, P.E.: On scattering theory for canonical differential operators, Izv. Akad. Nauk Armyan. SSR, 11 (1976), 291-313.

[O] Orlov, S.A.: Nested matrix disks analytically depending on a parameter, and theorems on the invariance of ranks of radii of limiting disks, Izv. Akad. Nauk SSSR Ser. Mat. 40 (1976), 593-644. [English Trans.: Math. USSR Izv. 10 (1976), 565-613].

[T] Titchmarsh, E.C.: The Theory of Functions (2nd edition), Oxford University Press, London, 1960.

[Y] Youla, D.C.: The deterministic multichannel spectral estimation problem-solution, Preprint.

[Z1] Faddeyev, L.D.: The inverse problem in the quantum theory of scattering, J. Math. Phys. 4 (1963), 72-104.

[Z2] Faddeev, L.D.: Inverse problem of quantum scattering theory, II, J. Soviet Math. 5 (1976), 334-396.

Department of Theoretical Mathematics
The Weizmann Institute of Science
Rehovot 76100, ISRAEL

Operator Theory:
Advances and Applications, Vol. 12

MINIMAL DIVISORS OF RATIONAL MATRIX FUNCTIONS WITH PRESCRIBED ZERO AND POLE STRUCTURE

I. Gohberg, M.A. Kaashoek, L. Lerer and L. Rodman

Necessary and sufficient conditions are given in order that a rational matrix function is a minimal divisor of another one. These conditions are expressed in terms of zero and pole structure of the given functions. In connection with this a description is obtained of all rational matrix functions with prescribed zero and pole data.

INTRODUCTION

1. This paper concerns minimal divisors of rational matrix functions. This concept has its origin in network and systems theory where it appears naturally in problems of analysis and synthesis of cascade connections.

To introduce minimal divisors and minimal factorization we use the following local version of the Smith-McMillan form. Let $W(\lambda)$ be a $n \times n$ rational matrix function which is regular, i.e., $\det W(\lambda)$ does not vanish identically. For each given point $\lambda_0 \in \mathbb{C}$ such a function can be represented as

$$(0.1) \qquad W(\lambda) = E_{\lambda_0}(\lambda) D_{\lambda_0}(\lambda) F_{\lambda_0}(\lambda) ,$$

where $E_{\lambda_0}(\lambda)$ and $F_{\lambda_0}(\lambda)$ have no poles and are invertible at the point λ_0, the middle term $D_{\lambda_0}(\lambda)$ is a diagonal matrix function of the form

$$(0.2) \qquad D_{\lambda_0}(\lambda) = \operatorname{diag}((\lambda-\lambda_0)^{\nu_i(\lambda_0)})_{i=1}^{n} ,$$

and

$$(0.3) \qquad \nu_1(\lambda_0) \geq \cdots \geq \nu_r(\lambda_0) > 0 \geq \nu_{r+1}(\lambda_0) \geq \cdots \geq \nu_n(\lambda_0)$$

are integers. The matrix $D_{\lambda_0}(\lambda)$, which is uniquely defined by the function W and the point λ_0, is called the *Smith-McMillan form* of W at λ_0. The integers

$$z(W;\lambda_0) = \sum_{j=1}^{r} \nu_j(\lambda_0) , \quad p(W;\lambda_0) = -\sum_{j=r+1}^{n} \nu_j(\lambda_0)$$

are referred to as the *zero* and the *pole multiplicity* of W at λ_0, respectively. Obviously, $z(W;\lambda_0) = p(W^{-1};\lambda_0)$. The number $p(W;\lambda_0)$ can also be described as the largest multiplicity of λ_0 as a pole of any minor of W.

The zero and pole multiplicities enjoy a sublogarithmic property. Namely, if W is factored as

(0.4) $$W(\lambda) = W_1(\lambda)W_2(\lambda) ,$$

where W_1 and W_2 are $n \times n$ rational matrix functions, then

(0.5) $$z(W;\lambda_0) \leq z(W_1;\lambda_0) + z(W_2;\lambda_0) ; \quad p(W;\lambda_0) \leq p(W_1;\lambda_0) + p(W_2;\lambda_0) .$$

Note that

(0.6) $$z(W;\lambda_0) = z(W_1;\lambda_0) + z(W_2;\lambda_0)$$

if and only if

(0.7) $$p(W;\lambda_0) = p(W_1;\lambda_0) + p(W_2;\lambda_0) ,$$

and in this case the factorization (0.4) is said to be *minimal at* λ_0 and W_2 is called a *(right) minimal divisor of* W *at* λ_0. If (0.6) (or, equivalently, (0.7)) holds for each $\lambda_0 \in \mathbb{C}$, one says that the factorization (0.4) is (globally) *minimal* and W_2 is a (global) *minimal divisor* of W. Roughly speaking, the minimality of the factorization (0.4) at λ_0 means that there is no pole-zero cancellation at this point.

Minimal factorizations (mainly in the global framework) have been studied recently by various authors (see [8], section 8.3 and the references there). We mention also the monograph [1] in which a geometric approach is developed to deal with factorization problems. In the present paper, in which we look for a description of minimal divisors, the emphasis is on divisors and not so much on factorizations. Roughly speaking the following problem is solved: Given two rational matrix functions $W(\lambda)$ and $W_2(\lambda)$, find necessary and sufficient conditions (stated in terms of W and W_2 only) in order that W_2 is a minimal divisor of W.

One of the important characteristics of a rational matrix function is its root functions. Recall that a vector function $\varphi(\lambda)$, which is analytic at λ_0, is called a *right root function of rank* α of W at λ_0, if $\varphi(\lambda_0) \neq 0$ and for λ in some neighbourhood of λ_0

$$W(\lambda)\varphi(\lambda) = \sum_{j=\alpha}^{\infty} (\lambda-\lambda_0)^j \psi_j , \quad \psi_\alpha \neq 0 .$$

Similarly one defines left root functions. The relation of the root functions to the local Smith-McMillan form is reflected, among other things, by the fact that the rank of a root function may take the values $\nu_i(\lambda_0)$ $(i=1,2,\cdots,r)$ only and, conversely, for each integer $\nu_i(\lambda_0)$ there is a root function with rank $\nu_i(\lambda_0)$. If $\varphi(\lambda) = \sum_{j=0}^{\infty} (\lambda-\lambda_0)^j \varphi_j$ is a right root function of order α, then the chain $\varphi_0, \varphi_1, \cdots, \varphi_{\beta-1}$ $(\beta \leq \alpha)$ is called a *right Jordan chain* (of length β) of W at λ_0. Similarly one defines *left Jordan chains* using left root functions.

We prove in this paper that if W_2 is a minimal divisor of W at λ_0 then each right Jordan chain of W_2 at λ_0 is also a Jordan chain of W at λ_0. The same is true for the left Jordan chains of W_2^{-1} and W^{-1} $(=W_2^{-1}W_1^{-1})$. Therefore these properties are necessary for W_2 to be a minimal divisor of W at λ_0. Simple examples (see Example 2.2) show that they are not sufficient. The additional conditions which one needs for sufficiency (as well as for necessity) are given and analyzed in Chapters 2 and 3 for global and local minimal divisors, respectively. In the first chapter we answer the following question which arises naturally in this context: To what extent is the function W_2 determined by its right Jordan chains and the left Jordan chains of W_2^{-1}? In other words, we describe all rational matrix functions W which have a prescribed system of right Jordan chains for W and left Jordan chains for W^{-1}. Here we assume that the rational matrix functions W and W_2 are analytic at infinity and take value I there. The general case of regular rational matrix functions can be reduced to this case by a Möbius transformation.

2. The solutions of the problems considered in this paper are stated mainly in the language of zero pairs of the given matrix functions and their inverses. In order to explain these notions and their connections with the root functions and Jordan chains we first recall that any $n \times n$ rational matrix function W which takes value I at infinity can be represented as

$$(0.8) \qquad W(\lambda) = I + C(\lambda I - A)^{-1}B ,$$

where C, A and B are matrices of sizes $n \times p$, $p \times p$ and $p \times n$, respectively. In this case the inverse matrix function $W(\lambda)^{-1}$ can be written as follows

$$W(\lambda)^{-1} = I - C(\lambda I - A^{\times})^{-1}B \,, \tag{0.9}$$

where $A^{\times} = A - BC$.

The right hand side of (0.8) is called a *realization* of $W(\lambda)$. Realizations for which the size $m \times m$ of A is as small as possible are said to be *minimal* and the corresponding number m is called the *Mc-Millan degree* of W. This number can be expressed in terms of zero and pole multiplicities:

$$m = \sum_{\lambda \in \mathbb{C}} z(W;\lambda) = \sum_{\lambda \in \mathbb{C}} p(W;\lambda) \,.$$

Now let (0.8) be a minimal realization of W. Assume that λ_0 is an eigenvalue of the matrix $A^{\times}$ and let

$$x_{i0}, x_{i1}, \cdots, x_{i,\nu_i-1} \quad (i=1,2,\cdots,r_0) \tag{0.10}$$

be the Jordan chains of $A^{\times}$ which form the part of a Jordan basis of $A^{\times}$ corresponding to the eigenvalue λ_0. The number r_0 is equal to the integer r in (0.3) and the lengths ν_i of these chains coincide with the integers $\nu_i(\lambda_0)$ in (0.1)-(0.3). Furthermore the right root functions of W at λ_0 can be described in the following way: Put $\varphi_{ij} = Cx_{ij}$ $(i=1,\cdots,r_0;$ $j=0,1,\cdots,\nu_i-1)$ and let $q = |\nu_n(\lambda_0)|$. For each $i = 1,\cdots,r$ and a suitable choice of the vectors $\varphi_{i,\nu_i},\cdots,\varphi_{i,\nu_i+q-1}$ the vector function

$$\varphi_i(\lambda) = \sum_{j=0}^{\nu_i+q-1} (\lambda-\lambda_0)^j \varphi_{ij}$$

is a right root function of rank ν_i of W at λ_0. In particular, the vectors $\varphi_{i0},\cdots,\varphi_{i,\nu_i-1}$ $(i=1,\cdots,r_0)$ form right Jordan chains of W at λ_0. If $\nu_{i_0-1} > \nu_{i_0} = \nu_{i_0+1} = \cdots = \nu_{i_0+k-1} > \nu_{i_0+k}$, then all root functions of rank ν_{i_0} are linear combinations of $\varphi_i(\lambda)$ $(i=i_0,i_0+1,\cdots,i_0+k-1)$ modulo vector functions with a zero of order $\geq q + \nu_{i_0}$ at λ_0.

It is clear now that the information about right Jordan chains of W at λ_0 is contained in the pair of matrices $(C|_{\lambda_0}, A^{\times}|_{\lambda_0})$, where $A^{\times}|_{\lambda_0}$

denotes the restriction of the matrix $A^{\times}$ to the subspace spanned by the vectors x_{ij} $(i=1,\cdots,r_0;\ j=0,1,\cdots,\nu_i-1)$ and the matrix $C_{|\lambda_0}$ is formed by the vectors $\varphi_{ij} = Cx_{ij}$ as follows

$$C_{|\lambda_0} = [\varphi_{10}\cdots\varphi_{1,\nu_1-1}\ \ \varphi_{20}\cdots\varphi_{2,\nu_2-1}\cdots\varphi_{r_0 0}\cdots\varphi_{r_0,\nu_{r_0}-1}] .$$

The pair $(C_{|\lambda_0}, A^{\times}{}_{|\lambda_0})$ will be referred to as a *right zero pair of* W *at* λ_0.

Analogously, starting with left Jordan chains y_{ij} $(i=1,\cdots,k_0;\ j=0,1,\cdots,\mu_i-1)$ of A corresponding to an eigenvalue λ_0 of A we obtain a similar description of the left root functions and left Jordan chains of W^{-1} at λ_0 via the (row) vectors $\psi_{ij} = y_{ij}B$. Denote by ${}_{\lambda_0|}A$ the restriction of A to the (left) invariant subspace N_{λ_0} of A spanned by the vectors y_{ij} $(i=1,\cdots,k_0;\ j=0,1,\cdots,\mu_i-1)$, and put

$${}_{\lambda_0|}B = [\psi_{10}^T\cdots\psi_{1,\mu_1-1}^T\ \ \psi_{20}^T\cdots\psi_{2,\mu_2-1}^T\cdots\psi_{k_0 0}^T\cdots\psi_{k_0,\mu_{k_0}-1}^T]^T ,$$

where the superscript T stands for transpose. The pair of matrices $({}_{\lambda_0|}A, {}_{\lambda_0|}B)$ will be called a *left zero pair* of W^{-1} at λ_0.

In this way the problem of describing all rational matrix functions W with prescribed systems of right Jordan chains for W and left Jordan chains for W^{-1} is transformed into the problem of producing all rational matrix functions with given right zero pair of W and left zero pair of W^{-1}. Note that in [3] another inverse problem is solved. There a full description is given of all rational matrix functions with prescribed right zero pairs for both functions W and W^{-1}.

The property that at the point λ_0 the right Jordan chains of W_2 are right Jordan chains of W means that there is a minimal realization (0.8) of W such that

$$C = [\ C_2 \quad * \ ] , \quad A^{\times} = \begin{bmatrix} A_2^{\times} & * \\ 0 & * \end{bmatrix} ,$$

where $W_2(\lambda) = I + C_2(\lambda I - A_2)^{-1}B_2$ is a minimal realization of $W_2(\lambda)$ and $A_2^{\times} = A_2 - B_2C_2$. Similarly, the property that the left Jordan chains of W_2^{-1} at λ_0 are left Jordan chains of W^{-1} at λ_0 means that there is a minimal realization of W (possible different from the previous one) such that

$$A = \begin{bmatrix} * & * \\ 0 & A_2 \end{bmatrix}, \quad B = \begin{bmatrix} * \\ B_2 \end{bmatrix}.$$

In Chapters 2 and 3 the final solutions of our problems are stated in this terminology. The problem of minimal divisors was studied in [10] as the synthesis problem for cascade connections and there for the global version the solution is stated in terms of certain spaces generated by Toeplitz matrices whose entries are the coefficients of the Laurent expansions of W and W_2 at each pole and zero.

I. RATIONAL MATRIX FUNCTIONS WITH PRESCRIBED ZERO AND POLE PAIRS

In this chapter we solve the following problem: Find a rational matrix function (in short, r.m.f.) with value I at infinity when its right zero pair and a left zero pair of its inverse are known. As we shall see, a solution of this problem does not exist always; and when it exists the solution may not be unique. In general, the solution of this problem depends on the existence of invertible solutions of a certain Lyapunov equation.

I.1 Zero Pairs and Pole Pairs

In this section we shall present the basic definitions and facts on zero pairs and pole pairs of r.m.f.'s, which will be frequently used in the sequel. Consider an $n \times n$ r.m.f. $W(\lambda)$ with $W(\infty) = I$, and let

$$(1.1) \qquad W(\lambda) = I + C(\lambda - A)^{-1}B$$

be one of its minimal realizations. Let $(C_{|\lambda_0}, A^{\times}{}_{|\lambda_0})$ be the (right) zero pair of $W(\lambda)$ at $\lambda_0 \in \mathbb{C}$ as defined in the introduction, and let M_{λ_0} be the root subspace of $A^{\times}$ corresponding to λ_0, i.e., M_{λ_0} is the subspace spanned by the vectors x_{ij} $(i=1,\cdots,r;\ j=0,1,\cdots,\nu_i-1)$ in the notations of the introduction (see (0.10)). Put $p = \dim M_{\lambda_0}$. Any pair of matrices (X,T), where the sizes of X and T are $n \times p$ and $p \times p$, respectively, which is *similar* to $(C_{|\lambda_0}, A^{\times}{}_{|\lambda_0})$ (i.e., there exists an invertible linear transformation $S : M_{\lambda_0} \to \mathbb{C}^p$ such that $XS = C_{|\lambda_0}$ and $S^{-1}TS = A^{\times}{}_{|\lambda_0}$) is also called a (right) *zero pair* of $W(\lambda)$ at λ_0. It is easily seen that a zero pair of $W(\lambda)$ at λ_0 does not depend on the choice of the minimal realization (1.1). This follows from the well known fact that a minimal realization of $W(\lambda)$ is unique up to similarity, i.e., if $W(\lambda) = I + \widetilde{C}(\lambda-\widetilde{A})^{-1}\widetilde{B}$ is another minimal realization, then $\tilde{C}F = C$, $F^{-1}\tilde{A}F = A$, $F^{-1}\tilde{B} = B$ for some invertible matrix F. Note that a zero pair (X,T) of $W(\lambda)$ at λ_0 is non-trivial if and only if the zero multiplicity of W at λ_0 is positive; in that case λ_0 is called a *zero* (or *eigenvalue*) of W, and the size p of the matrix T coincides with the zero multiplicity of W at λ_0. Observe that a zero pair (X,T) of $W(\lambda)$ at λ_0 is *minimal*, i.e., $\bigcap_{i=0}^{p-1} \mathrm{Ker}(XT^i) = (0)$, where $p \times p$ is the size of T. This property follows from the minimality of the realization (1.1).

Analogously, let $({}_{\lambda_0|}A, {}_{\lambda_0|}B)$ be the (*right*) *pole pair* of W at λ_0, which by definition is the left zero pair of W^{-1} at λ_0 as defined in the introduction, and let $q \times q$ be the size of ${}_{\lambda_0|}A$. Any pair of matrices (U,Y), where the sizes of U and Y are $q \times q$ and $q \times n$, respectively, which is *similar* to $({}_{\lambda_0|}A, {}_{\lambda_0|}B)$ (i.e. $U = S^{-1}\,{}_{\lambda_0|}AS$, $Y = S^{-1}\,{}_{\lambda_0|}B$ for some invertible linear transformation $S : \mathbb{C}^q \to N_{\lambda_0}$) is also called a (*right*) *pole pair* of W at λ_0. Again this definition does not depend on the choice of the minimal realization (1.1). A pole pair (U,Y) of W at λ_0 is non-trivial if and only if λ_0 is a *pole* of $W(\lambda)$, i.e., the pole multiplicity of W at λ_0 is positive; in that case the size of U coincides with the pole multiplicity of W at λ_0. A pole pair (U,Y) of W at λ_0 is *minimal*, which means that $\sum_{i=0}^{q-1} \mathrm{Im}(U^iY) = \mathbb{C}^q$, where $q \times q$ is the size of U.

The zero and pole pairs can be described also in terms of the singular parts of the Laurent expansions of W and W^{-1} at λ_0, as follows. A pair of matrices (X,T) will be called (*right*) *admissible* if the sizes of X and T are $n \times m$ and $m \times m$, respectively (m may depend on X,T, but n is fixed). It is not difficult to see that an admissible pair (X,T) is a zero pair of $W(\lambda)$ at λ_0 if and only if the following conditions hold:

i) λ_0 is the only eigenvalue of T;

ii) the pair (X,T) is minimal;

iii) for some $m \times n$ matrix Y the function $W(\lambda)^{-1} - X(\lambda-T)^{-1}Y$ is analytic at λ_0.

Note that i) and iii) together imply that $X(\lambda-T)^{-1}Y$ is the singular part of the Laurent expansion of $W(\lambda)^{-1}$ in a neighbourhood of λ_0. Analogously, a *left admissible* pair (U,Y) (which means that U (resp. Y) is an $m \times m$ (resp. $m \times n$) matrix) is a pole pair of $W(\lambda)$ at λ_0 if and only if the following holds:

i') λ_0 is the only eigenvalue of U;

ii') the pair (U,Y) is minimal;

iii') for some $n \times m$ matrix X the function $W(\lambda) - X(\lambda-U)^{-1}Y$ is analytic at λ_0.

This description allows us to define a zero (resp. pole) pair of an arbitrary regular r.m.f. $W(\lambda)$ at $\lambda_0 \in \mathbb{C}$ as a right (resp. left) admissible pair (X,T) (resp. (U,Y)) for which i)-iii) (resp. i')-iii')) is satisfied. Further, it follows that (U,Y) is a pole pair of a regular r.m.f. $W(\lambda)$ at λ_0 if and only if (Y^*,U^*) is a zero pair of the r.m.f. $(W(\bar{\lambda}))^{*-1}$ at $\bar{\lambda}_0$.

We define now zero and pole pairs for an r.m.f. with respect to the whole complex plane. Let $W(\lambda)$ be an $n \times n$ regular r.m.f., and let $\lambda_1,\cdots,\lambda_s$ be all the different zeros of $W(\lambda)$ in $\mathbb{C}$. An admissible pair (X,T) which is similar to the admissible pair

$$([X_1\ X_2\ \cdots\ X_s],\operatorname{diag}(T_1,\cdots,T_s)) ,$$

where (X_i,T_i) is a zero pair of $W(\lambda)$ at λ_i, $i = 1,\cdots,s$, is called a *zero pair* of $W(\lambda)$. Note that a zero pair of $W(\lambda)$ is minimal and it is unique up to similarity. Assuming $W(\infty) = I$, an admissible pair (X,T) is a zero pair of $W(\lambda)$ if and only if for some matrix Y the function $W(\lambda)^{-1}$ has a minimal realization $W(\lambda)^{-1} = I + X(\lambda-T)^{-1}Y$.

We pass now to the pole pairs. Let $\mu_1,\cdots,\mu_t$ be all the different poles in $\mathbb{C}$ of the regular r.m.f. $W(\lambda)$. A left admissible pair (U,Y) is called a *pole pair* of $W(\lambda)$ if it is similar to the left admissible pair

$$(\operatorname{diag}(U_1,\cdots,U_t) , \begin{bmatrix} Y_1 \\ \vdots \\ Y_t \end{bmatrix}) ,$$

where (U_i,Y_i) is a pole pair of $W(\lambda)$ at μ_i, $i = 1,\cdots,t$. A pole pair of $W(\lambda)$ is a minimal left admissible pair, and it is unique up to similarity. Assuming $W(\infty) = I$, a left admissible pair (U,Y) is a pole pair of $W(\lambda)$ if and only if for some matrix X the function $W(\lambda)$ has a minimal realization $W(\lambda) = I + X(\lambda-U)^{-1}Y$.

I.2 Main Results

In the following theorem we give a description of all r.m.f.'s with I at infinity for which a zero pair and a pole pair are prescribed in advance.

THEOREM 1.1 *Let* (C,A_z) *and* (A_p,B) *be a minimal admissible pair and a left minimal admissible pair, respectively. Then* (C,A_z) *is a zero pair and* (A_p,B) *is a pole pair of some* $n \times n$ *rational matrix function* $W(\lambda)$ *with* $W(\infty) = I$ *if and only if there exists an invertible matrix* S *such that*

(1.2) $$SA_z - A_pS = -BC .$$

In this case $W(\lambda)$ *is given by the formula*

(1.3) $$W(\lambda) = I + CS^{-1}(\lambda-A_p)^{-1}B .$$

There is one-to-one correspondence between invertible matrices S *for which* (1.2) *holds and the rational matrix functions* $W(\lambda)$ *for which* (C,A_z) *is a zero pair and* (A_p,B) *is a pole pair.*

Proof. Assume (C,A_z) and (A_p,B) are zero and pole pairs for some $W(\lambda)$, respectively. As observed in Section I.1, there exist unique matrices $\widetilde{C}$ and $\widetilde{B}$ such that

(1.4) $$W(\lambda) = I + \widetilde{C}(\lambda-A_p)^{-1}B , \quad W(\lambda)^{-1} = I + C(\lambda-A_z)^{-1}(-\widetilde{B})$$

are minimal realizations. The first equality (1.4) gives the minimal realization $(\widetilde{C},A_p-B\widetilde{C},-B)$ for $W(\lambda)^{-1}$, and since any two minimal realizations of $W(\lambda)^{-1}$ are similar, there exists a unique invertible S such that

$$\widetilde{C} = CS^{-1} , \quad A_p - B\widetilde{C} = SA_zS^{-1} , \quad B = S\widetilde{B} .$$

Substitute $\widetilde{C}$ in the second equality from the first, and get (1.2). Next, substitute $\widetilde{C}$ in (1.4) by CS^{-1} to obtain the formula (1.3). This argument shows also that there is a unique invertible S for which (1.2) and (1.3) hold.

Assume now (1.2) holds for some invertible S. Define $W(\lambda)$ by (1.3). We claim this is a minimal realization for $W(\lambda)$. Indeed, we have to check that

$$\bigcap_{i=0}^{m-1} \operatorname{Ker}(CS^{-1}A_p^i) = (0) , \quad \sum_{i=0}^{m-1} \operatorname{Im}(A_p^iB) = \mathbb{C}^m ,$$

where m is the size of A_p. The second equality is just the minimality of (A_p,B). To check the first equality, observe that $S^{-1}A_pS = A_z + S^{-1}BC$ because of (1.2), and therefore it is sufficient to check that

$$(1.5) \qquad \bigcap_{i=0}^{m-1} \operatorname{Ker}(C(A_z+S^{-1}BC)^i) = (0) .$$

It is well-known that a right admissible pair (X,T) is minimal if and only if the rank of $\begin{bmatrix} \lambda-T \\ X \end{bmatrix}$ is equal to the rank of T for all $\lambda \in ¢$. So

$$\operatorname{rank} \begin{bmatrix} \lambda-A_z \\ C \end{bmatrix} = m , \quad \lambda \in ¢ ;$$

therefore also

$$\operatorname{rank} \begin{bmatrix} \lambda-A_z-S^{-1}BC \\ C \end{bmatrix} = \operatorname{rank} \left(\begin{bmatrix} I & -S^{-1}B \\ 0 & I \end{bmatrix} \begin{bmatrix} \lambda-A_z \\ C \end{bmatrix} \right) = m , \quad \lambda \in ¢ ,$$

and (1.5) follows. So indeed (1.3) is a minimal realization. Consequently, (A_p,B) is a pole pair for $W(\lambda)$ and (CS^{-1},A_p-BCS^{-1}) is a zero pair for $W(\lambda)$. But $A_p - BCS^{-1} = SA_zS^{-1}$, so the pair (C,A_z) is a zero pair for $W(\lambda)$ as well.

As the realization (1.3) is minimal, it is easily seen that the equality

$$I + CS^{-1}(\lambda-A_p)^{-1}B = I + CS^{-1}(\lambda-A_p)^{-1}B , \quad \lambda \in ¢ ,$$

holds for two invertible matrices S_1 and S_2 satisfying (1.2) if and only if $S_1 = S_2$. □

In particular, a necessary condition in order that (C,A_z) is a zero pair and (A_p,B) is a pole pair of some r.m.f. $W(\lambda)$ with $W(\infty) = I$ is that the sizes of A_z and A_p are equal. The following simple example shows that even if A_z and A_p have the same size, equation (1.2) may not have any (invertible) solution.

EXAMPLE. Let

$$(C,A_z) = \left(\begin{bmatrix} 1 \\ 0 \end{bmatrix}, 0 \right) ; \qquad (A_p,B) = (0,[1 \quad 0])$$

(so C is 2×1, $A_z = A_p$ is 1×1 and B is 1×2). These pairs are minimal, but the equation (1.2) does not have any solution S. Hence there is no 2×2 r.m.f. $W(\lambda)$ with $W(\infty) = I$ whose zero pair is (C,A_z) and pole pair is (A_p,B).

In view of Theorem 1.1 the following remarks about equation (1.2) are in order. Let (C,A_z) and (A_p,B) be as in Theorem 1.1 and let the spectra of A_z and A_p are disjoint: $\sigma(A_z) \cap \sigma(A_p) = \emptyset$. Then there is a unique matrix S satisfying the equation (1.2); it is given by the well-known formula

$$(1.6) \qquad S = \frac{1}{2\pi i} \int_\Gamma (\lambda - A_p)^{-1} BC(\lambda - A_z)^{-1} d\lambda ,$$

where Γ is a suitable positively oriented contour such that $\sigma(A_p)$ (resp. $\sigma(A_z)$) is inside (resp. outside) Γ. So in case $\sigma(A_z) \cap \sigma(A_p) = \emptyset$, there is an r.m.f. $W(\lambda)$ with zero pair (C,A_z) and pole pair (A_p,B) if and only if the integral (1.6) is invertible; then such $W(\lambda)$ is unique and given by (1.3) with S taken from (1.6). Also, in this case one can drop the assumption of minimality of (C,A_z) and (A_p,B) from Theorem 1.1. Indeed, if S is an invertible matrix such that $SA_z - A_pS = -BC$ for a right (resp. left) admissible pair (C,A_z) (resp. (A_p,B)), then the realization (1.3) is minimal (this follows from the general fact that a realization $I + \widetilde{C}(\lambda - \widetilde{A})^{-1}\widetilde{B}$ is minimal whenever $\sigma(\widetilde{A}) \cap \sigma(\widetilde{A} - \widetilde{B}\widetilde{C}) = \emptyset$). Hence (C,A_z) and (A_p,B) are zero pair and pole pair of $W(\lambda)$, respectively; in particular, they are minimal.

Consider now the case when $\sigma(A) \cap \sigma(A) \neq \emptyset$. Then the homogeneous equation

$$(1.7) \qquad XA_z - A_pX = 0$$

has non-zero solutions X. Assume, in addition, that (1.2) has an invertible solution S. Then, obviously, any matrix of the form $S + \alpha X$, where α is a complex number different from the zeros of the polynomial $\det(S+\lambda X)$, is also an invertible solution of (1.2). In this way all invertible solutions of (1.2) are produced. We remark that all non-zero solutions X of (1.7) can be obtained in the following way: Let M be an A_p-invariant subspace and N be an A_z-coinvariant subspace (i.e., a subspace for which there exists an A_z-invariant direct complement), such that $A_{p|M}$ is similar to $QA_zQ : N \to N$, where Q is a projection on N along an A_z-invariant subspace. (Here and in the sequel $Z_{|L}$ stands for the restriction of a $p \times q$ matrix Z, considered as the linear transformation $\mathbb{C}^q \to \mathbb{C}^p$, to the subspace $L \subset \mathbb{C}^q$.) Then there exists a matrix $X : \mathbb{C}^{m_1} \to \mathbb{C}^{m_2}$ (here m_1 and m_2 are the size of A_z and A_p, respectively) such that $\operatorname{Im} X = M$, $\operatorname{Ker} X = \operatorname{Ker} Q$, the linear transformation $X_{|N} : N \to M$ is invertible and $A_{p|M} \cdot X_{|N} = X_{|N} \cdot QA_zQ$. The matrix X is a solution of (1.7), and every solution of (1.7) can be obtained in this way for some M and N as above.

Let now $W(\lambda)$ be an $n \times n$ r.m.f. with minimal realization

$$W(\lambda) = I + C(\lambda I-A)^{-1}B .$$

Using Theorem 1.1, we shall describe all rational matrix functions which have the same zero pair and the same pole pair as W. To this end introduce the following notation. If T is an $m \times m$ matrix (where m is the size of A) such that $\operatorname{Im} T$ is A-invariant and $I - T$ is invertible, we define the linear transformations

$$A_T = A|_{\operatorname{Im} T} ; \quad B_T = TB : \mathbb{C}^n \to \operatorname{Im} T , \quad C_T = C(I-T)^{-1}|_{\operatorname{Im} T} : \operatorname{Im} T \to \mathbb{C}^n ,$$

considered as matrices in some fixed basis in $\operatorname{Im} T$. Denote

$$(1.8) \qquad G_T(\lambda) = I + C_T(\lambda I-A_T)^{-1}B_T .$$

THEOREM 1.2 *In the previous notations the general form of all rational matrix functions* $W_1(\lambda)$ *with* $W_1(\infty) = I$ *and with the same pole and zero pair as* $W(\lambda)$ *is given by*

$$(1.9) \qquad W_1(\lambda) = G_T(\lambda)W(\lambda) ,$$

where T *is some solution of the equation*

$$(1.10) \qquad TA^{\times} - AT = 0 \quad (A^{\times} = A - BC)$$

such that $I - T$ *is invertible. In this case the realization* (1.8) *is minimal.*

<u>Proof</u>. Let $(C,A^{\times})$ and (A,B) be the zero and pole pairs, respectively, of $W(\lambda)$. By Theorem 1.1 we know that the general form of $W_1(\lambda)$ is

$$W_1(\lambda) = I + CS_0^{-1}(\lambda-A)^{-1}B ,$$

where S_0 is invertible and satisfies $S_0A^{\times} - AS_0 = -BC$. In particular:

$$(1.11) \qquad (\lambda-A)^{-1}S_0 - S_0(\lambda-A^{\times})^{-1} = (\lambda-A)^{-1}BC(\lambda-A^{\times})^{-1} .$$

Now compute

$$W_1(\lambda)W(\lambda)^{-1} = [I+CS^{-1}(\lambda-A)^{-1}B][I-C(\lambda-A^{\times})^{-1}B] =$$
$$= I + CS_0^{-1}(\lambda-A)^{-1}(I-S_0)B ,$$

which is easily seen to be equal to G_{I-S_0}. So (1.9) and (1.10) hold with $T = I - S_0$.

Conversely, assume that $W_1(\lambda) = G_T(\lambda)W(\lambda)$ for some T for which $S_0 := I - T$ is invertible and (1.10) holds. Then obviously $\operatorname{Im} T$ is A-invariant. Further,

$$S_0A^{\times} - AS_0 = -BC , \quad G_T(\lambda) = I + CS_0^{-1}(\lambda-A)^{-1}(I-S_0)B ,$$

and a computation of $G_T(\lambda)W(\lambda) = [I+CS_0^{-1}(\lambda-A)^{-1}(I-S_0)B][I+C(\lambda-A)^{-1}B]$, shows that $W_1(\lambda) = I + CS_0^{-1}(\lambda-A)^{-1}B$. By Theorem 1.1, $W_1(\lambda)$ has the same zero and pole pairs as $W(\lambda)$.

Finally, let us check that the realization (1.8) is minimal. To this end observe that

$$G_T(\lambda) = I + C(I-T)^{-1}(\lambda-A)^{-1}TB$$

and the pair $(C(I-T)^{-1},A)$ is minimal (cf. the proof of Theorem 1.1). Now the pair $(A^{\times},B)$ is minimal as well; therefore, taking into account equalities $A^jT = T(A^{\times})^j$ for $j = 0,1,2,\cdots$, we get

$$\operatorname{Im}[TB,ATB,\cdots,A^{m-1}TB] = T\operatorname{Im}[B,A^{\times}B,\cdots,(A^{\times})^{m-1}B] = \operatorname{Im} T ,$$

where m is the size of A. So the realization (1.8) is indeed minimal. □

We observe that the determinant of $G_T(\lambda)$ is identically 1. Indeed,

$$\det[I+C(I-T)^{-1}(\lambda-A)^{-1}TB] = \det[I+(\lambda-A)^{-1}TBC(I-T)^{-1}] =$$
$$= \det(\lambda-A)^{-1} \cdot \det(\lambda-(A-TBC(I-T)^{-1})) .$$

Using (1.10) we obtain

$$A - TBC(I-T)^{-1} = (I-T)(A^{\times}+(I-T)^{-1}BC-(I-T)^{-1}TBC)(I-T)^{-1} =$$
$$= (I-T)A(I-T)^{-1} .$$

Hence

$$\det G_T(\lambda) = \det(\lambda-A)^{-1} \cdot \det(\lambda-(I-T)A(I-T)^{-1}) = 1 .$$

Besides the minimal realization (1.8), other minimal realizations of $G_T(\lambda)$ are available. For instance, (1.8) implies (using the properties that $TA^\times = AT$ and $I - T$ is invertible) that

$$G_T(\lambda) = I - C(I-T)^{-1}T(\lambda-A^\times)^{-1}B \ .$$

This realization is not necessarily minimal. Passing to the minimal realization, one gets

$$G_T(\lambda) = I + \widetilde{C}(\lambda-\widetilde{A})^{-1}\widetilde{B} \ ,$$

where

$$\widetilde{C} = C(I-T)^{-1}T|_{\operatorname{Im} Q} : \operatorname{Im} Q \to \mathbb{C}^n \ ,$$

Q being a projection along $\operatorname{Ker} T$;

$$\widetilde{B} = QB : \mathbb{C}^n \to \operatorname{Im} Q \ ; \quad \widetilde{A} = QA^\times Q|_{\operatorname{Im} Q} : \operatorname{Im} Q \to \operatorname{Im} Q \ .$$

In particular, A_T and $\widetilde{A}$ are similar.

We conclude this section with two more observations. First, the Lyapunov equation (1.2) can be considered as a particular case of the matrix quadratic equation

$$(1.12) \qquad XM_1X + XM_2 + M_3X = M_4$$

(with $M_1 = 0$). Using the description of solutions of (1.12) in terms of angular subspaces, which is given in [1], Section 5.3, we obtain that, with (C,A_z) and (A_p,B) as in Theorem 1.1, there exists an r.m.f. $W(\lambda)$ with $W(\infty) = I$ and zero (resp. pole) pair (C,A_z) (resp. (A_p,B)) if and only if A_z and A_p are of the same size (say, m) and the $2m \times 2m$ matrix

$$\begin{bmatrix} A_p & -BC \\ 0 & A_z \end{bmatrix}$$

has an m-dimensional invariant subspace $\mathcal{M}$ such that

$$\mathcal{M} \cap \{\begin{bmatrix} x \\ 0 \end{bmatrix} \mid x \in \mathbb{C}^m\} = \mathcal{M} \cap \{\begin{bmatrix} 0 \\ x \end{bmatrix} \mid x \in \mathbb{C}^m\} = (0) \ .$$

Writing

$$= \{\begin{bmatrix} Sx \\ x \end{bmatrix} \mid x \in \mathbb{C}^m\}$$

for the unique invertible matrix S, the function $W(\lambda)$ is given again by (1.3).

Secondly, Theorem 1.1 can be regarded also as a criterium for existence of invertible solutions for Lyapunov equations. Consider, for instance, the scalar case $(n = 1)$. In this case evidently a minimal admissible pair (C,A_z) is a zero pair, and a left minimal admissible pair (A_p,B) is a pole pair of some scalar rational function with 1 at infinity if and only if the sizes of A_z and A_p coincide, and $\sigma(A_z) \cap \sigma(A_p) = \emptyset$. Using this observation and Theorem 1.1, we obtain the following.

COROLLARY 1.3 *Let* (C,A_z) *be a minimal admissible pair with row vector* C, *let* (A_p,B) *be a minimal left admissible pair with column vector* B, *and let the sizes of* A_z *and* A_p *coincide. Then the equation* $SA_z - A_pS = -BC$ *has an invertible solution if and only if* $\sigma(A_z) \cap \sigma(A_p) = \emptyset$.

Combining this result with Theorem 1 in [6] we conclude that under the assumptions of Corollary 1.3 the equation $SA_z - A_pS = -BC$ is solvable if and only if $\sigma(A_z) \cap \sigma(A_p) = \emptyset$ and in this case the (unique) solution S is invertible (see also Corollary 21 in [9] and the remarks thereafter).

I.3 Main Results in Terms of Coprime Matrix Fractions

Let $W(\lambda)$ be an $n \times n$ regular r.m.f.. As well-known (see, e.g., [8]), $W(\lambda)$ admits a decomposition

$$(1.13) \qquad W(\lambda) = D_L(\lambda)^{-1}N_L(\lambda) ,$$

where $N_L(\lambda)$ and $D_L(\lambda)$ are $n \times n$ matrix polynomials which are *left coprime*, i.e., there exist $n \times n$ matrix polynomials $X_1(\lambda)$ and $X_2(\lambda)$ such that

$$N_L(\lambda)X_1(\lambda) + D_L(\lambda)X_2(\lambda) = I , \quad \lambda \in \mathbb{C} .$$

Equivalently, N_L and D_L are left coprime if and only if they do not have a common left eigenvector corresponding to the same zero (see, e.g., [2]). Recall that a row (resp. column) vector $x_0 \neq 0$ is called a *left* (resp. *right*) *eigenvector* of an $n \times n$ matrix polynomial $M(\lambda)$ corresponding to its zero λ_0 if $x_0M(\lambda_0) = 0$ (resp. $M(\lambda_0)x_0 = 0$). The decomposition (1.13) is called *left coprime matrix fraction description* of the r.m.f. $W(\lambda)$, with *left numerator* N_L and *left denominator* D_L. Note that in the left coprime matrix fraction (1.13) the matrix polynomials $N_L(\lambda)$ and $D_L(\lambda)$ are uniquely determined up to multiplication from the left by the same matrix polynomial which is *unimodular*, i.e., with constant non-zero determinant.

The dual notion of a *right coprime matrix fraction* $W(\lambda) = N_R(\lambda)D_R(\lambda)^{-1}$ where $N_R(\lambda)$ and $D_R(\lambda)$ are matrix polynomials, is defined by the property that $N_R(\lambda)$ and $D_R(\lambda)$ are *right coprime*, i.e.

$$Y_1(\lambda)N_R(\lambda) + Y_2(\lambda)D_R(\lambda) = I \qquad (\lambda \in \mathbb{C})$$

for some matrix polynomials $Y_1(\lambda)$ and $Y_2(\lambda)$ (or, equivalently, N_R and D_R have no common right eigenvectors corresponding to the same zero). In this case $N_R(\lambda)$ is the *right numerator* of $W(\lambda)$ and $D_R(\lambda)$ is its *right denominator*.

In the sequel the following characterization of coprime matrix fractions will be used frequently (see Lemma 4.1 in [3]):

PROPOSITION 1.4 *A matrix fraction* $W(\lambda) = N_2(\lambda)^{-1}N_1(\lambda)$ *(resp.* $W(\lambda) = N_2(\lambda)N_1(\lambda)^{-1}$*), where* N_1 *and* N_2 *are regular* $n \times n$ *matrix polynomials, is left (resp. right) coprime if and only if the zero (resp. pole) pair of* $W(\lambda)$ *coincides with the zero (resp. left zero) pair of* $N_1(\lambda)$.

By definition, a *left zero pair* of a regular r.m.f. $V(\lambda)$ is a pole pair of $V(\lambda)^{-1}$.

In this section we study the following problem: Given regular $n \times n$ matrix polynomials D_R and N_L, find (if possible) regular $n \times n$ matrix polynomials N_R and D_L in such a way that

$$W(\lambda) := N_R(\lambda)D_R(\lambda)^{-1} = D_L(\lambda)^{-1}N_L(\lambda)$$

are right and left coprime matrix fraction descriptions of some r.m.f. $W(\lambda)$ with $W(\infty) = I$.

Let us explain how this problem is related to the inverse problem treated in the preceding section. Two regular $n \times n$ matrix polynomials $L_1(\lambda)$ and $L_2(\lambda)$ have the same zero (resp. left zero) pair if and only if $L_1(\lambda) = U(\lambda)L_2(\lambda)$ (resp. $L_1(\lambda) = L_2(\lambda)U(\lambda)$) for some unimodular matrix polynomial $U(\lambda)$ (see, e.g. Chapter 7 in [4]). Hence, taking into account Proposition 1.4, the knowledge of a left numerator of an r.m.f. $W(\lambda)$ amounts to the knowledge of the zero pair of $W(\lambda)$. Analogously, the knowledge of a right denominator of an r.m.f. of $W(\lambda)$ amounts to the knowledge of the pole pair of $W(\lambda)$.

The solution of the above mentioned problem is given by the following theorem.

THEOREM 1.5 *Let* D_R *and* N_L *be given* $n \times n$ *regular matrix polynomials. Then there exists an* $n \times n$ *rational matrix function* $W(\lambda)$ *with* $W(\infty) = I$ *such that* D_R *is a right denominator of* $W(\lambda)$ *and* N_L *is a left numerator of* $W(\lambda)$ *if and only if the equation*

$$(1.14) \qquad SA_z - A_pS = -BC ,$$

where (C,A_z) *is the right zero pair of* N_L *and* (A_p,B) *is the left zero pair of* D_R*, has an invertible solution* S*. In this case one may take*

$$(1.15) \qquad W(\lambda) = I + CS^{-1}(\lambda-A_p)^{-1}B = I - C(\lambda-A_z)^{-1}S^{-1}B .$$

The correspondence given by (1.15) between the set of invertible solutions S *of (1.14) and the set of rational matrix functions* $W(\lambda)$ *with the required properties, is one-to-one.*

Proof. Assume there exists $W(\lambda)$ with the required properties, and consider the coprime matrix fraction descriptions of $W(\lambda)$

$$W(\lambda) = N_R(\lambda)D_R(\lambda)^{-1} = D_L(\lambda)^{-1}N_L(\lambda)$$

with some matrix polynomials $N_R(\lambda)$ and $D_L(\lambda)$. Then by Proposition 1.4, the pair (C,A_z) is a zero pair of $W(\lambda)$ and (A_p,B) is a pole pair of $W(\lambda)$. Now use Theorem 1.1 to deduce the existence of a unique invertible solution of (1.14) for which (1.15) holds.

Conversely, assume that S is an invertible solution of (1.14). By Theorem 1.1 the r.m.f. given by (1.15) has zero pair (C,A_z) and pole pair (A_p,B). Write

$$(1.16) \qquad W(\lambda) = N_R(\lambda)D_R(\lambda)^{-1} = D_L(\lambda)^{-1}N_L(\lambda)$$

for some r.m.f.'s N_R and D_L. We claim that actually N_R and D_L are matrix polynomials. Indeed, take a left coprime matrix fraction $W(\lambda) = M(\lambda)^{-1}L(\lambda)$. Then (C,A_z) is the zero pair of $L(\lambda)$. As the matrix polynomials $L(\lambda)$ and $N_L(\lambda)$ have the same zero pairs, we have $L(\lambda) = U(\lambda)N_L(\lambda)$

for some unimodular matrix polynomial $U(\lambda)$ (see, e.g., Chapter 7 in [4]). Now

$$D_L(\lambda) = N_L(\lambda)W(\lambda)^{-1} = N_L(\lambda)L(\lambda)^{-1}M(\lambda) = U(\lambda)^{-1}M(\lambda)$$

is a matrix polynomial. Analogously one checks that N_R is a matrix polynomial. Now the coprimeness of (1.16) follows from Proposition 1.4. □

All remarks made in the preceding section in connection with Theorem 1.1 and invertible solutions of (1.2), apply also in connection with Theorem 1.5.

The problem which was described and solved in this section can be stated purely in terms of matrix polynomials, as follows. Given regular $n \times n$ matrix polynomials $L_1(\lambda)$ and $L_2(\lambda)$, find regular matrix polynomials $M_1(\lambda)$ and $M_2(\lambda)$ with the following properties:

1) $M_1(\lambda)M_2(\lambda) = L_1(\lambda)L_2(\lambda)$;
2) M_1 and L_1 are left coprime;
3) M_2 and L_2 are right coprime;
4) $M_1(\lambda)^{-1}L_1(\lambda)$ takes value I at infinity.

The solution to this problem is given by Theorem 1.5 (with $D_L = L_1$; $N_R = L_2$). If the matrix polynomials L_1 and L_2 are monic (i.e. with leading coefficient I) of the same degree, say k, and the scalar polynomials $\det L_1(\lambda)$ and $\det L_2(\lambda)$ have no common zeros, then this problem amounts to existence of a monic spectral divisor of a matrix polynomial from one side provided there exists a monic spectral divisor of the same matrix polynomial from the other side (see, e.g., [4] for the definition and basic properties of monic spectral divisors of matrix polynomials). In view of Theorem 1.5, a criterium for existence of such monic spectral divisor amounts to invertibility of the matrix

$$\int_\Gamma \begin{bmatrix} L_2(\lambda)^{-1}C_1(\lambda-A_1)^{-1} \\ \vdots \\ \lambda^{k-1}L_2(\lambda)^{-1}C_1(\lambda-A_1)^{-1} \end{bmatrix} d\lambda ,$$

where (C_1, A_1) is a zero pair for $L_1(\lambda)$ and Γ is a suitable contour such that the zeros of $\det L_2$ (resp. of $\det L_1$) are inside (resp. outside) Γ.

I.4 Rational Matrix Functions with Prescribed Right and Left Zero and Pole Pairs

Until now we have considered the problem of finding r.m.f.'s with I at infinity when their (right) zero and (right) pole pairs are given. An analogous problem can be solved also for the case when right zero and pole pairs, as well as the left ones, are given. By definition, a left zero pair of W is a pole pair for $W(\lambda)^{-1}$, and a left pole pair for $W(\lambda)$ is a zero pair for $W(\lambda)^{-1}$.

THEOREM 1.6 *Let* (C_1,A_z), (C_2,A_p) *(resp.* (A_p,B_1), (A_z,B_2)*) be right (resp. left) minimal admissible pairs. Then there exists a rational* $n\times n$ *matrix function* $W(\lambda)$ *with* $W(\infty)=I$ *for which* (C_1,A_z), (C_2,A_p), (A_p,B_1) *and* (A_z,B_2) *are a zero pair, a left pole pair, a pole pair and a left zero pair, respectively, if and only if there exists an invertible matrix* S_1 *such that*

$$S_1A_z - A_pS_1 = -B_1C_1 ,$$

and for some invertible matrices F *and* R *the following equalities hold:*

$$FA_z = A_zF \ ; \quad RA_p = A_pR \ ; \quad B_2 = FS_1^{-1}B_1 \ ; \quad C_2 = C_1S_1^{-1}R .$$

In such case one may take

$$W(\lambda) = I + C_1S_1^{-1}(\lambda-A_p)^{-1}B_1 . \tag{1.17}$$

The proof of Theorem 1.6 is obtained easily from Theorem 1.1 using similarity of minimal realizations of r.m.f.'s and the equality

$$S_2A_p - A_zS_2 = -B_2C_2$$

with $S_2 = -FS_1^{-1}R$, which follows from the conditions of the theorem.

In general, a function $W(\lambda)$ from Theorem 1.6 is not unique, as the following example shows.

EXAMPLE 1.2 Set

$$C_1 = C_2 = \begin{bmatrix}1\\ \alpha\end{bmatrix} ; \quad B_1 = B_2 = [-\alpha \ 1] ; \quad A_z = A_p = 0 .$$

The conditions of Theorem 1.6 are satisfied with $F=1$, $R=1$ and an arbitrary non-zero number $S_1 \neq 0$. In this case formula (1.17) gives

$$W_{S_1}(\lambda) = I + \frac{1}{S_1\lambda}\begin{bmatrix} -\alpha & 1 \\ -\alpha^2 & \alpha \end{bmatrix}.$$

Every function $W_{S_1}(\lambda)$ (independently of S_1) has $(\begin{bmatrix}1\\ \alpha\end{bmatrix},0)$, $(\begin{bmatrix}1\\ \alpha\end{bmatrix},0)$, $(0,[-\alpha\ \ 1])$ and $(0,[-\alpha\ 1])$ as its zero pair, left pole pair, pole pair and left zero pair, respectively.

II. MINIMAL DIVISIBILITY

In this chapter we shall study the minimal divisibility of r.m.f.'s in connection with their zero and pole pairs. Recall that an $n \times n$ r.m.f. W_2 with $W_2(\infty) = I$ is called a *right* (resp. *left*) *minimal divisor* of an $n \times n$ r.m.f. W with $W(\infty) = I$ if the McMillan degrees of W_2 and WW_2^{-1} (resp. of W_2 and $W_2^{-1}W$) add up to the McMillan degree of W. Since W_2 is a left minimal divisor of W if and only if $(W_2(\bar{\lambda}))^*$ is a right minimal divisor of $(W(\bar{\lambda}))^*$, in the sequel we shall consider only right minimal divisibility.

II.1 Minimal Divisors

Let $W(\lambda)$ and $W_2(\lambda)$ be $n \times n$ r.m.f.'s with value I at infinity. In this section we give necessary and sufficient conditions in terms of zero pairs and pole pairs of $W(\lambda)$ and $W_2(\lambda)$ in order that W_2 is a minimal divisor of W.

To this end we need the notions of restrictions and compressions of admissible pairs.

Let (X,T) and $(\tilde{X},\tilde{T})$ be two (right) admissible pairs. Let the sizes of matrices X, T, $\tilde{X}$, $\tilde{T}$ be $n \times p$, $p \times p$, $n \times q$, $q \times q$, respectively. The pair $(\tilde{X},\tilde{T})$ is called a *restriction* of (X,T) if there exist a T-invariant subspace M and a bijective linear transformation $\Phi : M \to \mathbb{C}^q$ such that

$$\text{(2.1)} \qquad \Phi T_{|M} = \tilde{T}\Phi , \qquad X_{|M} = \tilde{X}\Phi .$$

In this case we say that $(\tilde{X},\tilde{T})$ is the $\{\Phi,M\}$-*restriction* of (X,T). Similarly, a left admissible pair $(\tilde{U},\tilde{Y})$ is a *compression* of a left admissible pair (U,Y) if there exists an U-coinvariant (i.e. such that one of its direct complements is U-invariant) subspace N and a bijective linear transformation $\Psi : \mathbb{C}^r \to N$, where $r \times r$ is the size of $\tilde{U}$, such that

(2.2) $\Pi U_{|N}\Psi = \Psi\tilde{U}$, $\Pi Y = \Psi Y$,

where Π is a projection onto N and the kernel of Π is U-invariant. In this case $(\tilde{U},\tilde{Y})$ will be called $\{\Psi,\Pi\}$-*compression* of (U,Y).

Note that if the right admissible pair (X,T) is minimal, then for any $\{\Phi,M\}$-restriction $(\tilde{X},\tilde{T})$ of (X,T) the similarity Φ and the subspace M are uniquely defined by $(\tilde{X},\tilde{T})$ and (X,T) (see the proof of Lemma 7.12 in [4]). For the left admissible pairs the following holds: let (U,Y) be a minimal left admissible pair and let $(\tilde{U},\tilde{Y})$ be its $\{\Psi,\Pi\}$-compression. Let Π' be any projection with $\mathrm{Ker}\,\Pi' = \mathrm{Ker}\,\Pi$. Then $(\tilde{U},\tilde{Y})$ is also $\{\Psi',\Pi'\}$-compression of (U,Y), where $\Psi' = \Pi'_{|\mathrm{Im}\,\Pi} \cdot \Psi : \mathbb{C}^q \to \mathrm{Im}\,\Pi'$. Conversely, if $(\tilde{U},\tilde{Y})$ is a $\{\Psi',\Pi'\}$-compression of (U,Y), then $\mathrm{Ker}\,\Pi' = \mathrm{Ker}\,\Pi$ and $\Psi' = \Pi'_{|\mathrm{Im}\,\Pi} \cdot \Psi$.

The following theorem describes minimal divisibility of r.m.f.'s in terms of restrictions and compressions of their pole and zero pairs.

THEOREM 2.1 *Let*

(2.3) $W(\lambda) = I + C(\lambda-A)^{-1}B$

and

(2.4) $W_2(\lambda) = I + C_2(\lambda-A_2)^{-1}B_2$

be minimal realizations. Then W_2 *is a right minimal divisor of* W *if and only if* (A_2,B_2) *is a compression of* (A,B), *say* $\{\Psi,\Pi\}$-*compression,* $(C_2,A_2^\times)$ *is a restriction of* $(C,A^\times)$ (*where* $A_2^\times = A_2 - B_2C_2$, $A^\times = A - BC$), *say* $\{\Phi,M\}$-*restriction, and*

(2.5) $\Pi_{|M} = \Psi\Phi$.

Using the description given above of all $\{\Psi,\Pi\}$ for which (A_2,B_2) is a $\{\Psi,\Pi\}$-compression of (A,B), it is not difficult to see that the equality (2.5) does not depend on the choice of $\{\Psi,\Pi\}$.

Proof. Assume $W_2(\lambda)$ is a minimal divisor of $W(\lambda)$. According to [1], Theorem 4.8 this means that there exists a direct sum decomposition $\mathbb{C}^m = M \dotplus M^\times$, where m is the size of A, such that

$$AM \subset M \ , \quad A^\times M^\times \subset M^\times$$

and

$$W_2(\lambda) = I + C_{|M^\times}(\lambda-\Pi_0 A_{|M^\times})^{-1}\Pi_0 B$$

is a minimal realization. Here Π_0 is the projection of $\mathbb{C}^m$ along M onto $M^\times$. So there exists an invertible $S : \mathbb{C}^{m_2} \to M^\times$ (m_2 being the size of A_2) such that

$$SA_2S^{-1} = \Pi_0 A_{|M^\times} \ ; \quad SB_2 = \Pi_0 B \ ; \quad C_2S^{-1} = C_{|M^\times} .$$

In particular, $SA_2^\times S^{-1} = \Pi_0 A^\times{}_{|M^\times}$, and $(S_2, A_2^\times)$ is the $\{S^{-1}, M^\times\}$-restriction of $(C, A^\times)$. Also, (A_2, B_2) is a $\{S, \Pi_0\}$-compression of (A,B). Equality (2.5) is obviously satisfied with $\Pi = \Pi_0$, $S = \Psi$, $S^{-1} = \Phi$.

Now, conversely, assume $(C_2, A_2^\times)$ is the $\{\Phi, M\}$-restriction of $(C, A^\times)$, (A_2, B_2) is a $\{\Psi, \Pi\}$-compression of (A,B), and (2.5) holds. Since

$$\Psi\Phi = \Pi_{|M} : M \to \operatorname{Im} \Pi$$

is invertible, we have $\mathbb{C}^m = \operatorname{Ker} \Pi \dot{+} M$. Let Π_0 be the projection of $\mathbb{C}^m$ along $\operatorname{Ker} \Pi$ onto M. Note that $A(\operatorname{Ker} \Pi) \subset \operatorname{Ker} \Pi$ and $A^\times M \subset M$. So by Theorem 4.8 in [1] the r.m.f.

$$I + C_{|M}(\lambda-\Pi_0 A_{|M})^{-1}\Pi_0 B \tag{2.6}$$

is a minimal divisor of $W(\lambda)$. Using the equality $\Psi\Phi\Pi_0 = \Pi$ and the properties of restriction and compression one sees that

$$\Phi^{-1}A_2\Phi = \Pi_0 A_{|M} \ ; \quad \Phi^{-1}B_2 = \Pi_0 B \ ; \quad C_2\Phi = C_{|M} .$$

Hence the r.m.f. (2.6) is just $W(\lambda)$. □

Theorem 2.1 was stated in terms of minimal realizations of $W(\lambda)$ and $W_2(\lambda)$. This result can be also formulated starting with pole pairs and zero pairs of $W(\lambda)$ and $W_2(\lambda)$, as follows. Let $W(\lambda)$ and $W_2(\lambda)$ be r.m.f.'s with I at infinity such that $(\widetilde{C}, \widetilde{A}_z)$ and $(\widetilde{A}_p, \widetilde{B})$ are the zero pair and the pole pair of $W(\lambda)$, respectively, and $(\widetilde{C}_2, \widetilde{A}_{2z})$, $(\widetilde{A}_{2p}, \widetilde{B}_2)$ are the zero pair and the pole pair of $W_2(\lambda)$, respectively. By Theorem 1.1 there exist invertible matrices S and S_2 such that

$$S\widetilde{A}_z - \widetilde{A}_p S = -\widetilde{B}\widetilde{C} \ ; \quad S_2\widetilde{A}_{2z} - \widetilde{A}_{2p}S_2 = -\widetilde{B}_2\widetilde{C}_2$$

and

$$(2.7)\qquad W(\lambda) = I + \widetilde{C}S^{-1}(\lambda-\widetilde{A}_p)^{-1}\widetilde{B}\ ,\qquad W_2(\lambda) = I + \widetilde{C}_2S_2^{-1}(\lambda-\widetilde{A}_{2p})^{-1}\widetilde{B}_2\ .$$

Moreover, the realizations (2.7) are minimal. Then $W_2(\lambda)$ is a (right) minimal divisor of $W(\lambda)$ if and only if $(\widetilde{C}_2,\widetilde{A}_{2z})$ is the $\{\Omega,N\}$-restriction of $(\widetilde{C},\widetilde{A}_z)$; $(\widetilde{A}_{2p},\widetilde{B}_2)$ is a $\{\Psi,\Pi\}$-compression of $(\widetilde{A}_p,\widetilde{B})$ and $\Pi_{|SN} = \Psi S_2\Omega S^{-1}$. This statement follows immediately from Theorem 2.1 using the following observation: $(\widetilde{C}_2,\widetilde{A}_{2z})$ is the $\{\Omega,N\}$-restriction of $(\widetilde{C},\widetilde{A}_z)$ if and only if $(\widetilde{C}_2S_2^{-1},S_2\widetilde{A}_{2z}S_2^{-1})$ is the $\{S_2\Omega S^{-1},SN\}$-restriction of $(\widetilde{C}S^{-1},S\widetilde{A}_zS^{-1})$.

II.2 Minimal Divisors with Prescribed Zero and Pole Pairs

In this section we solve the following problem: given an r.m.f. $W(\lambda)$ (with $W(\infty) = I$) and given a right minimal pair (X,T) and a left minimal pair (U,Y) : when does there exist a right minimal divisor $W_2(\lambda)$ (with $W_2(\infty) = I$) of $W(\lambda)$ such that (X,T) is a zero pair of W_2 and (U,Y) is a pole pair of W_2? If such $W_2(\lambda)$ exists, describe all of them.

We need the following definition. Given right and left admissible pairs $(\widetilde{X},\widetilde{T})$ and $(\widetilde{U},\widetilde{Y})$, respectively, given a $\{\Phi,M\}$-restriction (X,T) of $(\widetilde{X},\widetilde{T})$ and given a $\{\Psi,\Pi\}$-compression (U,Y) of $(\widetilde{U},\widetilde{Y})$, the linear transformation $Z = \Psi^{-1}(\Pi_{|M})\Phi^{-1} : \mathbb{C}^q \to \mathbb{C}^r$, where q and r are the sizes of T and U, respectively, will be called the *indicator* of the pairs $(\widetilde{X},\widetilde{T})$, (X,T), $(\widetilde{U},\widetilde{Y})$, (U,Y). It is easily seen that the indicator does not depend on the choice of $\{\Psi,\Pi\}$.

THEOREM 2.2 *Let* (X,T) *and* (U,Y) *be right and left minimal pairs, respectively, and let*

$$W(\lambda) = I + C(\lambda-A)^{-1}B$$

be a minimal realization of $W(\lambda)$. *Then there exists a minimal divisor* $W_2(\lambda)$ *of* $W(\lambda)$ *with pole pair* (U,Y) *and zero pair* (X,T) *and such that* $W_2(\infty) = I$ *if and only if* (X,T) *is a restriction of* $(C,A-BC)$, (U,Y) *is a compression of* (A,B) *and the indicator* Z *of* $(C,A-BC)$, (X,T), (A,B), (U,Y) *is invertible. In this case the minimal divisor* $W_2(\lambda)$ *is unique and is given by the formula*

$$(2.8)\qquad W_2(\lambda) = I + XZ^{-1}(\lambda-U)^{-1}Y\ .$$

Using the definition, one checks easily that the indicator Z satisfies the Lyapunov equation

$$(2.9) \qquad ZT - UZ = -YX .$$

Proof. The proof is obtained by combining Theorem 1.1 with Theorem 2.1. Let us give the details.

Assume the conditions of the theorem hold true. In view of Theorem 1.1 the r.m.f. $W_2(\lambda)$ given by (2.8) has pole pair (U,Y) and zero pair (X,T). Now apply the statement at the end of the preceding section (with $\tilde{C} = C$, $\tilde{B} = B$, $\tilde{A}_p = A$, $\tilde{A}_z = A - BC$, $S = I$, $\tilde{C}_2 = X$, $\tilde{A}_{2z} = T$, $\tilde{A}_{2p} = U$, $\tilde{B}_2 = Y$, $S = Z$) to deduce that $W_2(\lambda)$ is a minimal divisor of $W(\lambda)$.

Conversely, assume there exists a minimal divisor $W_2(\lambda)$ $(W_2(\infty) = I)$ of $W(\lambda)$ with zero pair (X,T) and pole pair (U,Y). By Theorem 1.1 $W_2(\lambda)$ is given by (2.8), where Z is an invertible matrix satisfying (2.9). Now apply again the same statement to verify that (X,T) is a restriction of $(C,A-BC)$ and (U,Y) is a compression of (A,B). □

By taking $(X,T) = (C,A-BC)$ and $(U,Y) = (A,B)$ in Theorem 2.2, we conclude, that among all r.m.f.'s with value I at infinity whose pole pair and zero pair are (A,B) and $(C,A-BC)$, respectively, there is exactly one minimal divisor of W; namely, the function W itself.

We illustrate Theorem 2.2 with simple examples.

EXAMPLE 2.1 For

$$W(\lambda) = \begin{bmatrix} \frac{\lambda-1}{\lambda-2} & 0 \\ 0 & \frac{\lambda}{\lambda-1} \end{bmatrix}$$

write the minimal realization

$$W(\lambda) = I + C(\lambda-A)^{-1}B ,$$

with

$$A = \begin{bmatrix} 2 & 0 \\ 0 & 1 \end{bmatrix}, \quad C = B = \begin{bmatrix} 1 & 0 \\ 0 & 1 \end{bmatrix}, \quad A^{\times} = A - BC = \begin{bmatrix} 1 & 0 \\ 0 & 0 \end{bmatrix}.$$

Take

$$(X,T) = \left(\begin{bmatrix} 1 \\ 0 \end{bmatrix},1\right) ; \quad (U,Y) = ([0\ 1],1) .$$

Then (X,T) is the $\{\Phi,M\}$-restriction of $(C,A^{\times})$ with $M = \mathrm{Span}\{\begin{bmatrix}1\\0\end{bmatrix}\}$, $\Phi : M \to \mathbb{C}$ defined by $\Phi\begin{bmatrix}1\\0\end{bmatrix} = 1$. Also, (U,Y) is a $\{\Psi,\Pi\}$-compression of (A,B) with $\Pi = \begin{bmatrix}0&0\\0&1\end{bmatrix}$ and $\Psi : \mathbb{C} \to \mathrm{Im}\,\Pi$ defined by $\Psi(1) = \begin{bmatrix}0\\1\end{bmatrix}$. Using Theorem 1.1 one finds easily that all 2×2 r.m.f.'s with I at infinity whose zero pair is (X,T) and pole pair is (U,Y) are given by the formula

$$W_s(\lambda) = \begin{bmatrix} 1 & \frac{s}{\lambda-1} \\ 0 & 1 \end{bmatrix},$$

where $s \neq 0$ is a complex parameter. However, the indicator of $(C,A^{\times})$, (X,T), (A,B), (U,Y) is zero. So by Theorem 2.2 none of the functions $W_s(\lambda)$ is a minimal divisor of $W(\lambda)$. In fact

$$W(\lambda)W_s(\lambda)^{-1} = \begin{bmatrix} \frac{\lambda-1}{\lambda-2} & \frac{-s}{\lambda-2} \\ 0 & \frac{\lambda}{\lambda-1} \end{bmatrix}$$

has McMillan degree 2. □

EXAMPLE 2.2 Let

$$W_\beta(\lambda) = I + \frac{1}{\lambda\beta}\begin{bmatrix} 0 & 1 \\ 0 & 0 \end{bmatrix}, \quad \beta \neq 0$$

(cf. Example 1.2). Then all functions $W_\beta(\lambda)$ have the same zero pairs, the same pole pairs, the same left zero pairs and the same left pole pairs. The function $W_1(\lambda)$ has minimal realization

$$W_1(\lambda) = I - \begin{bmatrix}1\\0\end{bmatrix}\lambda^{-1}[0 \quad 1],$$

so (in notation of Theorem 2.2) $C = \begin{bmatrix}1\\0\end{bmatrix}$, $A = 0$, $B = [0 \quad 1]$. Theorem 2.2 shows that $W_1(\lambda)$ is the only minimal divisor of $W_1(\lambda)$ among all functions $W_\beta(\lambda)$. □

The description of minimal factorization given in [1] (Theorem 4.8) was used in order to prove Theorem 2.1 and hence Theorem 2.2. Conversely, it is easy to recover Theorem 4.8 in [1] using Theorems 1.1 and 2.1.

III. MINIMAL DIVISIBILITY AT A POINT

In this chapter we study minimal divisibility of r.m.f.'s at a point (or at a given set of points) in terms of their zero and pole pairs at this point, as well as in terms of coprime matrix fraction descriptions.

III.1 Locally Minimal Divisors

Given a set Λ in the complex plane, we say that a regular $n \times n$ r.m.f. $W_2(\lambda)$ is a *(right) minimal divisor with respect to* Λ of a regular $n \times n$ r.m.f. $W(\lambda)$ if W_2 is a right minimal divisor of W at each point $\lambda_0 \in \Lambda$. For an r.m.f. $W(\lambda)$ a *(right)* Λ*-zero pair* of $W(\lambda)$ is defined as any (right) admissible pair which is similar to

$$([C_1\ C_2\ \cdots\ C_r],\mathrm{diag}(A_1,A_2,\cdots,A_r)) ,$$

here (C_i,A_i) is a zero pair of $W(\lambda)$ at λ_i $(i = 1,\cdots,r)$ and $\lambda_1,\cdots,\lambda_r$ are all different zeros of $W(\lambda)$ in the set Λ. A *(right)* Λ*-pole pair* of $W(\lambda)$ is defined analogously.

THEOREM 3.1 *Let* $W(\lambda)$ *and* $W_2(\lambda)$ *be regular rational matrix functions, and let* $\Lambda \subset \mathbb{C}$. *If* $W_2(\lambda)$ *is a right minimal divisor of* $W(\lambda)$ *with respect to* Λ, *then the* Λ*-zero pair (resp.* Λ*-pole pair) of* $W_2(\lambda)$ *is a restriction (resp. compression) of the* Λ*-zero pair (resp.* Λ*-pole pair) of* $W(\lambda)$. *Conversely, if the* Λ*-zero pair (resp.* Λ*-pole pair) of* W_2 *is a restriction (resp. compression) of the* Λ*-zero pair (resp.* Λ*-pole pair) of* W, *and if any point in* Λ *is not simultaneously a zero and a pole of* $W_2(\lambda)$, *then* $W_2(\lambda)$ *is a right minimal divisor of* $W(\lambda)$ *with respect to* Λ.

Proof. We prove first the direct statement. Write a left coprime matrix fraction $W_2 = D_2^{-1}N_2$ and a right coprime matrix fraction $W_1 = N_1D_1^{-1}$, where $W_1 = WW_2^{-1}$. (For simplicity the variable λ is omitted). The minimality of the factorization $W = W_1W_2$ at each $\lambda_0 \in \Lambda$ implies, in view of Proposition 1.4, that

$$p(D_1^{-1};\lambda_0) + p(D_2^{-1};\lambda_0) = p(W;\lambda_0) \tag{3.1}$$

for each $\lambda_0 \in \Lambda$. As D_i, $i = 1,2$, are matrix polynomials, $z(D_i^{-1};\lambda_0) = 0$ and the left-hand side of (3.1) is equal to $p((D_2D_1)^{-1};\lambda_0)$. Here we use the fact (which can be easily proved by taking determinants) that

$$p(V_1V_2;\mu_0) + z(V_1V_2;\mu_0) = p(V_1;\mu_0) + z(V_1;\mu_0) + p(V_2;\mu_0) + z(V_2;\mu_0)$$

for any r.m.f.'s V_1 and V_2 and any $\mu_0 \in \mathbb{C}$. Further,

$$p((D_2D_1)^{-1};\lambda_0) \leq p((D_2D_1)^{-1}N_2;\lambda_0) \quad (\lambda_0 \in \Lambda) , \tag{3.2}$$

since $W = N_1D_1^{-1}D_2^{-1}N_2$ and N_1 is a matrix polynomial. As the opposite inequality is obvious, we have actually the equality in (3.2). Hence the factorization $\widetilde{W} := (D_2D_1)^{-1}\cdot N_2$ is minimal at λ_0. Since D_2D_1 is a matrix polynomial, it is evident from the definition of a Jordan chain that every Jordan chain of $\widetilde{W}$ at λ_0 is also a Jordan chain of N_2 at λ_0. But $z(\widetilde{W};\lambda_0) = z(N_2;\lambda_0)$ $(\lambda_0 \in \Lambda)$ in view of the minimality of the factorization $\widetilde{W} = (D_2D_1)^{-1}\cdot N_2$ at λ_0. Thus, in fact, the Jordan chains of N_2 and $\widetilde{W}$ at λ_0 coincide. On the other hand, in view of Proposition 1.4 the Jordan chains of W_2 and N_2 (and, consequently, of $\widetilde{W}$) at λ_0 coincide. Further, since N_1 is a matrix polynomial, every Jordan chain of $\widetilde{W}$ (and, consequently, of W_2) at $\lambda_0 \in \Lambda$ is also a Jordan chain of $W = N_1\widetilde{W}$. Now from the construction of a zero pair of an r.m.f. given in the Introduction it is not difficult to deduce that a Λ-zero pair of W_2 is a restriction of a Λ-zero pair of W.

The statement about the Λ-pole pairs follows from the part of Theorem 3.1 which we already proved by taking into account that (U,Y) is a Λ-pole pair for an r.m.f. $V(\lambda)$ if and only if (Y^*,U^*) is a $\bar{\Lambda}$-zero pair for the r.m.f. $(V(\bar{\lambda}))^{*-1}$ (here $\bar{\Lambda} = \{\bar{\lambda} \mid \lambda \in \Lambda\}$).

We prove now the converse statement of Theorem 3.1. Assume that $\lambda_0 \in \Lambda$ is not a pole of W_2, i.e., $p(W_2;\lambda_0) = 0$. In such case the minimality of the factorization $W = W_1W_2$ $(W_1 := WW_2^{-1})$ means that $p(W;\lambda_0) = p(W_1;\lambda_0)$. Let $W = D_L^{-1}N_L$ and $W_2 = D_{L2}^{-1}N_{L2}$ be left coprime matrix fractions. In view of Proposition 1.4 the zero pairs of N_L and N_{L2} coincide with the zero pairs of W and W_2, respectively. By the assumptions of the theorem the Λ-zero pair of N_{L2} is a restriction of the Λ-zero pair of N_L and therefore the r.m.f. $N_LN_{L2}^{-1}$ is analytic in Λ (see, e.g., [5]). Hence

$$p(W_1;\lambda_0) = p(D_L^{-1}N_LN_{L2}^{-1}D_{L2};\lambda_0) \leqslant p(D_L^{-1}) .$$

Applying Proposition 1.4 to $W^{-1} = N_L^{-1}D_L$ we conclude that $p(W;\lambda_0) = p(D_L^{-1};\lambda_0)$ and therefore

$$p(W_1;\lambda_0) \leqslant p(W;\lambda_0) .$$

On the other hand,

$$p(W;\lambda_0) \leqslant p(W_1;\lambda_0) + p(W_2;\lambda_0) = p(W_1;\lambda_0) ,$$

which implies the minimality of the factorization $W = W_1W_2$ at λ_0.

If λ_0 is not a zero of W_2 (i.e., $z(W_2;\lambda_0) = 0$), then, using right coprime matrix fraction descriptions of W and W_2, one proves that $z(W;\lambda_0) = z(W_1;\lambda_0)$. □

Note that the converse statement of Theorem 3.1 is not true in general if one omits the assumption that no point in Γ is a simultaneous zero and pole of $W_2(\lambda)$. This can be demonstrated by the following example.

EXAMPLE 3.1 Let

$$W(\lambda) = \begin{bmatrix} \frac{\lambda-1}{\lambda-2} & 0 \\ 0 & \frac{\lambda}{\lambda-1} \end{bmatrix}, \quad W_2(\lambda) = \begin{bmatrix} 1 & \frac{1}{\lambda-1} \\ 0 & 1 \end{bmatrix}.$$

Then $W(\lambda) = I + C(\lambda-A)^{-1}B$, with

$$A = \begin{bmatrix} 2 & 0 \\ 0 & 1 \end{bmatrix}, \quad B = C = \begin{bmatrix} 1 & 0 \\ 0 & 1 \end{bmatrix},$$

is a minimal realization of W. Also, $W_2(\lambda) = I + C_2(\lambda-A_2)^{-1}B_2$, with

$$C_2 = \begin{bmatrix} 1 \\ 0 \end{bmatrix}, \quad A_2 = 1, \quad B_2 = [0 \quad 1],$$

is a minimal realization of W_2. Now clearly $(C_2, A_2-B_2C_2)$ is a restriction of $(C, A-BC)$, and (A_2, B_2) is a compression of (A,B). However, the factorization

$$\begin{bmatrix} \frac{\lambda-1}{\lambda-2} & 0 \\ 0 & \frac{\lambda}{\lambda-1} \end{bmatrix} = \begin{bmatrix} \frac{\lambda-1}{\lambda-2} & -\frac{1}{\lambda-2} \\ 0 & \frac{\lambda}{\lambda-1} \end{bmatrix} \begin{bmatrix} 1 & \frac{1}{\lambda-1} \\ 0 & 1 \end{bmatrix}$$

is not minimal at $\lambda_0 = 1$. □

We pass now to the characterization of locally minimal divisors in terms of matrix fraction descriptions of the given r.m.f.'s. To this end we need the following definitions and facts about local divisibility. Let L and M be regular r.m.f.'s which are analytic in the given domain Λ. If the quotient ML^{-1} (resp. $L^{-1}M$) is also analytic in Λ, we say that L is a *right* (resp. *left*) *Λ-local divisor* of M. The following fact is useful (see [5]): L is a right (resp. left) Λ-local divisor of M if and only if a right (resp. left) Λ-zero pair of L is a restriction (resp. compression) of a right (resp. left) Λ-zero pair of M. (The definition of a left Λ-zero pair of M is analogous to the definition of a right Λ-zero pair given in the beginning of this section.) Let M_1, M_2 and L be regular r.m.f.'s which are analytic in Λ and let L be a right (resp. left) Λ-local divisor of both M_1 and M_2. The r.m.f. L is called a *greatest right* (resp. *left*) *Λ-local divisor* of M_1 and M_2 if any other right (resp. left) Λ-local divisor of both M_1 and M_2 is in turn a right (resp. left) Λ-local divisor of L. If L is a greatest right (resp. left) Λ-local divisor of M_1 and M_2, the integer $z(L;\lambda_0)$ $(\lambda_0 \in \Lambda)$ is called the *common right* (resp. *left*) *zero multiplicity of* M_1 *and* M_2 *at* λ_0 and the integer $\sum_{\lambda\in\Lambda} z(L;\lambda)$ will be referred to as the *common right* (resp. *left*) *Λ-zero multiplicity* of M_1 and M_2.

Now let

$$W = N_R D_R^{-1} \ , \quad W_2 = N_{R2} D_{R2}^{-1} \tag{3.3}$$

be right coprime matrix fraction descriptions of the given r.m.f.'s W and W_2. In view of Proposition 1.4 the pole pairs of W and W_2 coincide with the left zero pairs of D_R and D_{R2}, respectively. Therefore Theorem 3.1 implies, in particular, that if W_2 is a minimal divisor of W with respect to Λ, then, necessarily, D_{R_2} is a left Λ-local divisor of D_R, i.e., $A(\lambda) := D_{R2}^{-1}D_R$ is analytic in Λ. The following proposition reveals the extra condition which should be added to the analyticity of $A(\lambda)$ in order to obtain a criterium of minimality of the factorization $W = W_1W_2$ $(W_1 := WW_2^{-1})$ in Λ.

PROPOSITION 3.2 *Let a right denominator* D_{R_2} *of* W_2 *be a left Λ-local divisor of a right denominator* D_R *of* W. *Then* W_2 *is a right minimal divisor of* W *with respect to* Λ *if and only if the common right Λ-zero multiplicity of* WD_R $(= N_R)$ *and* W_2D_R $(= N_{R_2}A)$ *is equal to the sum of zero multiplicities of* $W_2D_{R_2}$ $(= N_2)$ *taken over all zeros of* $W_2D_{R_2}$ *in* Λ.

Proof. Let $W_1 = N_{R1}D_{R1}^{-1}$ be a right coprime matrix fraction description of the r.m.f. $W_1 := WW_2^{-1}$. Then

$$N_R = N_{R1}F \ , \quad N_{R_2}A = D_{R1}F \ , \tag{3.4}$$

where $F := N_{R1}^{-1}N_R = D_{R1}^{-1}N_{R_2}A$. Since N_{R1} and D_{R1} are right coprime, there exist matrix polynomials L and M such that

$$LN_{R1} + MD_{R1} = I \ .$$

Multiplying this equality by F from the right and using (3.4) we obtain the equality

(3.5) $LN_R + MN_{R_2}A = F$,

which shows, in particular, that F is analytic in Λ. We claim that F is a greatest right Λ-local divisor of N_R and $N_{R_2}A$. Indeed, equalities (3.4) show that F is a right Λ-local divisor of both N_R and $N_{R_2}A$. Moreover, if F_1 is another right Λ-local divisor of both N_R and $N_{R_2}A$, we have from (3.5)

$$FF_1^{-1} = LV_1 + MV_2 ,$$

where the r.m.f.'s $V_1 := N_R F_1^{-1}$ and $V_2 := N_{R_2}AF^{-1}$ are analytic in Λ. So, FF_1^{-1} is analytic in Λ, which proves that F is a greatest right Λ-local divisor of N_R and $N_{R_2}A$.

Now Proposition 1.4 implies that

$$p(W_i;\lambda_0) = z(D_{Ri};\lambda_0) \quad (i = 1,2) ; \quad p(W;\lambda_0) = z(D_R;\lambda_0) .$$

Using the equality

$$z(D_R;\lambda_0) - z(D_{R_2};\lambda_0) = z(A;\lambda_0) ,$$

which follows from the definition of A, we see that the factorization $W = W_1W_2$ is minimal at λ_0 if and only if

(3.6) $z(D_{R_1};\lambda_0) = z(A;\lambda_0)$.

But the second equality in (3.4) implies that

$$z(N_{R_2};\lambda_0) + z(A;\lambda_0) = z(D_{R_1};\lambda_0) + z(F;\lambda_0) ,$$

and therefore (3.6) is equivalent to the equality

$$z(N_{R_2};\lambda_0) = z(F;\lambda_0) ,$$

which proves the Proposition in the case $\Lambda = \{\lambda_0\}$. Now using the inequalities $p(W;\lambda_0) \leq p(W_1;\lambda_0) + p(W_2;\lambda_0)$ $(\lambda_0 \in \mathbb{C})$ one deduces easily the statement of the Proposition in the general case. □

We remark that an analogue of Proposition 3.2 can be stated in terms of left numerators of W and W_2.

III.2 Locally Minimal Divisors with Prescribed Local Pairs

In this section we solve the following problem: Given an $n \times n$ regular r.m.f. W, given a set $\Lambda \in \mathbb{C}$ and given right and left admissible pairs (X,T) and (U,Y), respectively; when does there exist an $n \times n$ regular r.m.f. W_2 with (X,T) as Λ-zero and (U,Y) as Λ-pole pair such that W_2 is a right minimal divisor of W with respect to Λ? By Theorem 3.1 a necessary condition for existence of such a W_2 is that (X,T) is a restriction of a Λ-zero pair of W and (U,Y) is a compression of a Λ-pole pair of W. Example 2.1 shows that for the case $\Lambda = \mathbb{C}$ (and $W_2(\infty) = I$) these conditions are not sufficient. However, it turns out that for the case when $\Lambda \neq \mathbb{C}$ the necessary conditions mentioned above are sufficient as well; moreover, there is such a W_2 with the additional properties that $W_2(\infty) = I$ and W_2 has poles and zeros outside Λ in at most one prescribed point. All these assertions are contained in the following theorem.

THEOREM 3.3 *Let a right admissible pair* (X,T) *be a restriction of a* Λ*-zero pair of an* $n \times n$ *regular rational matrix function* $W(\lambda)$ *and let a left admissible pair* (U,Y) *be a compression of a* Λ*-pole pair of* $W(\lambda)$, *where* $\Lambda \neq \mathbb{C}$ *is a set in the complex plane. Then for every* $\lambda_0 \in \mathbb{C} \setminus \Lambda$ *there exists a rational matrix function* $W_2(\lambda)$ *with* $W_2(\infty) = I$ *such that* (X,T) *is a* Λ*-zero pair of* W_2, (U,Y) *is a* Λ*-pole pair of* W_2 *and* W_2 *is a right minimal divisor of* W *with respect to* $\mathbb{C} \setminus \{\lambda_0\}$.

We need two lemmas for the proof of Theorem 3.3.

LEMMA 3.4 *Let* $\hat{V}(\lambda)$ *be a regular* $n \times n$ *rational matrix function (not necessarily with* $\hat{V}(\infty) = I$), *and assume that* $\mu_0 \in \mathbb{C}$ *is not a pole nor a zero of* $\hat{V}(\lambda)$. *Then there exists an* $n \times n$ *rational matrix function* $V(\lambda)$ *with the poles and zeros (if any) at* μ_0 *and* ∞ *only such that* $[V(\lambda)\hat{V}(\lambda)]_{\lambda=\infty} = I$.

<u>Proof</u>. Let R be a positive number so large that $\hat{V}(\lambda)$ does not have poles and zeros in the set $\{\lambda \in \mathbb{C} \mid |\lambda-\mu_0| \geq R\}$. Consider the Wiener-Hopf factorization of $\hat{V}(\lambda)$ with respect to the contour $\{\lambda \in \mathbb{C} \mid |\lambda-\mu_0| = R\}$:

$$\hat{V}(\lambda) = E(\lambda)\mathrm{diag}((\lambda-\mu_0)^{\kappa_i})_{i=1}^{n} F(\lambda) ,$$

where $E(\lambda)$ and $F(\lambda)$ are r.m.f.'s such that $F(\lambda)$ has no zeros and poles in the set $\{\lambda \in \mathbb{C} \mid |\lambda-\mu_0| \geq R\} \cup \{\infty\}$, $E(\lambda)$ has no zeros and poles in the set $\{\lambda \in \mathbb{C} \mid |\lambda-\mu_0| \leq R\}$, and $\kappa_1 \leq \cdots \leq \kappa_n$ are integers. Because of the choice of R it is easily seen that the only poles and zeros of $E(\lambda)$ are at infinity. Now put

$$V(\lambda) = F(\infty)^{-1}\operatorname{diag}((\lambda-\mu_0)^{-\kappa_i})_{i=1}^{n} E(\lambda)^{-1}$$

to satisfy the conditions of the lemma. □

LEMMA 3.5 *Let* $\mu_0 \in \mathbb{C}$ *be a zero (resp. a pole) of a regular rational matrix function* $W(\lambda)$ *and let* $E(\lambda)$ *and* $F(\lambda)$ *be rational matrix functions which are analytic and invertible at* μ_0. *If* (X,T) *(resp.* (U,Y)*) is a zero (resp. a pole) pair of* $W(\lambda)$ *at* μ_0*, then*

$$(3.7) \qquad \left(\frac{1}{2\pi i}\int_\Omega F(\lambda)^{-1}X(\lambda-T)^{-1}d\lambda\ ,\ T\right)$$

(resp.

$$(3.8) \qquad \left(U\ ,\ \frac{1}{2\pi i}\int_\Omega (\lambda-U)^{-1}YF(\lambda)^{-1}d\lambda\right))$$

is a zero (resp. a pole) pair of $\widetilde{W}(\lambda) := E(\lambda)W(\lambda)F(\lambda)$ *at* μ_0*, where* Ω *is a small circle around* μ_0. *In particular, the zero pairs at* μ_0 *of* W *and* EW *coincide, as well as their pole pairs at* μ_0.

<u>Proof</u>. Write left coprime matrix fraction $W = D_L^{-1}N_L$. Then by Proposition 1.4 (X,T) is a zero pair of N_L at μ_0. Further, the factorization $\widetilde{W} = ED_L^{-1}\cdot N_LF$ is minimal at μ_0. Indeed,

$$z(\widetilde{W};\mu_0) = z(W;\mu_0) = z(N_L;\mu_0) = z(N_LF;\mu_0)\ ;$$

also $z(ED_L^{-1};\mu_0) = p(D_LE^{-1};\mu_0) = 0$ because both D_L and E^{-1} are analytic at μ_0. So

$$z(\widetilde{W};\mu_0) = z(ED_L^{-1};\mu_0) + z(N_LF;\mu_0)\ ,$$

and the factorization $\widetilde{W} = ED_L^{-1}\cdot N_LF$ is minimal at μ_0. By Theorem 3.1, a zero pair (C,A) at μ_0 of N_LF is a restriction of a zero pair (C_1,A_1) at μ_0 of $\widetilde{W}$. But the sizes of A and A_1 coincide (in fact, the size of A (resp. of A_1) is equal to $z(N_LF;\mu_0)$ (resp. to $z(\widetilde{W};\mu_0)$). So, actually (C,A) and (C_1,A_1) are similar, and N_LF and $\widetilde{W}$ have the same zero pairs at μ_0. Now formula (3.7) follows from Theorem 2.3 in [7]. Formula (3.8) is obtained from (3.7) using the fact that (U,Y) is a pole pair of $W(\lambda)$ at μ_0 if and only if (Y^*,U^*) is a zero pair of $(W(\bar{\lambda}))^{*-1}$ at $\bar{\mu}_0$. □

Proof of Theorem 3.3. Taking into account Lemma 3.4 and the last statement of Lemma 3.5, it is sufficient to prove the existence of an r.m.f. $\hat{W}_2(\lambda)$ (not necessarily with $\hat{W}_2(\infty) = I$) whose zero (resp. pole) pair is (X,T) (resp. (U,Y)) and which is a right minimal divisor of W (with respect to $\mathbb{C}$).

We shall prove this by induction on the number p of different points where W has a pole and/or a zero. Consider the case $p = 1$. In view of Lemma 3.5 we can assume that W is in the Smith-McMillan form at λ_1, where λ_1 is the only zero and/or pole of W:

$$W(\lambda) = \mathrm{diag}((\lambda-\lambda_1)^{\nu_1},\cdots,(\lambda-\lambda_1)^{\nu_n}) ,$$

where $\nu_i \geq 0$ for $i = 1,\cdots,k$ and $\nu_i \leq 0$ for $i = k+1,\cdots,n$. By putting

$$\hat{W}_2(\lambda) = \begin{bmatrix} \hat{W}_{21}(\lambda) & 0 \\ 0 & \hat{W}_{22}(\lambda) \end{bmatrix} ,$$

where $\hat{W}_{21}(\lambda)$ (resp. $\hat{W}_{22}(\lambda)$) is a suitable r.m.f., we reduce the construction of $\hat{W}_2(\lambda)$ to the case when either all ν_i are nonnegative or all ν_i are nonpositive. In the former case there exists a matrix polynomial $\hat{W}_2(\lambda)$ with the desired properties (see, e.g., Chapter 7 in [4]). Analogously, in the latter case there exists $\hat{W}_2(\lambda)$ with the desired properties which is the inverse of a polynomial in λ.

Consider now the case of $p > 1$, assuming that for r.m.f.'s W with $p - 1$ different poles and zeros the existence of $\hat{W}_2$ has been proved. Let $\lambda_1,\cdots,\lambda_p$ be all the different poles and zeros of W. Write $W = V_1V_2$, where the r.m.f. V_1 (resp. V_2) has its poles and zeros in λ_1 (resp. in $\{\lambda_2,\cdots,\lambda_p\}$). Existence of such V_1 and V_2 follows, for instance, from the (global) Smith-McMillan form of W:

$$W(\lambda) = L_1(\lambda)\mathrm{diag}(p_1(\lambda),\cdots,p_n(\lambda))L_2(\lambda) ,$$

where L_1 and L_2 are unimodular matrix polynomials, and $p_i(\lambda)$, $i = 1,\cdots,n$ are scalar rational functions. Let (C,A) (resp. (G,B)) be a Λ-zero (resp. Λ-pole) pair of W. For a square matrix S denote by $M(S)$ (resp. $N(S)$) the maximal S-invariant subspace such that $\sigma(S_{|M(S)}) \subset \{\lambda_2,\cdots,\lambda_p\}$ (resp. $\sigma(S_{|N(S)}) \subset \{\lambda_1\}$), and let Π_S be the projection on $M(S)$ along $N(S)$. As (X,T) is a restriction of (C,A), the pair $(X_{|M(T)},T_{|M(T)})$ is a restriction of $(C_{|M(A)}, A_{|M(A)})$. Also, $(\Pi_U U_{|M(U)}, \Pi_U Y)$ is a compression of $(\Pi_G G_{|M(G)}, \Pi_G B)$. By the induction hypothesis, there exists an r.m.f. $\hat{V}_2$ which is a right minimal divisor of V_2 (with respect to $\mathbb{C}$) and whose zero (resp. pole) pair is $(X_{|M(T)}, T_{|M(T)})$ (resp. $(\Pi_U U_{|M(U)}, \Pi_U Y)$). Denote

$$\tilde{T} = T_{|N(T)}, \quad \tilde{X} = \frac{1}{2\pi i}\int_{\Omega} \hat{V}_2(\lambda) X_{|N(T)}(\lambda-\tilde{T})^{-1} d\lambda ,$$

$$\tilde{U} = (I-\Pi_U)U_{|N(U)}, \quad \tilde{Y} = \frac{1}{2\pi i}\int_{\Omega} (\lambda-\tilde{U})^{-1}(I-\Pi_U)Y\hat{V}_2(\lambda) d\lambda ,$$

where Ω is a small circle around λ_1. Decompose $W\hat{V}_2^{-1} = Q_1 Q_2$, where Q_1 (resp. Q_2) is an r.m.f. with all zeros and poles in $\{\lambda_1\}$ (resp. in $\{\lambda_2, \cdots, \lambda_p\}$); then, by the case $p = 1$, which we already considered, there exists an r.m.f. $\hat{V}_1(\lambda)$ which is a right minimal divisor of $Q_1(\lambda)$ and whose zero (resp. pole) pair is $(\tilde{X}, \tilde{T})$ (resp. $(\tilde{U}, \tilde{Y})$). (Lemma 3.5 ensures that $(\tilde{X}, \tilde{T})$ is a restriction of a zero pair of Q_1 at λ_1, and that $(\tilde{U}, \tilde{Y})$ is a compression of a pole pair of Q_1 at λ_1). Now put $\hat{W}_2 = \hat{V}_1 \hat{V}_2$. By Lemma 3.5, (X,T) (resp. (U,Y)) is a zero (resp. pole) pair of $\hat{W}_2$. It is easy to see that $\hat{W}_2$ is a right minimal divisor of W. Indeed, let $H_1(\lambda) = Q_1(\lambda)\hat{V}_1(\lambda)^{-1}$ and write

$$W = Q_2 H_1 \cdot \hat{W}_2 . \tag{3.9}$$

As Q_2 and $\hat{V}_2$ are analytic and invertible at λ_1, we have $z(W;\lambda_1) = z(Q_1;\lambda_1)$. Further, $z(Q_1;\lambda_0) = z(H_1;\lambda_0) + z(\hat{V}_1;\lambda_0) = z(Q_2 H_1;\lambda_0) + z(\hat{V}_1\hat{V}_2;\lambda_0)$, and the factorization (3.9) is minimal at λ_1. Using the representation $W = V_1 H_2 \hat{V}_1^{-1} \cdot \hat{W}_2$, where $H_2 = V_2 \hat{V}_2^{-1}$, one checks the minimality of (3.9) in $\lambda_2, \cdots, \lambda_p$ analogously. □

REFERENCES

1. Bart, H., Gohberg, I., Kaashoek, M.A.: Minimal factorization of matrix and operator functions, Operator Theory: Advances and Applications, Vol. 1, Birkhauser, Basel, 1979.

2. Gohberg, I., Kaashoek, M.A., Lerer, L., Rodman, L.: Common multiples and common divisors of matrix polynomials, II. Vandermonde and resultant matrices, Linear and Multilinear Algebra 12 (1982), 159-203.

3. Gohberg, I., Kaashoek, M.A., van Schagen, F.: Rational matrix and operator functions with prescribed singularities, Integral Equations and Operator Theory 5 (1982), 673-717.

4. Gohberg, I., Lancaster, P., Rodman, L.: Matrix polynomials, Academic Press, New York etc., 1982.

5. Gohberg, I., Rodman, L.: Analytic matrix functions with prescribed local data, J. d'Analyse Mathematique 40 (1981), 90-128.

6. Hearon, J.Z.: Nonsingular solutions of TA - BT = C, Linear Algebra and Appl. 16 (1977), 57-65.

7. Kaashoek, M.A., van der Mee, C.V.M., Rodman, L.: Analytic operator functions with compact spectrum, I. Spectral nodes, linearization and equivalence, Integral Equations and Operator Theory 4 (1981), 504-547.

8. Kailath, T.: Linear system, Prentice-Hall, Englewood Cliffs, 1980.

9. Lancaster, P., Lerer, L., Tismenetsky, M.: Factored forms for solutions of AX - XB = C and X - AXB = C in companion matrices, Linear Algebra and Appl., to appear.

10. Vanderwalle, J., Dewilde, P.: A local I/O structure theory for multivariable systems and its application to minimal cascade realization, IEEE Trans. on Circuits and Systems, CAS-25 (1978), 279-289.

I. Gohberg,
School of Mathematical Sciences,
Tel-Aviv University,
Tel-Aviv,
Israel

M.A. Kaashoek,
Wiskundig Seminarium,
Vrije Universiteit,
1007 MC Amsterdam,
The Netherlands

L. Lerer,
Department of Mathematics,
Technion-Israel Institute of Technology,
Haifa,
Israel

L. Rodman,
School of Mathematical Sciences,
Tel-Aviv University,
Tel-Aviv,
Israel

Operator Theory:
Advances and Applications, Vol. 12

THE LINEAR-QUADRATIC OPTIMAL CONTROL PROBLEM--THE OPERATOR THEORETIC VIEWPOINT[1)]

E. A. Jonckheere and L. M. Silverman

The celebrated linear-quadratic optimal control problem is examined in the light of some operator-theoretic techniques. Special attention is devoted to the case where the quadratic integrand of the cost functional is not necessarily positive semidefinite, as this directly opens the road to a fundamental "positivity" issue in general system theory. In essence, the operator theoretic approach to the linear-quadratic problem identifies the crucial role played by two bounded self-adjoint operators defined on appropriate Hilbert spaces of controls. Central in the problem of the existence of a stabilizing solution to the Algebraic Riccati Equation is the positive definiteness of a Wiener-Hopf operator, while the existence of an antistabilizing solution involves the positive definiteness of a perturbed Wiener-Hopf operator. Further, the spectra of the Wiener-Hopf and the perturbed Wiener-Hopf operators reveal a great deal about the "fine" structure of the linear-quadratic problem. Finally, the spectra of both operators can be determined using the factorization ideas of Gohberg and Krein.

INTRODUCTION

Linear-quadratic optimal control, as originally formulated in the pioneering work of Kalman [16], is well known to be one of the most fundamental problems of modern system theory. In addition to its interest in its own right, infinite-time linear-quadratic optimal control with positive semidefinite integrand of the integral performance criterion provides a systematic design procedure for linear, stationary full-state feedback gain, guaranteeing closed-loop stability and other desirable robustness properties. The significance of linear-quadratic optimization to control system design was discovered by Kalman [18]; some latest developments are due to Jameson and Kreindler [6], and Safonov and Athans [25]. When the quadratic integrand and the terminal weighting are no longer restricted to be positive semidefinite, the range of interpretation

1) This research was supported by AFOSR Grant 80-0013.

and application of linear-quadratic control becomes even broader. Indeed, beyond the "academic" problem of "the existence of a lower bound to the cost functional" or "the boundedness of the infimum," it is the extremely important issue of positivity of functionals which is involved, and which is central in positive realness and in the stability of nonlinear feedback systems, as the way has been indicated by Popov [22], Zames [30], Brockett [3], and Willems [27]. In addition to this, there is plenty of room left to reinterpret all of the aforementioned features in the stochastic framework, see Kailath [15], but in this paper we shall remain within the confines of control, although the translation from control to filtering is sometimes not quite straightforward.

Beside a hard core of well-established and well-understood results, there are still several topics in linear-quadratic control which have not yet been completely elucidated, despite more than two decades of intensive research effort. One will recognize, among other things, the linkage between the so-called time-domain and frequency-domain properties of linear-quadratic control, along which several asserted relationships, e.g., Willems [27, Theorem 4] and Anderson [1, Theorem 2], were subsequently found to be wrong; see Willems [28] and Anderson [2]. Also, Kalman [17] formulated some outstanding invariant theoretic issues in linear-quadratic control, and Khargonekar [20] showed how to derive canonic forms for linear-quadratic problems.

The purpose of this paper is to develop a new avenue of approach to linear-quadratic problems. The analysis relies on the Hilbert space formulation of the linear-quadratic problem, due to Jonckheere and Silverman [11-14]. In this approach, the cost is shown to be related, in a crucial way, to several bounded, self-adjoint operators defined over appropriate Hilbert spaces of control. The positivity of these operators is tightened to the positivity of quadratic functionals and the spectra of these operators have already been shown to be the missing chain links in the previous attempts to clearly connect time-domain and frequency-domain properties of linear-quadratic control; see Jonckheere and Silverman [11]. As a spinoff, new, deeper insights into the linear-quadratic problem are gained, and this sheds light on some of the outstanding problems listed in the previous paragraph.

BASIC DEFINITIONS AND RESULTS

Consider the continuous-time, linear, finite-dimensional system

$$\dot{x}(t) = A\,x(t) + B\,u(t) \;, \tag{1a}$$

$$x(a) = \xi \;, \quad a \leq t \leq b \;, \tag{1b}$$

where A and B are time-invariant matrices whose sizes are consistent with $x(t) \in \mathbb{R}^n$ and $u(t) \in \mathbb{R}^r$; the pair (A, B) is controllable, and we further assume that the matrix A is asymptotically stable, i. e., $\mathrm{Re}\,\lambda(A) < 0$. Here we invoke feedback invariance to justify the (apparent) restriction that A is asymptotically stable; for a thorough exposition of the feedback invariance properties of linear-quadratic control, see Khargonekar [20]. Together with the plant (1), define the integral performance index with quadratic integrand

$$I\left(\xi, u_{[a,b]}\right) := \int_a^b w[x(t), u(t)]\,dt \;, \tag{2a}$$

$$w(x, u) := x^T Q x + 2x^T S u + u^T R u \;. \tag{2b}$$

The overall weighting matrix $\begin{pmatrix} Q & S \\ S^T & R \end{pmatrix}$ is time-invariant, symmetric, but not necessarily positive semidefinite. $u_{[a,b]}$ is a control signal in $L^2_{\mathbb{R}^r}[a, b]$, which guarantees the existence of a unique, absolutely continuous solution x(t) to (1).

The linear-quadratic problem consists in finding, <u>if it exists,</u> the infimum of (2) for all $u_{[a,b]} \in L^2_{\mathbb{R}^r}[a, b]$ subject to the "dynamical constraints" (1).

It will prove convenient in the sequel to express the quadratic integrand w as the scalar product of two artificially defined outputs. To this end, we factor the overall weighting matrix as

$$W = \begin{pmatrix} C^T \\ D^T \end{pmatrix} (E \;\; F) \;, \tag{3a}$$

$$C, E \in \mathbb{R}^{m \times n} \;, \tag{3b}$$

$$D, F \in \mathbb{R}^{m \times r} \;. \tag{3c}$$

The outputs are then defined as

$$y(t) := Cx(t) + Du(t) \quad , \tag{4a}$$

$$z(t) := Ex(t) + Fu(t) \quad , \tag{4b}$$

which themselves define the transfer matrices

$$\hat{J}(s) := D + C(sI - A)^{-1}B \quad , \tag{5a}$$

$$\hat{K}(s) := F + E(sI - A)^{-1}B \quad . \tag{5b}$$

In the case of a positive semidefinite quadratic integrand, it is clear that one has the freedom to take (C D) = (E F), hence y = z. With this system theoretic notation, (2a) can be rewritten

$$I\left(\xi, u_{[a,b]}\right) = \int_a^b y^T(t)\, z(t)\, dt \quad . \tag{6}$$

In this approach to linear-quadratic optimization, we consider $I(\xi, u_{[a,b]})$ as defining, for each initial state $\xi \in \mathbb{R}^n$, a functional: $L^2_{\mathbb{R}^r}[a,b] \to \mathbb{R}$. Of paramount importance is the dependency of this functional on the end points a and b, and this motivates the use of the subscript [a, b] to specify over what time interval a function or an operator is defined. With this notation, we rewrite (6) as

$$I\left(\xi, u_{[a,b]}\right) = \left(y_{[a,b]}, z_{[a,b]}\right) . \tag{7}$$

The cost functional is now manipulated with the objective of uncovering the underlying operators. Combining (1) and (4) yields

$$y_{[a,b]} = C\, e^{A(\cdot - a)}\xi + J_{[a,b]}\, u_{[a,b]} \, , \tag{8a}$$

$$z_{[a,b]} = E\, e^{A(\cdot - a)}\xi + K_{[a,b]}\, u_{[a,b]} \, . \tag{8b}$$

$C\, e^{A(\cdot - a)}\xi$ is a mapping in $L^2_{\mathbb{R}^m}[a,b]$, $t \mapsto C\, e^{A(t-a)}\xi$. $J_{[a,b]}$ is the (bounded) Volterra convolution operator: $L^2_{\mathbb{R}^r}[a,b] \to L^2_{\mathbb{R}^m}[a,b]$, $u(t) \mapsto (J_{[a,b]}u_{[a,b]})(t) := Du(t) + \int_a^t C e^{A(t-\beta)} Bu(\beta)\, d\beta$; this operator can be thought of as being associated with $\hat{J}(s)$. The mapping $E\, e^{A(\cdot - a)}\xi$ and the operator $K_{[a,b]}$ are defined in similar obvious ways.

Combining (7) and (8) and using classical properties of the $L^2_{\mathbb{R}^m}[a,b]$-scalar product yield

$$I\left(\xi, u_{[a,b]}\right) = \left(e^{A(\cdot - a)}\xi, C^T E\, e^{A(\cdot - a)}\xi\right)$$
$$+ \left(e^{A(\cdot - a)}\xi, C^T K_{[a,b]} u_{[a,b]}\right)$$
$$+ \left(E^T J_{[a,b]} u_{[a,b]}, e^{A(\cdot - a)}\xi\right)$$
$$+ \left(u_{[a,b]}, J^*_{[a,b]} K_{[a,b]} u_{[a,b]}\right) . \tag{9}$$

It should be noted that the right hand side does no longer depend onthe particular factorization (3) of W. Indeed, by (3), $C^T E = Q$. Further, $C^T K_{[a,b]}$ is defined as the Volterra operator associated with $C^T \hat{K}(s) = C^T E(sI-A)^{-1}B + C^T F = Q(sI-A)^{-1}B + S$, that is, $C^T K_{[a,b]}: L^2_{\mathbb{R}^r}[a,b] \to L^2_{\mathbb{R}^n}[a,b]$, $u_{[a,b]} \mapsto (C^T K_{[a,b]} u_{[a,b]})(t) = \int_a^t Qe^{A(t-\beta)}Bu(\beta)\,d\beta + Su(t)$. Likewise, $E^T J_{[a,b]}$ is defined as the Volterra operator associated with $E^T \hat{J}(s) = E^T C(sI-A)^{-1}B + E^T D = Q(sI-A)^{-1}B + S = C^T \hat{K}(s)$. Hence, $E^T J_{[a,b]} = C^T K_{[a,b]}$. Finally, defining

$$R_{[a,b]} := J^*_{[a,b]} K_{[a,b]} , \tag{10a}$$

it is easily seen that this is a bounded, self-adjoint operator

$$R_{[a,b]}: L^2_{\mathbb{R}^r}[a,b] \to L^2_{\mathbb{R}^r}[a,b] , \tag{10b}$$

with kernel

$$R_{[a,b]}(\alpha,\beta) = R\delta(\alpha-\beta) + \int_{\max\{\alpha,\beta\}}^b B^T e^{A^T(t-\alpha)} Q e^{A(t-\beta)} B\, dt$$
$$+ \begin{cases} B^T e^{A^T(\beta-\alpha)} S, & \text{for } a \le \alpha \le \beta \le b , \\ S^T e^{A(\alpha-\beta)} B , & \text{for } a \le \beta \le \alpha \le b . \end{cases} \tag{10c}$$

Using the above, the functional (9) can now be written in its definitive format:

$$I\left(\xi, u_{[a,b]}\right) = \left(e^{A(\cdot-a)}\xi,\ Q\, e^{A(\cdot-a)}\xi\right) + 2\left(e^{A(\cdot-a)}\xi,\ C^T K_{[a,b]} u_{[a,b]}\right) + \left(u_{[a,b]},\ R_{[a,b]} u_{[a,b]}\right). \tag{11}$$

Clearly, the existence of an infimum requires $R_{[a,b]} \geq 0$, and, further, the optimal control involves $R^{-1}_{[a,b]}$. The operator $R_{[a,b]}$ is henceforth crucial. But it remains to relieve the properties of this operator from their dependency on the end points a and b. This is accomplished by examining the functional (11) in some limiting situations. First, as $b \to \infty$, it is quite clear that everything is well behaved, and we get

$$I\left(\xi, u_{[a,\infty)}\right) = \left(e^{A(\cdot-a)}\xi,\ Q\, e^{A(\cdot-a)}\xi\right) + 2\left(e^{A(\cdot-a)}\xi,\ C^T K_{[a,\infty)} u_{[a,\infty)}\right) + \left(u_{[a,\infty)},\ R_{[a,\infty)} u_{[a,\infty)}\right). \tag{12}$$

$C^T K_{[a,\infty)}$ is the Volterra operator associated with $C^T\hat{K}(s) = Q(sI-A)^{-1}B+S$ and defined over $[0,\infty)$, that is, $C^T K_{[a,\infty)} : L^2_{\mathbb{R}^r}[a,\infty) \to L^2_{\mathbb{R}^n}[a,\infty)$, $u_{[a,\infty)} \mapsto (C^T K_{[a,\infty)} u_{[a,\infty)})(t) = \int_a^t Q e^{A(t-\beta)} Bu(\beta)d\beta + Su(t)$. $R_{[a,\infty)}$ is the bounded, self-adjoint operator

$$R_{[a,\infty)} : L^2_{\mathbb{R}^r}[0,\infty) \to L^2_{\mathbb{R}^r}[a,\infty) \tag{13a}$$

with kernel

$$R_{[a,\infty)}(\alpha,\beta) = R\delta(\alpha-\beta) + \int_{\max\{\alpha,\beta\}}^{\infty} B^T e^{A^T(t-\alpha)} Q e^{A(t-\beta)} B\, dt + \begin{cases} B^T e^{A^T(\beta-\alpha)} S, & \text{for } a \leq \alpha \leq \beta < \infty \\ S^T e^{A(\alpha-\beta)} B, & \text{for } a \leq \beta \leq \alpha < \infty. \end{cases} \tag{13b}$$

Some elementary manipulations show that this kernel can be further rewritten as

$$R_{[a,\infty)}(\alpha,\beta) =$$

$$R\delta(\alpha-\beta) + \begin{cases} B^T e^{A^T(\beta-\alpha)}(S+YB), & \text{for } a \leq \alpha \leq \beta < \infty \\ (S^T + B^T Y)\, e^{A(\alpha-\beta)} B, & \text{for } a \leq \beta \leq \alpha < \infty, \end{cases} \tag{13c}$$

where $Y = Y^T \in \mathbb{R}^{n\times n}$ is the (unique) solution to the Lyapunov equation $A^T Y + YA = -Q$. From this last writing, it is quite evident that the operator $R_{[a,\infty)}$ has its kernel depending only on the difference of arguments ($R_{[a,\infty)}(\alpha,\beta) = R_{[a,\infty)}(\alpha-\beta)$) and is hence a Wiener-Hopf or convolution operator [11].

Consider now the other extreme situation, i. e., the case where $a \to -\infty$. Define

$$y_{(-\infty,b]} := J_{(-\infty,b]}\, u_{(-\infty,b]},$$

$$z_{(-\infty,b]} := K_{(-\infty,b]}\, u_{(-\infty,b]}.$$

$J_{(-\infty,b]}$ is the Volterra operator associated with $\hat{J}(s)$ and defined over $(-\infty, b]$, i. e., $J_{(-\infty,b]} : L^2_{\mathbb{R}^r}(-\infty,b] \to L^2_{\mathbb{R}^m}(-\infty,b]$, $u_{(-\infty,b]} \mapsto (J_{(-\infty,b]}u_{(-\infty,b]})(t) = \int_{-\infty}^{t} Ce^{A(t-\beta)}Bu(\beta)\,d\beta + Du(t)$. $K_{(-\infty,b]}$ is defined in a similar obvious way. With a slight abuse of notation, let

$$I\left(0, u_{(-\infty,b]}\right) := \left(y_{(-\infty,b]}, z_{(-\infty,b]}\right)$$

$$= \left(u_{(-\infty,b]}, J^*_{(-\infty,b]}K_{(-\infty,b]}u_{(-\infty,b]}\right). \tag{14a}$$

Defining

$$R_{(-\infty,b]} := J^*_{(-\infty,b]}K_{(-\infty,b]} \tag{15a}$$

yields

$$I\left(0, u_{(-\infty,b]}\right) = \left(u_{(-\infty,b]}, R_{(-\infty,b]}u_{(-\infty,b]}\right), \tag{14b}$$

which, owing to (11), justifies the notation. It is also easily checked that

$R_{(-\infty,b]}$ is a bounded, self-adjoint operator

$$R_{(-\infty,b]} : L^2_{\mathbb{R}^r}(-\infty, b] \to L^2_{\mathbb{R}^r}(-\infty, b] \tag{15b}$$

with kernel

$$R_{(-\infty,b]}(\alpha, \beta) = R\delta(\alpha-\beta) + \int_{\max\{\alpha,\beta\}}^{b} B^T e^{A^T(t-\alpha)} Q e^{A(t-\beta)} B \, dt$$

$$+ \begin{cases} B^T e^{A^T(\beta-\alpha)} S, & \text{for } -\infty < \alpha \le \beta \le b, \\ S^T e^{A(\alpha-\beta)} B, & \text{for } -\infty < \beta \le \alpha \le b. \end{cases} \tag{15c}$$

The operators $R_{[a,\infty)}$ and $R_{(-\infty,b]}$ are the central mathematical objects of concern in this approach, and their role is already brought to light by the following theorems:

THEOREM 1. Consider the problem (1)-(2), with A asymptotically stable and (A, B) controllable. The following statements are equivalent:

a) There exists a uniformly bounded, symmetric matrix-valued function $M(b-a) = M^T(b-a) \in \mathbb{R}^{n \times n}$ such that $I(\xi, u_{[a,b]}) \ge \xi^T M(b-a)\xi$, for all $b \ge a$, all $\xi \in \mathbb{R}^n$, and all $u_{[a,b]}$ subject to $x(b) = 0$.

b) $I(0, u_{[a,b]}) \ge 0$, for all $u_{[a,b]}$ subject to $x(b) = 0$.

c) $R_{[a,\infty)} \ge 0$.

d) $\inf \left\{ I(\xi, u_{[a,\infty)}) : u_{[a,\infty)} \in L^2_{\mathbb{R}^r}[a, \infty) \right\} = \xi^T P_+ \xi$ for some bounded matrix $P_+ = P_+^T$.

e) (In the case $R > 0$) There exists a stabilizing $(\mathrm{Re}\,\lambda(A - BR^{-1}(B^T P_+ + S^T)) \le 0)$ solution (namely P_+) to the algebraic Riccati equation.

PROOF. This result only requires straightforward modification of the discrete-time argument of Jonckheere and Silverman [11, Theorem 1].

THEOREM 2. Consider the problem (1)-(2), with A asymptotically stable and (A, B) controllable. The following statements are equivalent:

a) There exists a uniformly bounded, symmetric matrix-valued function $N(b-a) = N^T(b-a) \in \mathbb{R}^{n\times n}$ such that $I(\xi, u_{[a,b]}) \geq \xi^T N(b-a)\xi$, for all $b \geq a$, all $\xi \in \mathbb{R}^n$, and all $u_{[a,b]}$.

b) $I(0, u_{[a,b]}) \geq 0$ for all $u_{[a,b]}$.

c1) $R_{[a,b]} \geq 0, \quad \forall\, b \geq a$.

c2) $R_{(-\infty,b]} \geq 0$.

d) $\inf \{I(0, u_{(-\infty,b]}) : u_{(-\infty,b]} \in L^2_{\mathbb{R}^r}(-\infty, b] \text{ and } x(b) = \eta\}$
$= -\eta^T P_- \eta$ for some bounded matrix $P_- = P_-^T \leq 0$.

e1) (In the case $R > 0$) There exists an antistabilizing ($\mathrm{Re}\,\lambda\,(A - BR^{-1}(B^T P_- + S^T)) \geq 0$) solution (namely P_-) to the algebraic Riccati equation.

e2) (In the case $R > 0$) The Riccati differential equation
$-\dot{P} = A^T P + PA + Q - (PB + S)R^{-1}(B^T P + S^T), \; P(b) = 0$
has a global solution over the interval $(-\infty, b]$.

PROOF. Again by straightforward adaptation of the discrete-time argument of Jonckheere and Silverman [11, Theorem 2].

These two theorems have unified several seemingly unrelated issues. Theorem 1 deals with the so-called "zero terminal state case," while Theorem 2 refers to the less restrictive "free terminal state case." One will recognize in Statement a) of either theorem what is probably the most intuitive issue--"the existence of a lower bound to the cost." This issue is nothing else than the "well posedness" of the linear-quadratic problem, for the minimization of an integral cost, without apriori positive semidefiniteness assumption on the quadratic integrand, could potentially result in an optimal cost diverging toward $-\infty$. Statement b) is merely a more convenient rewriting of Statement a). A deeper but less intuitive issue (requiring $R > 0$) is the existence of some special solutions to the "Riccati equation," as asserted by Statement e). These special solutions can be constructed from some related infinization problems (Statement d). However, beyond these equivalent statements, the main point is that the crucial mathematical issue is the positivity of an operator, as asserted by Statement c). It is the point of view of this paper that the most natural way to express such things as "the existence of a lower bound to the cost" or "the existence of some solution to the Riccati equation" is the clear, mathematically well posed positivity of Hilbert space operators defined over appropriate Hilbert spaces of controls. In the next two sections, the structure--more precisely the spectrum--of each operator is examined, and the link between Hilbert space properties of operators and conventional properties of linear-quadratic optimization is made.

THE OPERATOR $R_{[a,\infty)}$

The operator $R_{[a,\infty)}$, intimately related to the "zero terminal state case" and the stabilizing solution of the algebraic Riccati equation, has been shown to be the Wiener-Hopf operator with kernel (13c). Obviously, this kernel can be given a Fourier transform:

$$\hat{R}(j\omega) := \int_{-\infty}^{+\infty} R_{[a,\infty)}(t)\, e^{-j\omega t}\, dt$$

$$= R + (S^T + B^T Y)(j\omega I - A)^{-1} B + B^T(-j\omega I - A^T)^{-1}(S + YB) \,. \qquad (16)$$

The Fourier transform of the kernel of a Wiener-Hopf operator is usually referred to as the symbol of the operator, and the Wiener-Hopf operator is uniquely characterized by its symbol [5]. $R_{[a,\infty)}$ is thus the Wiener-Hopf operator on the half-line $[a, \infty)$ with symbol $\hat{R}(j\omega)$, and we write this as $R_{[a,\infty)} = W^{\hat{R}}_{[a,\infty)}$. Some further manipulations on the symbol (16) yield

$$\begin{aligned}\hat{R}(j\omega) &= R + S^T(j\omega I - A)^{-1}B + B^T(-j\omega I - A^T)^{-1}S \\ &\quad + B^T(-j\omega I - A^T)^{-1}[(-j\omega I - A^T)Y + Y(j\omega I - A)](j\omega I - A)^{-1}B \\ &= R + S^T(j\omega I - A)^{-1}B + B^T(-j\omega I - A^T)^{-1}S \\ &\quad + B^T(-j\omega I - A^T)^{-1}Q(j\omega I - A)^{-1}B \,. \end{aligned} \qquad (17)$$

Hence the symbol $\hat{R}(j\omega)$ of the Wiener-Hopf operator $R_{[a,\infty)}$ is nothing else than the celebrated frequency-domain function of Popov; see Popov [22] and Willems [27]. That Popov's function plays a central role in the linear-quadratic problem is a fact that has been known for a while, but the natural interpretation of Popov's function as the symbol of the Wiener-Hopf operator of the stabilizing solution of the Riccati equation had not been formulated before the work of Jonckheere and Silverman [11].

Obviously, $\hat{R}(j\omega)$ can be analytically continued over the entire complex plane, save a finite set of points. Therefore, we shall sometimes write $\hat{R}(s)$, where $s \in \mathbb{C}$ is the "Laplace symbol."

Consistently with the overall system theoretic flavor of this approach, we rewrite the symbol as

$$\hat{R}(j\omega) = J^T(-j\omega) K(j\omega) , \tag{18}$$

which is easily proved from (3), (5), and (17). This "factorization result" will play a crucial role in the sequel.

Coming back to the main point of the positivity of the bounded self-adjoint operator $R_{[a,\infty)}$, probably the most natural procedure consists in looking at its spectrum ...

THEOREM 3. Consider the single input case $(r = 1)$ with $\operatorname{Re}\lambda(A) < 0$. Then

$$\operatorname{spec}(R_{[a,\infty)}) = \operatorname{closure}\{\hat{R}(j\omega) : \omega \in \mathbb{R}\} . \tag{19}$$

Further, this is an essential spectrum.

PROOF. Following Gohberg and Krein [5, Theorem 13.1], the spectrum of the Wiener-Hopf operator associated with $\hat{R}(j\omega)$ consists of the closure of $\{\hat{R}(j\omega)\}$ plus some eigenvalues outside this set and with finite dimensional eigenspaces.[2)] It thus remains to prove that there are no such eigenvalues. This is proved by contradiction. Assume λ is such an eigenvalue with $\hat{u}(s)$ the corresponding eigenvector. Let $(\cdot)_+ : L^2(\Pi) \to H^2(\Pi)$ denote the orthogonal projection onto the space of functions analytic in the ORHP. Then the eigenvalue equation can be written

$$(\hat{R}(s)\,\hat{u}(s))_+ = \lambda\,\hat{u}(s) . \tag{20}$$

Counting the number of stable (OLHP) poles in the left and the right hand sides, it follows that this eigenequation has a nontrivial solution if and only if $\hat{R}(s) = \text{constant} = \lambda$, in which case λ already lies in the closure of

---

[2)] Strictly speaking, sticking to Gohberg and Krein's formulation requires $R \neq 0$, which is not part of the hypothesis of the theorem. It is easy to circumvent this difficulty. Choose δ such that $R+\delta \neq 0$. Replace R by $R+\delta$. This merely shifts the spectrum, but does not change its structure.

$\{\hat{R}(j\omega)\}$. A contradiction. The proof is completed.[3)]

From this result, we directly derive the following:

COROLLARY 1. Consider the single input case $(r = 1)$ with Re $\lambda(A) < 0$. Then $R_{[a,\infty)} \geq 0$ (>0) if and only if $\hat{R}(j\omega) \geq 0$ (>0), $\forall \omega \in \mathbb{R}$. Consequently, any of the statements of Theorem 1 is verified if and only if $\hat{R}(j\omega) \geq 0$, $\forall \omega \in \mathbb{R}$.

Observe the deep insight provided by Theorem 3. It tells us that Popov's function is not only "related to" the equivalent statements of Theorem 1, but it also tells us that the ultimate significance of Popov's function is the fact that its image is the spectrum of an operator which occupies a central position in the LQ problem.

The connection between the condition $\hat{R}(j\omega) \geq 0$, $\forall \omega \in \mathbb{R}$ and the statements of Theorem 1 has been known for a while, see Willems [27], but this connection had been established via Parseval's theorem and an intricate string of arguments. We feel that the "spectrum of $R_{[a,\infty)}$" is the natural link between Popov's condition and the statements of Theorem 1.

We now turn our attention to the multi-input $(r \geq 1)$ case which is more involved.

THEOREM 4. Let Re $\lambda(A) < 0$. Then

$$\text{ess spec}\,(R_{[a,\infty)}) = \text{closure} \bigcup_{i=1}^{r} \left\{ \lambda_i(\hat{R}(j\omega)) : \omega \in \mathbb{R} \right\} \tag{21a}$$

and

$$\text{spec}\,(R_{[a,\infty)}) = \text{ess spec}\,(R_{[a,\infty)}) \cup \{\lambda_1, \lambda_2, \ldots, \lambda_N\}, \tag{21b}$$

---

3) In the discrete-time case of a Toeplitz operator, this result is usually referred to as Hartman and Wintner's theorem; see Douglas [4, Corollaries 1.5 and 2.4] and Reference [40] cited therein. Theorem 3 could have been recovered taking for granted Hartman and Wintner's result for Toeplitz operators, and making use of Rosenblum's isomorphism between Toeplitz and Wiener-Hopf operators [4, Introduction]. In the same vein, let $\int_{-\infty}^{+\infty} \lambda\, dE(\lambda)$ be the spectral decomposition of a bounded self-adjoint Toeplitz or Wiener-Hopf operator with continuous scalar-valued symbol. In the Toeplitz case, it can be shown that the spectral family $E(\cdot)$ is absolutely continuous; see Rosenblum [24]. Similar statements can be made for the Wiener-Hopf case. But this is not essential as far as the spectrum is concerned.

where the λ's are (at most finitely many) real eigenvalues, with finite dimensional eigenspaces, located outside ess spec $(R_{[a,\infty)})$ and between the connected components of ess spec $(R_{[a,\infty)})$.

PROOF. We still refer to Gohberg and Krein [5, Theorem 13.1]. According to that theorem, (21a) is trivial. Following the same theorem there may exist some isolated eigenvalues with finite dimensional eigenspaces and located outside the essential spectrum. It thus remains to prove (i) that the λ's are located between the connected components of the essential spectrum and (ii) that there are at most finitely many such points.

To prove Claim (i), it suffices to show that either situation

(i1) $\lambda < \inf$ ess spec $(R_{[a,\infty)})$,

(i2) sup ess spec $(R_{[a,\infty)}) < \lambda$,

is impossible.

To disprove (i1), consider the operator $R_{[a,\infty)} - \lambda I$. On the one hand, by (21b), this operator is not invertible. On the other hand, by (i1) and (21a), $\hat{R}(j\omega) - \lambda I > 0$, $\forall\, \omega \in \mathbb{R}$. Hence, by the spectral factorization result of Youla [29], there exists a spectral factor $\hat{G}$, stable $(\hat{G} \in H^{\infty}_{\mathbb{C}^{r\times r}}(\mathbb{I}))$ and minimum phase $(\hat{G}^{-1} \in H^{\infty}_{\mathbb{C}^{r\times r}}(\mathbb{I}))$, such that $\hat{R}(j\omega) - \lambda I = \hat{G}^{T}(-j\omega)\,\hat{G}(j\omega)$.

It then follows that $(R_{[a,\infty)} - \lambda I)^{-1} = (W^{\hat{R}-\lambda I}_{[a,\infty)})^{-1} = (W^{\hat{G}^*\hat{G}}_{[a,\infty)})^{-1}$
$= (W^{\hat{G}^*}_{[a,\infty)}\, W^{\hat{G}}_{[a,\infty)})^{-1} = (W^{\hat{G}}_{[a,\infty)})^{-1}(W^{\hat{G}^*}_{[a,\infty)})^{-1} = W^{\hat{G}^{-1}}_{[a,\infty)}(W^{\hat{G}^{-1}}_{[a,\infty)})^{*}$. Thus, $R_{[a,\infty)} - \lambda I$ has an inverse. A contradiction. Hence (i1) is impossible.

Consider now (i2). Remembering that the substitution $\hat{R}(j\omega) \to \hat{R}(j\omega) + \delta I$ merely shifts the spectrum without altering its structure (see [2)]), we can assume without loss of generality that

$$\sup\left\{ \lambda_i(\hat{R}(j\omega)) : \omega \in \mathbb{R};\ i=1,2,\dots,r \right\}$$
$$= \sup\left\{ |\lambda_i(\hat{R}(j\omega))| : \omega \in \mathbb{R};\ i=1,2,\dots,r \right\}.$$

With this convention, (i2) together with (21a) implies

$$\| W^{\hat{R}}_{[a,\infty)} \| > \sup\left\{ \lambda_i(\hat{R}(j\omega)) : \omega \in \mathbb{R};\ i=1,\dots,r \right\}. \tag{22}$$

$\|\cdot\|$ denotes the usual spectral norm. Let $W^{\hat{R}}_{(-\infty,+\infty)}$ denote the convolution operator on the full line with kernel $R(\alpha-\beta)$; in the frequency domain, it is just a multiplication by $\hat{R}(j\omega)$. We have

$$\|W^{\hat{R}}_{[a,\infty)}\| \leq \|W^{\hat{R}}_{(-\infty,+\infty)}\| =$$

$$\sup \{ |\lambda_i(\hat{R}(j\omega))| : \omega \in \mathbb{R};\ i=1,\ldots,r\} \quad . \tag{23}$$

The inequality is trivial. The equality is easily proved using Parseval's theorem. Comparing (22) and (23) yields the desired contradiction.

It remains to prove that there are at most finitely many eigenvalues. If λ is such an eigenvalue, the corresponding eigenequation can be written

$$\hat{R}(s)\,\hat{u}(s) = \lambda\,\hat{u}(s) + \hat{v}(s) \ ;$$

$\hat{u} \in H^2_{\mathbb{C}^r}(\Pi)$ is the eigenvector, and $\hat{v}(s)$ has its poles in the ORHP. Write $\hat{R}(s) = N(s)/d(s)$, where N is the numerator polynomial matrix and d the denominator polynomial. It is easily seen that the eigenequation canbe rewritten

$$\hat{u}(s) = \frac{d(s)}{\det\,(N(s) - \lambda\,d(s)\,I)}\ \mathrm{adj}\,(N(s) - \lambda\,d(s)\,I)\,\hat{v}(s) \ .$$

Some "pole/zero cancellation" must occur, since $\hat{u}(s)$ and $\hat{v}(s)$ have their poles in the OLHP and ORHP, respectively. In particular, let z_λ be any root of $\det\,(N(s) - \lambda\,d(s)\,I)$ in the ORHP, i.e.,

$$\det\,(N(z_\lambda) - \lambda\,d(z_\lambda)\,I) = 0, \qquad \mathrm{Re}\,z_\lambda > 0 \ . \tag{24a}$$

The pole/zero cancellation requires

$$\mathrm{adj}\,(N(z_\lambda) - \lambda\,d(z_\lambda)\,I)\,\hat{v}(z_\lambda) = 0 \ . \tag{24b}$$

From nowon, the rest of the proof is only sketched since the detailed argument is long. Following the same procedure as Jonckheere and Silverman [12, Section IV], it is possible to "eliminate the z_λ's between (24a) and (24b)." The elimination procedure constructs a "Bezoutian" matrix, polynomial in λ. Further, any λ simultaneously verifying (24a) and (24b) is a zero of this polynomial matrix (the converse is not necessarily true). Hence there are at most finitely many such λ's. The proof is complete.

The substance of this multi-input theorem is the same as the single-input Theorem 3. The only discrepancy between the single- and the multi-input cases is that in the latter some eigenvalues may exist in addition to the essential spectrum. (We have examples of this situation.) If the essential spectrum is connected, there are no such eigenvalues. Further, in Khargonekar's canonic form of the LQ problem [20], these eigenvalues do not exist either. Therefore, we conjecture that the existence of some additional eigenvalues reflects some "pathology," which remains to be identified. In either case, because of their location, these additional eigenvalues are not relevant as far as the positivity or the invertibility of the operator $R_{[a,\infty)}$ is concerned, and we have the following multi-input

COROLLARY 2. Let $\mathrm{Re}\,\lambda(A) < 0$. Then $R_{[a,\infty)} \geq 0$ (>0) if and only if $\hat{R}(j\omega) \geq 0$ (>0), $\forall\, \omega \in \mathbb{R}$. Consequently any of the statements of Theorem 1 is verified if and only if $\hat{R}(j\omega) \geq 0$, $\forall\, \omega \in \mathbb{R}$.

The position of the spectrum of $R_{[a,\infty)}$ along the real axis is thus crucially related to whether or not the cost is bounded from below in the zero terminal state case. However, there are other links between the spectral properties of $R_{[a,\infty)}$ and the "conventional" properties of the linear-quadratic problem.

The problem of the existence and the uniqueness of the optimal control is amenable to the spectral analysis. To simplify the notation, rewrite (12) as

$$\begin{aligned} I(\xi, u_{[a,\infty)}) &= (\xi, Y\xi) \\ &+ 2(\xi, S_{[a,\infty)} u_{[a,\infty)}) \\ &+ (u_{[a,\infty)}, R_{[a,\infty)} u_{[a,\infty)}) \,, \end{aligned} \tag{25}$$

where $S_{[a,\infty)}: L^2_{\mathbb{R}^r}[a,\infty) \to \mathbb{R}^n$ is a compact operator. Clearly, if $R_{[a,\infty)} > 0$, the optimal control exists, is unique, and is given by

$$u^*_{[a,\infty)} = -R^{-1}_{[a,\infty)} S^*_{[a,\infty)} \xi \,. \tag{26}$$

Now, if $R_{[a,\infty)}$ is merely ≥ 0 and if its spectrum has a component of the form $[0, \delta)$ with 0 not an eigenvalue, then $R^{-1}_{[a,\infty)}$ is an unbounded operator, resulting in the optimal control to lie outside $L^2_{\mathbb{R}^r}[a,\infty)$.

Another singular situation occurs when $R_{[a,\infty)}$ is merely positive semidefinite with one of the components of its essential spectrum reduced to $\{0\}$. In this situation, since $R_{[a,\infty)} \geq 0$, the functional $I(\xi, u_{[a,\infty)})$ is still bounded from below; this and (25) yield $\text{Ker}(R_{[a,\infty)}) \subseteq \text{Ker}(S_{[a,\infty)})$. Hence the optimal control is defined up to any element of $\text{Ker}(R_{[a,\infty)})$.

The operator theoretic point of view also sheds some light on the structure of the solution P_+ of the algebraic Riccati equation. Assume $R > 0$ and $R_{[a,\infty)} \geq 0$. Then Statements d) and e) of Theorem 1 together with (25) and (26) yield

$$P_+ = Y - S_{[a,\infty)} R^{-1}_{[a,\infty)} S^*_{[a,\infty)} . \tag{27}$$

From this particular expression of the stabilizing solution of the Riccati equation, it is quite evident that the main computational burden is the inversion of the Wiener-Hopf operator $R_{[a,\infty)}$; this latter operation is known to be equivalent to a spectral factorization. But the complexity of this spectral factorization depends on the structure of the spectrum of $R_{[a,\infty)}$. This can be seen by taking the extreme situation where $r = 1$ and $\hat{R}(j\omega) = \lambda = \text{constant}, \forall \omega \in \mathbb{R}$. In this case $R_{[a,\infty)} = \lambda I$, and its inversion does not necessitate a spectral factorization. More generally, it can be proved that if any of the components of ess spec $(R_{[a,\infty)})$ is reduced to an eigenvalue (with infinite dimensional eigenspace) then a reduction of the amount of computations involved is possible.

Finally, singular or cheap control problems (i. e., $R = 0$ or $R \downarrow 0$) can also be looked at from the spectral theoretic point of view. Observe that

$$\text{spec}(R) \subseteq \text{spec}(R_{[a,\infty)}) ; \tag{28}$$

this is easily proved from Theorem 4 together with the fact that $\lim_{|\omega| \to \infty} \hat{R}(j\omega) = R$. According to this inclusion, if R is singular (singular control problem) then 0 is in the spectrum of $R_{[a,\infty)}$, resulting in the optimal control (26) to exist only in a very peculiar sense.

THE OPERATOR $R_{(-\infty,b]}$

We now come to Theorem 2 and the operator it involves. Taking a sharper look at the kernel (15c) of the operator $R_{(-\infty,b]}$, it is easily seen that it can be split as

$$R_{(-\infty,b]}(\alpha,\beta) = R\delta(\alpha-\beta) + \begin{cases} B^T e^{A^T(\beta-\alpha)}(S+YB), & -\infty < \alpha \le \beta \le b, \\ (B^TY+S^T)\, e^{A(\alpha-\beta)}B, & -\infty < \beta \le \alpha \le b, \end{cases}$$

$$-B^T e^{A^T(b-\alpha)} Y\, e^{A(b-\beta)} B \,. \tag{29a}$$

Hence the operator $R_{(-\infty,b]}$ can itself be split as

$$R_{(-\infty,b]} = W^{\hat{R}}_{(-\infty,b]} + H_{(-\infty,b]} \,; \tag{29b}$$

$W^{\hat{R}}_{(-\infty,b]}$ is the Wiener-Hopf operator: $L^2_{\mathbb{R}^r}(-\infty,b] \to L^2_{\mathbb{R}^r}(-\infty,b]$, with symbol $\hat{R}(j\omega)$, defined over the half-line $(-\infty,b]$; $H_{(-\infty,b]}$ is the bounded, self-adjoint, Hankel-like operator: $L^2_{\mathbb{R}^r}(-\infty,b] \to L^2_{\mathbb{R}^r}(-\infty,b]$, with kernel $H_{(-\infty,b]}(\alpha,\beta) = -B^T e^{A^T(b-\alpha)} Y\, e^{A(b-\beta)}B$. It is easily checked that this latter operator is compact. Hence the operator $R_{(-\infty,b]}$ of Theorem 2 has the perturbed Wiener-Hopf structure. Regarding its spectrum, we have the following

THEOREM 5. Let $\mathrm{Re}\,\lambda(A) < 0$. Then

$$\text{ess spec}\,(R_{(-\infty,b]}) = \text{ess spec}\,(R_{[a,\infty)})$$

$$= \text{closure} \bigcup_{i=1}^{r} \{\lambda_i(\hat{R}(j\omega)) : \omega \in \mathbb{R}\} \,. \tag{30}$$

PROOF. Since the compact perturbation does not change the essential spectrum [19, IV, Th. 5.35], the decomposition (29) yields $\text{ess spec}\,(R_{(-\infty,b]}) = \text{ess spec}\,(W^{\hat{R}}_{(-\infty,b]})$. It is readily seen that $W^{\hat{R}}_{(-\infty,b]}$ has exactly the same essential spectrum as $R_{[a,\infty)} = W^{\hat{R}}_{[a,\infty)}$. These facts, together with Theorem 4, yield the result.

The essential spectrum of $R_{(-\infty,b]}$ is thus easy to determine from Popov's function. But, outside the essential spectrum (30) of $R_{(-\infty,b]}$, there are, in general, some additional eigenvalues with finite-dimensional eigenspaces, and the location of these eigenvalues with respect to the essential spectrum cannot, in general, be predicted. This explains why Popov's frequency-domain condition $\hat{R}(j\omega) \ge 0,\ \forall\, \omega \in \mathbb{R}$, is necessary, but not in general sufficient, to guarantee the time-domain property $I(0, u_{(-\infty, b]}) \geq 0$, $\forall u_{(-\infty, b]}$.

There are, however, some situations where something can be said about the location of the additional eigenvalues with respect to the essential spectrum.

In the case where the quadratic integrand is positive semidefinite, we have $Q \geq 0$, hence $Y \geq 0$, and $H_{(-\infty, b]} \leq 0$. Further, in this case, $R_{(-\infty, b]} \geq 0$. Hence the additional eigenvalues, if any, are located in the interval $[0, \inf \operatorname{ess\,spec}(R_{(-\infty, b]}))$.

The so-called positive real case is characterized, in the system theoretic formulation (6), by one of the output, say z, equal to the input u. In this case, $Q = 0$, $Y = 0$, and $H_{(-\infty, b]} = 0$. Hence $R_{(-\infty, b]} = W^{\hat{R}}_{(-\infty, b]}$. Using an argument similar to that of Theorem 4, it follows that the spectrum of $R_{(-\infty, b]}$ is closure $\bigcup_{i=1}^{r} \{\lambda_i(\hat{R}(j\omega)) : \omega \in \mathbb{R}\}$, plus some eigenvalues between the connected components of the essential spectrum. As a corollary, Popov's condition is sufficient to guarantee $R_{(-\infty, b]} \geq 0$.

In the so-called scattering case, the integrand has the form $w(x, u) = u^T u - p^T p$, where p is an output of the form $Hx + Ju$. It follows that $Q = -H^T H \leq 0$, hence $Y \leq 0$, and $H_{(-\infty, b]} \geq 0$. It then follows that the additional eigenvalues of finite multiplicities in the spectrum of $R_{(-\infty, b]}$ are located to the right of the closure of $\bigcup_{i=1}^{r} \{\lambda_i(\hat{R}(j\omega)) : \omega \in \mathbb{R}\}$. Hence the classical frequency-domain condition of Popov is sufficient for positive semidefiniteness of $R_{(-\infty, b]}$.

The above perturbation analysis does not, however, resolve the problem of computing the troublesome eigenvalues of finite multiplicities responsible for the failure of $\hat{R}(j\omega) \geq 0$, $\forall \omega \in \mathbb{R}$ to provide a sufficient condition for the positive semidefinitess of $R_{(-\infty, b]}$. Apparently, the structure of $R_{(-\infty, b]}$ revealed by the splitting (29) is irrelevant to eigenvalue computation, unless the LQ problem is in Khargonekar's canonic form [20], [10]. The approach expanded upon here is based on the observation that the perturbed Wiener-Hopf operator $R_{(-\infty, b]} : L^2_{\mathbb{R}^r}(-\infty, b] \to L^2_{\mathbb{R}^r}(-\infty, b]$ can be "embedded" in a true Wiener-Hopf operator: $L^2_{\mathbb{R}^m}(-\infty, b] \to L^2_{\mathbb{R}^m}(-\infty, b]$, which is better suited to eigenvalue computation, but at the cost of an increase of dimensionality since $m \geq r$. This "embedding" is found by tracing back to the very definition (15a) of $R_{(-\infty, b]}$ and by remembering the "invariance" of the spectrum of an operator product under permutation of the factors:

$$\operatorname{spec}(R_{(-\infty, b]}) - \{0\} = \operatorname{spec}(K_{(-\infty, b]} J^*_{(-\infty, b]}) - \{0\} . \tag{31}$$

The operator

$$K_{(-\infty, b]} J^*_{(-\infty, b]} : L^2_{\mathbb{R}^m}(-\infty, b] \rightarrow L^2_{\mathbb{R}^m}(-\infty, b] \tag{32a}$$

is easily found to be the operator with kernel

$$K_{(-\infty, b]} J^*_{(-\infty, b]}(\alpha, \beta) =$$

$$FD^T \delta(\alpha - \beta) + \begin{cases} (EZ + FB^T)\, e^{A^T(\beta-\alpha)} C^T, & -\infty < \alpha \leq \beta \leq b, \\ E e^{A(\alpha-\beta)}(ZC^T + BD^T), & -\infty < \beta \leq \alpha \leq b; \end{cases} \tag{32b}$$

$Z = Z^T > 0$ is the unique solution of the Lyapunov equation $AZ + ZA^T = -BB^T$. From its kernel, it is evident that $K_{(-\infty, b]} \hat{J}^*_{(-\infty, b]}$ is a <u>Wiener-Hopf</u> operator; further, the Fourier transform of its kernel is found to be $\hat{K}(j\omega)\, \hat{J}^T(-j\omega)$; hence

$$K_{(-\infty, b]} J^*_{(-\infty, b]} = W^{\hat{K}(s)\hat{J}^T(-s)}_{(-\infty, b]} . \tag{32c}$$

Drawing upon the Wiener-Hopf property of $K_{(-\infty, b]} J^*_{(-\infty, b]}$, we could use Gohberg and Krein [5, Theorem 13.1] to derive

$$\operatorname{ess\,spec}(K_{(-\infty, b]} J^*_{(-\infty, b]}) =$$

$$\operatorname{closure} \bigcup_{i=1}^{m} \{ \lambda_i(\hat{K}(j\omega)\hat{J}^T(-j\omega)) : \omega \in \mathbb{R} \} . \tag{33}$$

This together with the "invariance of the spectrum of an operator product under commutation of the factors" would yield essentially the same result as Theorem 5. Also, observe that, in contrast to $R_{[a, \infty)}$, the Wiener-Hopf operator $K_{(-\infty, b]} J^*_{(-\infty, b]}$ is <u>not necessarily self-adjoint;</u> therefore, the argument utilized in Theorem 4 for locating the additional eigenvalues of a self-adjoint Wiener-Hopf operator brakes down, and nothing can be said, <u>at this point,</u> about the position of the additional eigenvalues of $K_{(-\infty, b]} J^*_{(-\infty, b]}$ with respect to its essential spectrum.

To pinpoint the additional eigenvalues of the spectrum of $K_{(-\infty,b]}J^*_{(-\infty,b]}$, one has to penetrate very deeply into the structure of $\hat{K}(s)J^T(-s)$; the mere "frequency-response" analysis of (33) does not provide enough insight. A much deeper insight into the spectrum of $K_{(-\infty,b]}J^*_{(-\infty,b]}$ is provided by the factorization "à la" Gohberg and Krein [5]. The spectrum of $K_{(-\infty,b]}J^*_{(-\infty,b]}$, its (frequency-domain) factorization interpretation, and its relation to the "boundedness of the infimum in the free terminal state case" constitute the culmination of the spectral theoretic approach to the LQ problem:

THEOREM 6. Let $\operatorname{Re}\lambda(A) < 0$. Then the operator $R_{(-\infty,b]}$ is positive semidefinite (and hence the other statements of Theorem 2 are verified) if and only if for all $\lambda < 0$ there exists a factorization

$$\hat{K}(j\omega)\hat{J}^T(-j\omega) - \lambda I = \hat{U}_\lambda(j\omega)\hat{V}_\lambda^T(-j\omega), \tag{34a}$$

where

$$\hat{U}_\lambda, \hat{U}_\lambda^{-1}, \hat{V}_\lambda, \hat{V}_\lambda^{-1} \in H^\infty_{\mathbb{C}\, m\times m}(\Pi) . \tag{34b}$$

(Actually these factors are rational and analytic in the ORHP, including at infinity.) Further, the spectrum of $R_{(-\infty,b]}$ (disregarding $\{0\}$ which escapes this analysis) is the set of those λ's at which the above factorability condition brakes down. Finally

$$\begin{aligned} &\text{ess spec}\,(R_{(-\infty,b]}) - \{0\} \\ &= \{\lambda : \det(\hat{K}(j\omega)\hat{J}^T(-j\omega) - \lambda I) = 0 \text{ for some } \omega \in \mathbb{R}\} - \{0\} . \end{aligned} \tag{35}$$

PROOF. To check whether the statements of Theorem 2 are verified, we have to look at the spectrum of either $R_{(-\infty,b]}$ or $K_{(-\infty,b]}J^*_{(-\infty,b]}$. This reduces to looking at the invertibility of $K_{(-\infty,b]}J^*_{(-\infty,b]} - \lambda I$. Observe that the latter is the Wiener-Hopf operator associated with $\hat{K}(s)\hat{J}^T(-s) - \lambda I$. By a known result of Gohberg and Krein [5] (see also Jonckheere and Silverman [12], [13] and Jonckheere and Delsarte [9]), the invertibility of a Wiener-Hopf operator is equivalent to the factorability of its symbol. This yields (34). Further, obviously, for any λ in the set (35), there is no factorization; and this set is easily seen to be the essential spectrum. The proof is complete.

The ability to factor the symbol $\hat{K}(s)\hat{J}^T(-s) - \lambda I$ as in (34) and for all $\lambda < 0$ can be thought of as the frequency-domain condition for positive semidefiniteness of the functional $I(0, u_{(-\infty, b]})$. This frequency-domain characterization had been left open since the publication of the counterexample [28].

With this characterization at hand, the hard problem is to pinpoint the additional eigenvalues of $R_{(-\infty, b]}$ located outside the essential spectrum. This is accomplished by going through the factorization (34), keeping λ as a symbol, and by finding the λ's at which something goes wrong. What goes wrong is the loss of a certain degree property [7], similar to the "degree dominance of the diagonal" in the factorization algorithm of Oono and Yasuura [21]. Another procedure, remote but equivalent to the factorization, consists in searching those λ's at which the operator with symbol $\hat{K}(s)\hat{J}^T(-s) - \lambda I$ has a nontrivial kernel. This procedure is expanded upon in [12], and the salient result is the fact that the critical λ's are among the zeros of a polynomial matrix. In view of these results, the computation of the spectrum of $R_{(-\infty, b]}$ should now be considered as a tractable problem.

We now come to some system theoretic interpretations of the spectrum of $R_{(-\infty, b]}$. First, to simplify the notation, observe that the constraint $x(b) = \eta$ of Statement d) of Theorem 2 can be rewritten $T_{(-\infty, b]}u_{(-\infty, b]} = \eta$, where $T_{(-\infty, b]}: L^2_{\mathbb{R}^r}(-\infty, b] \to \mathbb{R}^n$ is the (compact) reachability operator. With this notation, observe that

$$H_{(-\infty, b]} = -T^*_{(-\infty, b]} \, Y \, T_{(-\infty, b]} \, . \tag{36}$$

Now let $R_{(-\infty, b]} \geq 0$, $R > 0$, and $\hat{R}(j\omega) > 0$, $\forall \omega \in \mathbb{R}$. Then the optimal control of the infinization problem d) of Theorem 2 is easily found to be

$$u^*_{(-\infty, b]} = \left(W^{\hat{R}}_{(-\infty, b]}\right)^{-1} T^*_{(-\infty, b]} \left(T_{(-\infty, b]}(W^{\hat{R}}_{(-\infty, b]})^{-1} T^*_{(-\infty, b]}\right)^{-1} \eta ; \tag{37}$$

further,

$$P_- = Y - \left(T_{(-\infty, b]}(W^{\hat{R}}_{(-\infty, b]})^{-1} T^*_{(-\infty, b]}\right)^{-1} . \tag{38}$$

The conclusions are similar to those of the case of Theorem 1. The existence of the optimal control still requires $\hat{R}(j\omega) > 0$, $\forall \omega \in \mathbb{R}$. The main burden in the computation of P_- is the inversion of the Wiener-Hopf operator $W^{\hat{R}}_{(-\infty, b]}$, that is, the spectral factorization of $\hat{R}$. Should one of the eigenvalues of $\hat{R}(j\omega)$ be independent of ω, then a reduction of the amount of computation involved is possible.

The additional eigenvalues in the spectrum of $R_{(-\infty, b]}$ also admit some "classical" interpretations. For example, by Theorem 2, the existence of $P_- \leq 0$ is necessary and sufficient to guarantee the "well posedness" of the LQ problem. But this disregards round-off error problems. Canabal [23] has proved that the stronger condition $P_- < 0$ is necessary and sufficient for the solution P(t) of the Riccati differential equation to converge, as $t \to -\infty$, to the stabilizing solution P_+, for any terminal condition $P(b) \geq 0$, in spite of round-off errors. The condition $P_- < 0$ can be rephrased in the spectral theoretic setting as follows:

THEOREM 7. Let $\mathrm{Re}\,\lambda(A) < 0$, $R_{(-\infty, b]} \geq 0$, and $R > 0$. Then $P_- < 0$ if and only if 0 is not an eigenvalue of $R_{(-\infty, b]}$.

PROOF. By combining Statement d) of Theorem 2, (14), (29), and (36).

What is the ultimate interpretation of the additional eigenvalues in the spectrum of $R_{(-\infty, b]}$? The complete answer is not yet entirely clear. However, a clue is provided by the following:

THEOREM 8. Consider a single-input ($r = 1$), regulator problem ($w(x, u) \geq 0$, $\forall x$, $\forall u$) with $\mathrm{Re}\,\lambda(A) < 0$. Assume that the factorization (3) of the overall weighting matrix yields $\hat{J}(s) = \hat{K}(s)$ invertible and minimum phase. Then there are no eigenvalues with finite dimensional eigenspaces in the spectrum of $R_{(-\infty, b]}$ and $P_+ = 0$.

PROOF. From the definition (15a) of $R_{(-\infty, b]}$, we derive

$$R_{(-\infty, b]} = W^{\hat{J}(-s)}_{(-\infty, b]} \, W^{\hat{J}(s)}_{(-\infty, b]} \; .$$

By the minimum phase property of $\hat{J}(s)$, this operator has an inverse:

$$\begin{aligned} R^{-1}_{(-\infty, b]} &= \left(W^{\hat{J}(s)}_{(-\infty, b]}\right)^{-1} \left(W^{\hat{J}(-s)}_{(-\infty, b]}\right)^{-1} \\ &= W^{\hat{J}^{-1}(s)}_{(-\infty, b]} \, W^{\hat{J}^{-1}(-s)}_{(-\infty, b]} \\ &= W^{\hat{J}^{-1}(s)\hat{J}^{-1}(-s)}_{(-\infty, b]} \\ &= W^{\hat{R}^{-1}(s)}_{(-\infty, b]} \; . \end{aligned}$$

Thus, although $R_{(-\infty, b]}$ is not in general a Wiener-Hopf operator, its inverse is the Wiener-Hopf operator with symbol $\hat{R}^{-1}(j\omega)$. Hence, by an argument similar to that of Theorem 3, the spectrum of $R^{-1}_{(-\infty, b]}$ is essential and is the set $\{\hat{R}^{-1}(j\omega) : \omega \in \mathbb{R}\}$. Then the spectrum of $R_{(-\infty, b]}$ is essential and is the set $\{\hat{R}(j\omega) : \omega \in \mathbb{R}\}$. This completes the proof of the first claim.

To prove the second claim, observe that

$$\xi^T P_+ \xi = \inf \left\{ \int_a^\infty (Cx + Du)^2 dt : u \in L^2[a, \infty) \right\} .$$

Now, because J(s) is invertible and minimum phase, the output $Cx + Du$ can be set to zero by means of a stabilizing control. This completes the proof.

From this theorem, it is fair to conjecture that the number of additional eigenvalues in the spectrum of $R_{(-\infty, b]}$ is a measure of the complexity of the problem. Indeed, in the extreme situation of the above theorem, there are no additional eigenvalues and the solution to the optimal control problem does not require a Riccati equation. We further conjecture that there are at most n additional eigenvalues in the spectrum of $R_{(-\infty, b]}$. Actually, these facts can be proved in full generality and become much more transparent in Khargonekar's canonic form of the LQ problem [20], but this is postponed to a further paper [10].

CONCLUSIONS

In this paper, the infinite-horizon linear-quadratic problems have been looked at from the natural operator theoretic point of view. Such issues as the existence of special solutions to the "Riccati equation" and the positivity of functionals have been shown to be related to the spectra of underlying operators. These spectra are most naturally characterized in frequency-domain terms, and as such the link between time-domain and frequency-domain "positivity" properties have been established in what is probably the most natural way.

REFERENCES

1. Anderson, B.D.O.: Algebraic properties of minimal degree spectral factors, Automatica 9 (1973), 491-500.

2. Anderson, B.D.O.: Corrections to: Algebraic properties of minimal degree spectral factors, Automatica 11 (1975), 321-322.

3. Brockett, R.W.: Finite dimensional linear systems, New York: Wiley (1970).

4. Douglas, R.G.: Banach algebra techniques in the theory of Toeplitz operators, Regional Conf. Series, Amer. Math. Soc. 15 (1972).

5. Gohberg, I.C., Krein, M.G.: Systems of integral equations on a half line with kernels depending on the difference of arguments, Amer. Math. Soc. Transl. 14 (1960), 217-287.

6. Jameson, A., Kreindler, E.: Inverse problem of linear optimal control, SIAM J. Control 11 (1973), 1-19.

7. Jonckheere, E.A.: Continuous-time linear-quadratic spectral computations revisited, in preparation (1983).

8. Jonckheere, E.A.: On the existence of a negative semi-definite, antistabilizing solution to the discrete-time algebraic Riccati equation, IEEE Trans. Automatic Control AC-26 (1981), 707-712.

9. Jonckheere, E.A., Delsarte, P.: Inversion of Toeplitz operators, Levinson equations and Gohberg-Krein factorization--a simple and unified approach for the rational case, J. Mathematical Analysis and Applications 87 (1982), 295-310.

10. Jonckheere, E.A., Khargonekar, P.P.: Linear-quadratic optimal control--an algebro-spectral theory, in preparation (1983).

11. Jonckheere, E.A., Silverman, L.M.: Spectral theory of the linear-quadratic optimal control problem: discrete-time single-input case, IEEE Trans. Circuits and Systems CAS-25 (1978), 810-825.

12. Jonckheere, E.A., Silverman, L.M.: Spectral theory of the linear-quadratic optimal control problem: a new algorithm for spectral computations, IEEE Trans. Automatic Control AC-25 (1980), 880-888.

13. Jonckheere, E.A., Silverman, L.M.: Spectral theory of the linear-quadratic optimal control problem: analytic factorization of rational matrix valued functions, SIAM J. Control and Optimization 19 (1981), 262-281.

14. Jonckheere, E.A., Silverman, L.M.: Spectral theory of the linear-quadratic optimal control problem: review and perspectives, European Conference on Circuit Theory and Design, The Hague, The Netherlands (1981).

15. Kailath, T.: A view of three decades of linear filtering theory, IEEE Trans. Information Theory IT-20 (1974), 146-181.

16. Kalman, R. E.: Contributions to the theory of optimal control, Bol. Soc. Mat. Mex. 5 (1960), 102-119.

17. Kalman, R. E.: Canonical forms for quadratic optimization problems, unpublished notes, University of Florida (1975).

18. Kalman, R. E.: When is a linear system optimal? Trans. ASME J. Basic Engineering, D 86 (1964), 51-60.

19. Kato, T.: Perturbation theory for linear operators, New York: Springer-Verlag (1966).

20. Khargonekar, P. P.: Canonical forms for linear-quadratic optimal control problems, Ph.D. dissertation, University of Florida (1981).

21. Oono, Y., Yasuura, K.: Synthesis of finite passive 2n-terminal networks with prescribed scattering matrices, Mem. Kyushu Univ. (Engineering), Japan 14 (1954), 125-177.

22. Popov, V. M.: Hyperstability and optimality of automatic systems with several control functions, Rev. Roum. Sci. Tech., Elektrotek. et Energ. 9 (1964), 629-690.

23. Rodriguez-Canabal, J. M.: The geometry of the Riccati equation, Stochastics 1 (1973), 129-149.

24. Rosenblum, M.: The absolute continuity of Toeplitz's matrices, Pacific J. Math. 10 (1960), 987-996.

25. Safonov, M. G., Athans, M.: Gain and phase margin for multiloop LQG regulators, IEEE Trans. Automatic Control AC-22 (1977), 173-179.

26. Simon, B.: On the absorption of eigenvalues by continuous spectrum in regular perturbation problems, Amer. J. Math. 87 (1965), 709-718.

27. Willems, J. C.: Least squares stationary optimal control and the algebraic Riccati equation, IEEE Trans. Automatic Control AC-16 (1971), 621-634.

28. Willems, J. C.: On the existence of a nonpositive solution to the Riccati equation, IEEE Trans. Automatic Control AC-19 (1974), 592-593.

29. Youla, D. C.: On the factorization of rational matrices, IRE Trans. Information Theory IT-7 (1961), 172-189.

30. Zames, G.: On the input-output stability of time-varying nonlinear feedback systems - parts I and II, IEEE Trans. Automatic Control AC-11 (1966), 228-238 and 465-476.

E. A. Jonckheere,
Department of Electrical
Engineering,
University of Southern California,
Los Angeles, California 90089-0781,
U.S.A.

L. M. Silverman,
Department of Electrical
Engineering,
University of Southern California,
Los Angeles, California 90089-0781, U.S.A.

Operator Theory:
Advances and Applications, Vol. 12

ON THE STRUCTURE OF INVERTIBLE OPERATORS IN A NEST-SUBALGEBRA OF A VON NEUMANN ALGEBRA

Gareth J. Knowles and Richard Saeks

The topological structure of the collection of all invertible elements of certain operator algebras is investigated. The first such algebra A will be a nest subalgebra of a von Neumann algebra. It is shown that, the collection of invertible elements of A with inverse also in A, satisfying certain boundary conditions, are in the principal component of the identity when assigned the strong operator topology. The second $\mathbb{A}$ is a subalgebra of A. When assigned the uniform topology, it is shown that the invertible elements of $\mathbb{A}$ are in the principal component of the identity. The results are applied to a large variety of examples where $\mathbb{A}$ is shown to be extensive.

INTRODUCTION

The central theme of this paper is, an investigation into the topological properties of, the collection of all invertible elements of certain Banach algebras. The questions which the paper attempts to resolve have a lengthy history in both mathematics and systems theory. From mathematics, a simple case of these questions, is, to determine the structure of the invertible, infinite (scalar) upper triangular matrices. In the direction of systems theory, one would like to provide satisfactory criteria for stability.

The first algebra that we consider, is a nest subalgebra C of a von Neumann algebra B. It is shown that, the collection of invertible elements of C, denoted by $C \cap C^{-1}$, will form a connected component of the identity whenever the nest is discrete, and C is assigned strong operator topology. In section 3 the case of an arbitrary nest is investigated. To each operator in C is associated a bounded operator-valued analytic function on the upper half plane. Motivated by the concepts of classical system theory we restrict to those operators having a well defined limit at infinity from which it follows, that if such an operator is in $C \cap C^{-1}$ then it is path connected through $C \cap C^{-1}$ to the identity operator. Here, C is again assigned the strong operator topology.

In section 4, a Banach subalgebra $\mathbb{A}$, of C, which contains most of the differential and integral operators used in system theory, and, yet, is still quite tractable is formulated. It is shown that $\mathbb{A} \cap \mathbb{A}^{-1}$ is contained in the principal component of the identity in the uniform topology.

1. Preliminaries

Fix H a (separable) Hilbert space. Denote by $L(H)$ the corresponding collection of all bounded linear operators on H. Let N be a, totally ordered, lattice of (closed) subspaces of H closed under the operation of meet and join on any subset of N. Such an N is termed a complete nest of subspaces provided $\vee N = H$ and $\wedge N = (0)$. The corresponding (self-adjoint, orthogonal) projections, onto the members of N will be denoted by E. We shall need the following.

LEMMA 1.1. (Kadison) Let 0 and I be in E. There exists a spectral measure $E(\cdot)$ on the Borel subsets of $\mathbb{R}$ with support in [0,1], and an order preserving bijection $t_E \leftrightarrow E_t$ between [0,1] and E with $E = E([0,t_E])$.

The next result we will use now follows from this

LEMMA 1.2. A nest of projections can always be indexed by some subset of the real line. Indeed, allowing degeracy, it follows that $E = \{E_t\}_{t \in \mathbb{R}}$.

Let E be a nest of projections on H. An operator leaving each member of E invariant is said to be *causal*. The collection of all such causal operators is ultraweakly closed in $L(H)$, and will be denoted by Alg$E$, or just C if there is no confusion about the choice E. Causal operators arise naturally in a system theoretic setting, in that, unlike an arbitrary operator they are physically realizable. These operators correspond to the usual situation where the output of a system does not depend upon some future input.

Denote by u-lim, s-lim, ω-lim respectively, the uniform, strong, weak limits, should they exist. Let H_+ denote the upper half plane of $\mathbb{D}$ the open unit disc.

Let B be a von Neumann algebra, let E be a complete nest of projections contained in B, and let A = Alg$E \cap B$ denote the nest subalgebra of B relative to E. Then A is ultraweakly closed in $L(H)$. Writing E as $\{E_t\}_{t \in \mathbb{R}}$ we will need the following, which may be found in [6] for example.

LEMMA 1.3. Let A and B be as above. There exists a strongly continuous, one-parameter group representation of $\mathbb{R}$ as a group of unitary operators in B. This is given by

$$U_t = \int_{\mathbb{R}} e^{-its} d\, E(s)$$

Following [9] the notation n.s.v.a will denote a nest subalgebra of a von Neumann algebra.

Given an operator which is both causal and has a bounded inverse, it does not always follow that its inverse will be causal. For example, consider the bilateral shift of $\ell^2(\mathbb{Z})$ with its usual nest of subspaces obtained by truncation. Let A be an n.s.v.a. The collection of invertible elements of A whose inverse is also in A will be written $A \cap A^{-1}$. In particular, when $\mathcal{B}$ is $L(H)$, this is written $C \cap C^{-1}$. If A is an n.s.v.a. of $\mathcal{B}$, as $\mathcal{B}$ is a *-algebra it immediately follows that $A \cap A^{-1} = \mathcal{B} \cap C \cap C^{-1}$. An n.s.v.a. is said to be discrete if E is a discrete nest [9]. Since A is a Banach algebra in the uniform topology $A \cap A^{-1}$ is the disjoint union of connected components in this topology. We shall have occasion to use the following, which can be found for example in [5] Cor. 2.14.

LEMMA 1.4. The principal component (component containing the identity) of a Banach *-algebra contains all the unitary elements.

The diagonal of a nest algebra $\mathcal{D}$ is the von Neumann algebra $C \cap C^*$, denoted by $\mathcal{D}$. A member of $\mathcal{D}$ is said to be *memoryless* [6]. We shall need the following property of the Deddens operator D [4] given by $D = \int_0^1 e^t \, dE(t)$.

LEMMA 1.5. Let E be contained in $\mathcal{B}$. Then if $D = \int_0^1 e^t \, dE(t)$, where $E(\cdot)$ is the corresponding spectral measure, D is a positive memoryless operator in A with $\mathrm{Lat}(D) = \overline{E}$. Clearly then, D is in $A \cap A^{-1}$.

Let G be a locally compact group with Haar measure μ. Denote, by $H^\infty(\mu,\mathcal{B})$ the collection of all essentially bounded, integrable, holomorphic operator-valued functions from G to $\mathcal{B}$. In particular, when $\mathcal{B} = \mathbb{C}$ and μ is Lebesque measure on the circle T or $\mathbb{R}$, this is denoted by $H^\infty(\mathrm{T})$ or $H^\infty(\mathbb{R})$ respectively. The next result can be found in [13] 5.2.

LEMMA 1.6. Let μ be Lebesque measure on T. A $\mathcal{B}$-valued function $\Phi(\omega)$ is in $H^\infty(\mu,\mathcal{B})$ if and only if it has an extension into the disc $\mathbb{D}$ of the form

$$\Phi(\lambda) = \sum_{n=0}^{\infty} \lambda^n \phi_n \qquad ;\lambda \in \mathbb{D}$$

where ϕ_n are operators, the series converges in the (weak $\equiv$ strong $\equiv$ uniform) topology and satisfies $\|\Phi(\lambda)\| < M < \infty$ for all λ in $\mathbb{D}$. Furthermore $\Phi(e^{i\theta}) = \underset{r\uparrow 1^-}{\text{s-lim}}\,\Phi(re^{i\theta})$ a.e.

Following [13] denote by $E^\infty(\mathbb{R})$ the set $\{f \in H^\infty(\mathbb{R}) : \lim_{|z|\to\infty} f(z)$ exists on $H_+\}$

2. <u>Invertibility in a Discrete n.s.v.a.</u>

For the purposes of intelligibility this section is devoted to studying a special case of the algebras in section 3. The insight obtained of the techniques developed in sections 3 and 4, will hopefully, justify a small amount of duplication. Some of the material on discrete n.s.v.a.'s appears also in [2].

Given a discrete nest $\{E_n\}$ in a von Neumann algebra $\mathcal{B}$, an application of Lemma 1.3 provides the corresponding frequency group given by

$$U(\omega) = \sum_{-\infty}^{\infty} \bar{\omega}^n (E_n - E_{n-1}) \quad ; \omega \in T$$

Let $\mathcal{A}$ be the corresponding n.s.v.a. For each A in $\mathcal{B}$ define the bounded, measurable function $A(\omega)$ on T by

$$A(\omega) = U(\omega)\ A\ U(\omega)^{-1}$$

It follows that $A(\omega)$ has a bounded "harmonic" extension into the disc, given by

$$\rho(A(re^{it})) = \frac{1}{2\pi}\int_0^{2\pi} \rho(A(e^{it}))\ P_r(t-\theta)d\theta$$

where P_r is the usual Poisson kernal, and $\rho \in \mathcal{B}_*$ the pre-dual of $\mathcal{B}$. The next result is ([2] Theorem 2.9) and we note that the last part is also implicit in [1] and [10].

LEMMA 2.1. The following are equivalent.

(a) A is in $\mathcal{A}$.

(b) $A(\omega)$ is in $H^\infty(T,\mathcal{B})$.

(c) $A(\omega)$ is in $H^\infty(\mathbb{D},\mathcal{B})$.

(d) $\rho(A(\lambda))$ is in $H^\infty(\mathbb{D})$ for each ρ in $\mathcal{B}_*$.

LEMMA 2.2. For each λ, $A(\lambda) \in \mathcal{A}$. If $A \in \mathcal{A} \cap \mathcal{A}^{-1}$ then so is each $A(\lambda)$.

Furthermore

$$A(\lambda)^{-1} = A^{-1}(\lambda)$$

PROOF: It is easy to show [2] that for such $\lambda \in \mathbb{D}\backslash\{o\}$ the operator $E^m A(\lambda)E_n$ is given by

$$E^m A(\lambda)E_n = E^m\ U(\lambda)\ \ A\ U(\lambda)^{-1}\ E_n.$$

Here $U(\lambda)$ is the (possibly) unbounded, memoryless linear operator given by $U(\lambda) = \sum_{-\infty}^{\infty} \lambda^n(E_n - E_{n-1})$. Since $\mathcal{A}$ is strongly closed it now follows that $A(\lambda)$ is in $\mathcal{A}$. Since A^{-1} is in $\mathcal{A}$, it follows from Lemma 2.1 that A^{-1} has an extension into the disc $\mathbb{D}$ via the Poisson integral formula. Then $\rho(A^{-1}(re^{it}))$ $= \rho(A^{-1}(e^{it})) = \rho(A(e^{it})^{-1}$ μ-a.e.. From the uniqueness of extension the result now follows.

COROLLARY 2.3. If A is in $\mathcal{A} \cap \mathcal{A}^{-1}$ then $A(\lambda)$ is in $H\ (\mathbb{D},\mathcal{A} \cap \mathcal{A}^{-1})$.

PROOF: Since $A(\lambda)$ and $A^{-1}(\lambda)$ are in $H^\infty(\mathbb{D},\mathcal{B})$, and $A(0) = \text{s-}\lim_{r\downarrow 0^+} A(r)$,

$= \text{u-}\lim_{r \downarrow 0^+} A(r)$, the above follows from continuity of taking inverses.

THEOREM 2.4 [2]. In the strong operator topology $A \cap A^{-1}$ is path connected, for a discrete nest E.

PROOF: First observe that since $A(\lambda)$ is of the form $A(\lambda) = \sum_{n=0}^{\infty} \lambda^n \phi_n(A)$, for some operators $\phi_n(A)$. Thus, $A(o) = \text{s-}\lim_{r \downarrow 0^+} A(r)$ will show that $A(o)$ is memoryless, invertible in A. Then there exists a uniform path $\ell:[0,1] \to \mathcal{D} \cap \mathcal{B}$ with $\ell(1) = A(0)$, $\ell(0) = I$ and $\ell(t) \to \mathcal{D} \cap \mathcal{D}^{-1} \cap \mathcal{B}$, for $t \in [0,1]$, as $\mathcal{D} \cap \mathcal{B}$ is a von Neumann algebra. Since $U(\omega)$ is memoryless, unitary in A. It follows from Lemma 1.4 that there will exist a uniform path $h:[0,1] \to A \cap A^{-1} \cap \mathcal{D}$ satisfying $h(0) = U(\omega)$ and $h(1) = I$. From Lemma 1.6 pick ω with $\text{s-}\lim_{r \uparrow 1^-} A(r\omega) = A(\omega)$. Define a path $k:[0,1] \to A \cap A^{-1}$ by

$$k(t) = \begin{cases} \ell(3t) & 0 \leq t \leq 1/3 \\ A([3t-1]\omega) & 1/3 \leq t \leq 2/3 \\ h_\omega(3t-2)\ A\ h_\omega(3t-2)^{-1} & 2/3 \leq t \leq 1 \end{cases}$$

It now follows that $k(t)$ is a continuous path from A to I in $A \cap A^{-1}$, in the strong operator topology.

3. Invertibility Criteria for an n.s.v.a.

In this section we investigate the topology aspects of invertibility for an arbitrary n.s.v.a. The technical problem which arises, and which distinguishes those from a discrete n.s.v.a. is the problem of finding a suitable memoryless operator. It will be shown that, with a mild restriction, such an operator will exist. The operator will, however, correspond to a point on the circle, rather than in the disc.

Fix $\{E_t\}_{t \in \mathbb{R}}$ an arbitrary nest in a von Neumann algebra $\mathcal{B}$. Let $\{U_t\}_{t \in \mathbb{R}}$ denote the corresponding frequency group (Lemma 1.3). An operator A is in $E^\infty(\mathcal{B})$ provided that $t \to \rho(A(t))$ is in $E^\infty(\mathbb{R})$ for each ρ in $\mathcal{B}_*$, where $A(t) = U_t\, A\, U_t^{-1}$.

LEMMA 3.1. Let A be in $E^\infty(\mathcal{B})$. There exists an extension $A(z)$ of $A(t)$ into the upper half plane satisfying

(i) $A(z) = D^{-iz} A D^{iz}$ on a dense set, where D is the Deddens operator

(ii) $A(z)$ is in A

(iii) $\rho(A(t))$ is in $H^\infty(\mathbb{R})$ for each ρ in $\mathcal{B}_*$.

(iv) $A(t) = \text{s-}\lim_{s \uparrow 1^-} A(\gamma(s))$ whenever $\gamma:[0,1] \to H_+$ is a path converging non-tangentially to $t \in \mathbb{R}$.

PROOF:

(i) Since $A(t) = D^{-it} A D^{it}$ it follows as in [4] that $\tilde{A}(t)$ has a bounded extension to the upper half plane, given by $\tilde{A}(z) = D^{-iz} A D^{iz}$.

Define $A(z)$ by $A(z)\,h = \frac{1}{2\pi}\int_0^{2\pi} \frac{y}{(x-t)^2 + y^2} U_t A U_t^{-1} h\,dt$ [7]. Since U_t is a strongly continuous representation of $\mathbb{R}$, and H is separable, the above is well defined. For $r,s \in \mathbb{R}$ it is easy to see that

$$(I-E_s)\ A(z)\ E_r = (I-E_s)\ \tilde{A}(z)\ E_r \ldots\ldots(\dagger)$$

(ii) Since $(I-E_s)\ \tilde{A}(z)\ E_r$ is in A and A is strongly closed, it follows from $(\dagger)$ that $A(z)$ in Alg$E$, thus in A.

(iii) This follows from the definition of $E^{\infty}(\mathcal{B})$.

(iv) Define $B(\lambda)$ by

$$\rho(B(\lambda)) = \rho\left(A\left[\frac{i(1+\lambda)}{(1-\lambda)}\right]\right) \ ; \ \rho \in \mathcal{B}_*$$

It now follows that $\lambda \to B(\lambda)$ is in $H\ (\mathbb{D},\mathcal{B})$. Applying Lemma 1.6, it follows that $B(\omega) = \underset{t\ 1^-}{\text{s-lim}}\ B(\alpha(t))$ whenever $\alpha\colon [0,1] \to \mathbb{D}$ is any non-tangential path approaching ω. Since the map $\lambda \to \frac{i(1+\lambda)}{(1-\lambda)}$ is conformal the result follows

Define an operator M by

$$\rho(M) = \lim_{|z|\to\infty} \rho(a(z)) \ ; \ \rho \in \mathcal{B}_*;\ z \in H_+$$

That this exists and defines an operator in $\mathcal{B}$ follows directly from A being in $E^{\infty}(\mathcal{B})$.

LEMMA 3.2. With the notation as above, M is a memoryless operator in A.

PROOF. For each $\lambda = x+iy$ in H_+ the upper half plane and h in H

$$A(\lambda)h = \frac{1}{2\pi}\int_0^{2\pi} \frac{y}{(x-t)^2+y^2} U_t A U_t^{-1} h\,dt$$

From which it follows that, for each $s \in \mathbb{R}$

$$U_s^{-1}M = U_s^{-1}[\underset{|\lambda|\to\infty}{\text{s-lim}}\ A(\lambda)] = \underset{|\lambda|\to\infty}{\text{s-lim}}\ U_s^{-1}A(\lambda)$$

However,

$$U_s^{-1}A(\lambda)h = \frac{1}{2\pi}\int_0^{2\pi}\frac{y}{(x-t)^2+y^2} U_s^{-1}U_t A\, U_t^{-1}h\,dt$$

$$= \frac{1}{2} \int_{o}^{2\pi} \frac{y}{(x+s-r)^2+y^2} U_r^{-1} A U_r U_s^{-1} h \, dr$$

$$= A(\lambda+s) U_s^{-1} h$$

From this is deduced the equality $U_s M = MU_s$ for all $s \in \mathbb{R}$. Therefore, M is indeed a memoryless operator in $\mathcal{B}$, hence in $\mathcal{A}$.

LEMMA 3.3. The subalgebra $E^\infty(\mathcal{B})$ is a strongly closed subset of $\mathcal{A}$.

PROOF: Since $E^\infty(\mathcal{B}) \subseteq H^\infty(\mathbb{R},\mathcal{B})$ it follows that $E^\infty(\mathcal{B}) \subseteq \mathcal{A}$. To see that it is strongly closed, observe that, for a fixed h in H,

$$\|(A_n - A)h\| < \varepsilon \Rightarrow \|[A_n(z) - A(z)]\, h\| < \varepsilon \cdot \|A\|$$

for each z in $\overline{H}_+$, the closed upper half plane. This is shown to hold, by applying the above Poisson integral formula to the function $U_t A U_t^{-1} h$. The next result follows as in Corollary 2.3.

LEMMA 3.4. If A in $E^\infty(\mathcal{B})$ is invertible, with its inverse in $E^\infty(\mathcal{B})$ then A(z) is in $\mathcal{A} \cap \mathcal{A}^{-1}$ for all z in H_+.

THEOREM 3.5. Assign $\mathcal{A}$ the strong operator topology. With the assumptions of Lemma 3.4, A is in the principal component of $\mathcal{A} \cap \mathcal{A}^{-1}$.

PROOF: As above, define $B(\lambda)$ by

$$B(\lambda) = \begin{cases} A\left[\dfrac{i(+\lambda)}{1-\lambda}\right] & \lambda \in \mathbb{D}\backslash\{1\} \\ M = \underset{r\uparrow 1^-}{\text{s-lim}}\, B(r) & \lambda = 1 \end{cases}$$

An application of Theorem 2.4 will show that $B(re^{i\theta})$ converges strongly a.e. to $B(e^{i\theta})$. Pick such a θ and define $k:[0,1] \to \mathcal{A} \cap \mathcal{A}^{-1}$ by

$$k(t) = \begin{cases} B[(1-2t)e^{i\theta}] & 0 \le t \le 1/2 \\ B\,(2t-1) & 1/2 \le t \le 1 \end{cases}$$

This defines a strongly continuous path in $\mathcal{A} \cap \mathcal{A}^{-1}$ from $U_t A U_t^{-1}$ to M, where $t = i(1+e^{i\theta})(1-e^{i\theta})^{-1}$. Since $M \in \mathcal{D} \cap \mathcal{D}^{-1} \cap \mathcal{B}$ it follows that there is a path $\ell:[0,1] \to \mathcal{D} \cap \mathcal{D}^{-1} \cap \mathcal{B}$ from M to the identity operator. Joining the maps $k(\cdot)$ and $U_t^{-1}\ell(\cdot)U_t$ completes the proof.

4. Invertibility and the Uniform Topology

This section studies conditions on the boundary behavior of A(z) in $H^\infty(\mathbb{R},\mathcal{A})$ constructed above. It will be shown that there exists a viable uniform homotopy theory for a large subalgebra of causal operators. In order to motivate some of the results below, we begin by considering the following counterexample to uniform radial convergence on the unit disc. The following is essentially contained in [12].

EXAMPLE. Let $A = M_\phi$ acting on $\ell^2(\mathbb{Z}^+)$ by convolution, where ϕ is an arbitrary member of $\ell(\mathbb{Z}^+)$. Assuming M_ϕ to be causal with respect to the usual truncation structure on $\ell^2(\mathbb{Z}^+)$, it can be represented by the semi-infinite matrix

$$M_\phi = \begin{bmatrix} a_o & a_1 & a_2 & \cdots \\ & a_o & a_1 & \cdots \\ \bigcirc & & a_o & \cdots \\ & & \vdots & \ddots \end{bmatrix}$$

Thus $A(\lambda) = U_\lambda A U_\lambda^{-1}$ is given by

$$A(\lambda) = \begin{bmatrix} a_o & \lambda a_1 & \lambda^2 a_2 & \cdots \\ & a_o & \lambda a_1 & \cdots \\ \bigcirc & & a_o & \\ & & \vdots & \end{bmatrix}, \ \lambda \in \mathbb{D}$$

Suppose that, for some fixed θ_o, $A(re^{i\theta_o})$ converges to $A(e^{i\theta_o})$. The operator $A(e^{i\theta_o}) - A(re^{i\theta_o})$ is also a convolution operator with norm

$$\|A(re^{i\theta_o}) - A(e^{i\theta_o})\| = \operatorname*{ess\,sup}_{o \le \theta \le 2\pi} \left| \sum_{n=o}^{\infty} a_n [r^n e^{in\theta_o} - e^{in\theta_o}] \, e^{in\theta} \right|$$

$$= \sup_{o \le \theta \le 2} \left| \sum_{n=o}^{\infty} a_n (r^n - 1) \, e^{in\theta} \right|$$

which is independent of θ_o.

From this is obtained $\|A(re^{i\theta_o}) - A(e^{i\theta_o})\| = \|A(re^{i\gamma}) - A(e^{i\gamma})\|$ for $o \le \gamma \le 2\pi$. Showing that the level curves $A(re^{i\gamma})$, $o \le r \le 1$ converge uniformly to $A(e^{i\gamma})$ everywhere. However, it is easy to see that, the function $\gamma \to A(re^{i\gamma})$ is uniformly continuous for $o \le r \le 1$. From this it is deduced that the function $\gamma \to A(e^{i\gamma})$ is also uniformly continuous. This will be continuous exactly when $\hat{\phi}$ is continuous in T. Picking $\hat{\phi}$ with $\hat{\phi}$ not in $C(\mathrm{T})$ provides a counterexample to uniform radial convergence for any point on T. Indeed, the above will provide a suitable counterexample to uniform nontangential convergence anywhere to a point on the circle, by an obvious adaptation.

Fix $\mathcal{E}$ a discrete nest in the von Neumann algebra $\mathcal{B}$ define $\mathbb{A}$ $(=\mathbb{A}(\mathcal{E}))$ by

$$\mathbb{A} = \{B \in A\colon \omega \to U(\omega)BU(\omega^{-1}) \text{ is } C(\mathrm{T},\mathcal{B})\}$$

LEMMA 4.1. If for some θ_o.

$$A(e^{i\theta_o}) = \underset{r\uparrow 1^-}{\text{u-lim}}\; A(re^{i\theta_o}), \text{ then}$$

$$A(e^{i\theta}) = \underset{r\uparrow 1^-}{\text{u-lim}}\; A(re^{i\theta}) \text{ for } o \le \theta \le 2\pi.$$

Furthermore, the mapping $\omega \to A(\omega)$ is uniformly continuous from T to $\mathcal{B}$.

PROOF: Since $U(e^{i\gamma})A\,U(e^{i\gamma})^{-1} - U(re^{i\gamma})\,A\,U(re^{i\gamma})^{-1}$

$= U(e^{i(\gamma+\theta_o)})\,[U(e^{i\theta_o})\,A\,U(e^{i\theta_o})^{-1} - U(re^{i\theta_o})\,A\,U(re^{i\theta_o})^{-1}]\,U(e^{i(\gamma+\theta_o)})^{-1}$ the uniform convergence follows. Since this is independent of θ, it follows that the level curves $\omega \to A(r\omega)$ form a Cauchy sequence converging uniformly to $\omega \to A(\omega)$ in $C(\mathrm{T},\mathcal{B})$. The result follows.

LEMMA 4.2. $\mathbb{A}$ is a uniformly closed subalgebra of A consisting of all operators A in A with

$$A(e^{i\theta}) = \underset{r\uparrow 1^-}{\text{u-lim}}\; A(re^{i\theta}) \text{ for some } \theta$$

PROOF: If A is in $\mathbb{A}$ the function $\omega \to A(\omega)$ will be $C(\mathrm{T},A)$. Lemma 2.1(c) shows that $\mathbb{A}$ is the uniform closure of all $\mathcal{B}$-valued analytic trigonometric polynomials of the form $\sum_{n=o}^{N} \lambda^n \phi_n$ for some choice if "nth diagonals" ϕ_n. Therefore $\mathbb{A}$ is uniformly closed in A.

Suppose that $A(e^{i\theta}) = \underset{r\uparrow 1^-}{\text{u-lim}}\, A(re^{i\theta})$ for some fixed θ. Applying Lemma 4.1 shows that A will be in $\mathbb{A}$. The converse follows immediately, completing the proof.

THEOREM 4.3. In the following (i) $\Rightarrow$(ii) $\Rightarrow$(iii)

(i) A is on the principal component of $\mathbb{A} \cap \mathcal{L}(H)^{-1}$

(ii) A is in $\mathbb{A} \cap \mathbb{A}^{-1}$

(iii) A is in the principal component of $A \cap \mathcal{L}(H)^{-1}$

PROOF:

(i) $\Rightarrow$ (ii) Let $t \to A_t$ be a uniform path from A to I in $\mathbb{A} \cap \mathcal{L}(H)^{-1}$. Choosing M with

$$\max_{o \le t \le 1} \| A_t^{-1} \| < M < \infty$$

there will exist an $\varepsilon > o$ with $\|A_s - A_t\| < \frac{1}{M}$ for $|t-s| < \varepsilon$. Let A_t^{-1} be in $\mathbb{A}$ for some t in $[0,1]$. Then for s in $[0,1]$ with $|s-t| < \varepsilon$

$$||I - A_t^{-1}A_s|| \leq ||A_t^{-1}|| \cdot ||A_t - A_s|| < 1$$

Therefore $A_s^{-1} A_t$, hence A_s^{-1}, is in $\mathbb{A}$. A finite number of continuations beginning with I will complete the proof.

(ii) $\Rightarrow$ (iii) Let A be in $\mathbb{A} \cap \mathbb{A}^{-1}$. The function $r \to A(r)$ $(o \leq r \leq 1)$, is a uniformly continuous path from A to the memoryless operator A(o) in $\mathbb{A} \cap \mathbb{A}^{-1}$. Let A(o) = expV. P be the polar decomposition of A(o). Define the function ϕ on [0,1] by $\phi(s) = [\exp sV].(sP + (1-s)I]$. Then $\phi(s)$ $(o \leq s \leq 1)$ will be a path in $A \cap L(H)^{-1}$ from A(o) to I. Joining the above two paths will give a path in $A \cap L(H)^{-1}$ from A to I.

REMARK: It is worth observing that (i) and (ii) above are actually equivalent. If A is in $\mathbb{A}$ then $A(z) = U(z)A\,U(z)^{-1} = D^{-iz}AD^{iz}$ will also be in $\mathbb{A}$. That this holds follows from the observation that D^{iz} is in $\mathcal{D}'$. Furthermore $\mathcal{D}' \subset \mathbb{A}$, for if T is in $\mathcal{D}$ then $T(z) \equiv T, \forall z \in \overline{\mathbb{D}}$. Hence, it follows that, the path constructed above is actually in $\mathbb{A} \cap L(H)^{-1}$. An application of (ii) along this path will now reveal the following:

COROLLARY 4.4. $\mathbb{A} \cap \mathbb{A}^{-1}$ is a uniformly connected, uniformly closed subgroup of the principal component of $A \cap A^{-1}$.

We now consider the case of arbitrary n.s.v.a.'s. Given a nest E in $\mathcal{B}$ define $\mathbb{A}$ (= $\mathbb{A}(E)$) by

$$\mathbb{A}(E) = \{A \in A : t \to U_t\, A\, U_t^{-1} \text{ is in } C(\mathbb{R},A) \text{ and } \underset{|z|\to\infty}{\text{u-lim}}\, A(z) \text{ exists for } z \in H_+\}$$

Given A in $\mathbb{A}$, define B: $T \to A$ by $B(\lambda) = A\left[\frac{i(1+\lambda)}{1-\lambda}\right]$ $\lambda \in \overline{\mathbb{D}} \setminus \{-1\}$ and $B(-1) = \underset{|z|\to\infty}{\text{u-lim}}\, A(z)$. Then it is deduced that $\omega \to B(\omega)$ is in $C(T, \mathbb{A})$.

The following provides a Nyquist-like criterion

THEOREM 4.5. In the following (i) $\Rightarrow$ (ii) $\Rightarrow$ (iii)

(i) A is in the principal component of $\mathbb{A} \cap L(H)^{-1}$

(ii) A in in $\mathbb{A} \cap \mathbb{A}^{-1}$

(iii) A is in the principal component of $A \cap L(H)^{-1}$

PROOF:

(i) $\Rightarrow$ (ii). This will follow as in Theorem 4.3 once it has been established that $\mathbb{A}$ is uniformly closed. Pick a net $\{A_\alpha\}_{\alpha\in\Lambda}$ in $\mathbb{A}$ converging uniformly to A. Pick α_0 with $||A_\alpha - A|| < \varepsilon/4||A||$ for all $\alpha \geq \alpha_0$. It follows that $||A_{\alpha,t} - A_t|| < \varepsilon/4||A||$ for $t \in \mathbb{R}$. From the Poisson interval formula for defining the extension, it follows that $||A(z) - A_\alpha(z)|| < \varepsilon/2$ for $\alpha \geq \alpha_0$. Choose $\delta > o$ with $|s-t| < \delta \Rightarrow ||A_{\alpha,t} - A_{\alpha,s}|| < \varepsilon/2$ for $|s-t| < \delta$, $||A_s - A_t|| < \varepsilon$, showing continuity. An exactly similar argument shows that $\underset{|z|\to\infty}{\text{u-lim}}\, A(z)$ will exist. Thus $\mathbb{A}$ is norm closed in A.

(ii) ⇒ (iii). Let A be in $\mathbb{A} \cap \mathbb{A}^{-1}$, defining $B(\omega)$ as above $\omega \to B(\omega)$ is uniformly continuous from $\mathbb{T}$ to $\mathbb{A} \cap \mathbb{A}^{-1}$ in the norm topology. Define $\psi(r) = B(e^{i\pi r})$. It follows that ψ is a uniform path from A to $M = \underset{|z|\to\infty}{\text{u-lim}}\, A(z)$, a memoryless operator in $\mathbb{A} \cap \mathbb{A}^{-1}$. The proof can now be completed as in Theorem 4.3.

With an exactly similar argument to the above remarks following Theorem 4.3, the generalization to arbitrary nests can be obtained.

If it is worth noting that Theorem 4.3 will apply to a larger class of operators than Theorem 4.5 when the discrete nest is embedded in $\mathbb{R}$.

5. Sufficiency Conditions and their Applications

In this section sufficiency conditions are presented for an operator to be in $\mathbb{A}$. These are then applied to a variety of examples

(a) INFINITE MATRICES. Let $\{P_n\}_{n\in\mathbb{Z}}$ be a nest with rank $(P_n \ominus P_{n-1}) = 1$. A causal operator is one with an upper triangular (scalar) representation. Denote by $\phi_n(A)$ the "nth diagonal" of A given by

$$\phi_n(A) = \sum_{-\infty}^{\infty} (P_m - P_{j-1})\; A\; (P_{j+n} - P_{j+n-1})$$

Then each $\phi_n(A)$ defines a bounded linear operator with norm at most $\|A\|$ [2], where it is shown that A is in $\text{Alg}\{P_n\}$ if and only if $\phi_n(A) = 0$ for $n < o$.

LEMMA 5.1. If A in $\text{Alg}\{P_n\}$ satisfies $\{\|\phi_n(A)\|\}_{n=o}^{\infty} \in \ell^1(\mathbb{Z}^+)$ then A is in $\mathbb{A}$ $(\{P_n\})$, therefore A is in the principal component of $C \cap C^{-1}$ $(=\text{Alg}\{P_n\})$.

PROOF: The function $A(\lambda) = \sum_{n=o}^{\infty} \lambda^n \phi_n(A)$ is in $H^\infty(\mathbb{D},A)$. Next, observe that the partial sums of $\sum_{n=o}^{\infty} \phi_n(A)$, are uniformly convergent. From Lemma of [2] the partial sums converge ultraweakly to A. From the above it is deduced that the partial sums converge uniformly to A, showing that $A(\lambda)$ is the extension into the interior of $\mathbb{T}$ of the function $A(\omega) = U(\omega)\; A\; U(\omega)^{-1}$. Defining $g(\lambda) = \sum_{n=o}^{\infty} \lambda_n \|\phi_n\|$, the power series expansion converges on $\mathbb{D} \cup \{1\}$. That is, $g(\lambda)$ is continuous at 1. Giving

$$\|A(\lambda) - A\| = \|\sum_{n=o}^{\infty} (1 - \lambda^n)\; \phi_n\| \leq |g(\lambda) - g(1)|$$

Therefore, $A(r)$ converges uniformly to A as $r \uparrow 1^-$. An application of Lemma 4.1 completes the proof.

APPLICATION TO WEIGHTED SHIFTS. Let T_n be the weighted shift given by $T_n(e_j) = \omega_j^{(n)}\; e_{j+1}$. For a simple shift, for example T_1, it follows that T_1

is an $A \cap L(H)^{-1}$ exactly when $\omega_j^{(1)} \neq o$ for all j in $\mathbb{Z}$. Observing that $\phi_1(T_1) = T_1$ it follows that T_1 is in $\mathbb{A}$. It is deduced from this that $T_1 \in A \cap A^{-1}$ if and only if T_1 is path connected to I in $A \cap L(H)^{-1}$ (in the uniform topology). Let V denote the bilateral shift on $\ell^2(\mathbb{Z})$. A similar argument shows that putting $X = \sum_{j=1}^{n} T_j V^j$, X is in $A \cap A^{-1}$ if and only if X is in the principal component of $A \cap L(H)^{-1}$ $(A \cap A^{-1})$. Define $\tilde{\omega}_j = \|T_j\| = \sup_n |\omega_j^{(n)}|$. If $X = \sum_{j=o}^{\infty} T_j V^j$ with X in $A \cap A^{-1}$ and $\sum_{j=o}^{\infty} \tilde{\omega}_j < \infty$, then X is in the principal component of $A \cap L(H)^{-1}$ $(A \cap A^{-1})$.

(b) LIPSCHITZ OPERATORS. Let Ω_α, $o<\alpha<\infty$ be the Lipschitz algebra [11] consisting of function $\mathbb{D}$ satisfying

$$\lim \frac{f(e^{i\theta}) - f(e^{i\theta})}{|\theta - \gamma|^\alpha} = 0$$

uniformly as $|\theta - \gamma| \to 0$.

We will say that an operator T in $\mathcal{B}$ is in $\Omega_\alpha(\mathcal{B})$ if, for each $\rho \in B_*$, the map $\omega \to \rho[U(\omega)\, T\, U(\omega)^{-1}]$ is in Ω_α (where, as before, $U(\omega)$ is the strongly continuous, unitary representation of T corresponding to a nest of projections $\{E_n\}_{n\in\mathbb{Z}}$).

LEMMA 5.2. If A is in $\Omega_\alpha(\mathcal{B})$ then A is in $\mathbb{A}(\{E_n\})$. Consequently A is in the principal component of $\mathrm{Alg}\{E_n\} \cap \mathrm{Alg}\{E_n\}^{-1}$.

PROOF: Fix $o<\alpha<\infty$ and $\rho \in \mathcal{B}_*$ with $\|\rho\| \leq 1$. Since A is in $\Omega_\alpha(\mathcal{B})$ it follows that $A \in A = \mathrm{Alg}\{E_n\} \cap \mathcal{B}$. Putting $f_\rho(\omega) = \rho(A(\omega))$

$$|f_\rho(re^{i\theta}) - f_\rho(e^{i\theta})| \leq \sup_{-\delta\leq\gamma\leq\delta} |f_\rho(e^{i\gamma}) - f_\rho(e^{i(\gamma-\theta)})| + 2\|f_\rho\|_\infty \sup_{|\theta|>\delta} |P_r(\theta)|$$

where $P_r(\theta)$ is the Poisson kernal [8] p. 18. Pick $\varepsilon > o$ there exist an $\eta > o$ with $|\gamma| < \eta$ implying

$$|f_\rho(e^{i\gamma}) - f_\rho(e^{i(\gamma-\theta)})| < \varepsilon\, |\theta-\gamma|$$

This shows that $\dfrac{|f_\rho(e^{i\gamma}) - f_\rho(e^{i(\gamma-\theta)})|}{|\theta-\gamma|}$ is bounded for each ρ. An application of the boundness theorem shows that

$$|f_\rho(e^{i\gamma}) - f_\rho(e^{i(\gamma-\theta)})| < K|\theta-\gamma|$$

independently of ρ. Hence

$$|f_\rho(re^{i\theta}) - f_\rho(e^{i(\gamma-\theta)})| < K|\theta-\gamma| + 2\|A\| \sup_{|\theta|>\delta} |P_r(\theta)|$$

for all ρ in $\mathcal{B}_*$.

Therefore $\|A(re^{i\theta}) - A(e^{i\theta})\| = \sup_{\rho\varepsilon(\mathcal{B}_*)_1} |\rho[A(re^{i\theta})] - A(e^{i\theta})]| < \varepsilon$, giving $A(e^{i\theta}) = \text{u-}\lim_{r\uparrow 1^-} A(re^{i\theta})$. The second statement in the Lemma immediately follows.

(c) STRONG MEASURABILITY AND THE BOCHNER INTEGRAL. There we investigate the interaction between strong measurability, Bochner integrability and continuous boundary values. The results are then applied to a class of cross product algebras.

As before if $\lim_{|t|\to\infty} A(t)$ exists, define $B(\omega)$, $\omega \in \mathrm{T}$ by

$$B(\omega) = \begin{cases} A \quad \frac{i(\omega-1)}{\omega+1} & \omega \neq -1 \\ \lim_{|t|\to\infty} A(t) & \omega = -1 \end{cases}$$

LEMMA 5.3. Let $t \to U_t$ $(\omega \to U(\omega))$ be strongly measurable. For A in $L(H)$, $t \to U_t A U_t^{-1}$ $(\omega \to U(\omega) A U(\omega)^{-1})$ is uniformly continuous on $\mathbb{R}$ (T) in the uniform topology. If $\text{u-}\lim_{|t|\to\infty} A(t)$ exists then $B(\omega)$ is in $L^1(\mathrm{T}, L(H))$.

PROOF: As $t \to U_t$ it is strongly measurable representation of $\mathbb{R}$ as a group of unitary operators, an application of ([7] Theorem 10.2.1) will show that $t \to U_t^{-1} A U_t$ is uniformly continuous. From the assumption that $\text{u-}\lim_{|t|\to\infty} A(t)$ exists it follows that $B(\omega)$ is continuous from T to $L(H)$ in the uniform topology. Thus $B(\omega)$ is Bochner integrable.

Given $\{U_t\}_{t\,\mathbb{R}}$ as a strongly measurable representation of $\mathbb{R}$ as a group of unitary operators (thus strongly continuous [7]), let $E = \{E_s\}_{s\in\mathbb{R}}$ denote the corresponding nest of projections obtained via Stones theorem. Let C denote Alg$E$.

COROLLARY 5.4. If A is in $C \cap C^{-1}$ and $\text{u-}\lim_{|t|\to\infty} U_t A U_t^{-1}$ exists, then A is in the principal component of the identity in $C \cap L(H)^{-1}$ $(\equiv C \cap C^{-1})$.

PROOF: Defining $B(\omega)\colon \mathrm{T} \to \text{Alg}E$ as above, we have $k(t) = B(e^{it\pi})$ $0 \le t \le 1$ is a path in $C \cap C^{-1}$ from A to $B(-1)$. A similar argument to that contained in Lemma 3.2 shows $B(-1)$ is memoryless, hence path connected in $C \cap C^{-1}$ to the identity operator.

APPLICATION TO CROSSED PRODUCT ALGEBRAS. A cross product algebra can be viewed as a twisted group algebra [3] with a trivial twist. Indeed, the case we will consider will be even more special. Let A be a C^*-algebra and T a strongly Borel map $T\colon G \to \text{Aut}_1^*(A)$, from a locally compact, totally ordered, abelian group G to the group of isometric *-automorphisms of A. Denote by $L^1(A,G,T)$ the Banach space of all Bochner-integrable functions from

G to A with the norm

$$\|f\| = \int_G \|f_t\|_A d\mu(t)$$

where μ denotes Haar measure on G. Convolution given by

$$(f*g)_t = \int_G f_s\, T(s) g_{t-s} d\mu(s)$$

an involution

$$(f^*)_t = T(t)\, \overline{f}_{-t}.$$

Then $L^1(A,G,T)$ is the cross product, with respect to T, of G with A. Since we will be concerned with identity and invertibility the usual extension will be made to $A \oplus L^1(A,G,T)$ by

$$(a,f)(t) = at + f_t$$

The algebra A will be assumed, throughout the discussion below, to be a subalgebra of $C(T) \cap L^\infty(G)$, with the sup norm. The homomorphism T is given by

$$(T(s)a)(x) = a(x-s) \;,\; a \in A \;;\; xs \in G$$

In these circumstances an f in $L^1(A,G,T)$ induces a bounded linear operator on $L^2(G)$ by

$$(A_f u)(y) = \int_G f_t(y)\, u(y-t)\, d\mu(t)$$

for $u \in L^2(G)$, $f \in L^1(A,G,T)$. It is easy to see that $f \mapsto A_f$ is a contractive linear mapping and extends to $A \oplus L^1(A,G,T)$ by

$$(a,f) \to L_a + A_f \text{ where } L_a(u) = au$$

We now let G be $\mathbb{Z}$ or $\mathbb{R}$ and assign $L^2(G)$ its usual nest structure obtained by truncation. It can now be easily deduced from the above that $f(t) = o$ for $t < o$ precisely when A_f is causal. Given $X = L_a + A_f$ causal on $L^2(G)$ as above, let $X(\lambda)$ be the extension into $\mathbb{D}$ or H_+ respectively. It is easily seen that $X(\lambda)$ is also in $A \oplus L^1(A,G,T)$ acting on $L^2(G)$. Denote by C the algebra of all causal operators on $L^2(G)$, and R the subset of all causal operators in $A \oplus L^1(A,G,T)$.

THEOREM 5.5. If $X = L_a + A_f$ for some $a \in A$ and $f \in L^1(A,G,T)$ satisfies $X \in R \cap L(L^2(G))^{-1}$ and $\operatorname{u-lim}_{|t|\to\infty} X(t)$ exists then X is in the principal component of $C \cap C^{-1}$.

PROOF: When T(s) is restricted to act on A, by assumption it will be strongly measurable. An inspection of the proof of Theorem 5.3 reveals that this will be sufficient to ensure that $s \to T(s)\,[L_a + A_f]\,T(-s)$ is continuous. Since $\operatorname{u-lim}_{|t|\to\infty} X(t)$ exists defining

$$B(\omega) = \begin{cases} X\left[\dfrac{i(\omega-1)}{\omega+1}\right] & \omega \neq -1 \\ \underset{|t|\to\infty}{u\text{-}\lim} X(t) & \omega = -1 \end{cases}$$

it follows that $B(\omega) \in C(\mathrm{T},C)$. The completion of the proof follows as in Lemma 5.3. For applications of this result to stability see [12]

REFERENCES

1. Arveson, W.B.: On groups of automorphisms of operator algebras, Amer J. Math (1967) 578-642.

2. Ashton, G.J.: Frequency domain theory for certain nest subalgebras of von Neumann algebras (Preprint).

3. Busby, R.C. and Smith, H.A.: Representations of twisted group algebras, Trans. Amer. Math. Soc. 149 (1970) 503-537.

4. Dedden, J.: Another description of nest algebras, Lecture Notes in Mathematics No. 693, Springer-Verlag, New York, (1978) 77-86.

5. Douglas, R.G.: Banach space techniques in operator theory, Academic Press, New York.

6. Feintuch, A., and Saeks, R.: Systems theory: A Hilbert space approach, Academic Press, New York, 1982.

7. Hille, E., and Philips, R.S.: Functional analysis and semi-groups, Amer. Math. Soc., Colloqu. Publ. 32 (1957).

8. Hoffman, K.: Banach spaces of analytic functions, Prentice-Hall, Inc., NJ. (1962).

9. Gilfeather, F., and Larson, D.R.: Nest-subalgebras of von Neumann algebras: commutants modulo compacts and distance estimates, J. Operator Theory 7 (1982) 279-302.

10. Loebl, R., and Muhly, P.S.: Analyticity and flows in von Neumann algebras, J. Functional Analy. 29 (1978) 214-252.

11. Matheson, E.L.: Closed ideals in rings satisfying a Lipschitz condition, Springer-Verlag, Lecture Notes in Mathematics 604, 67-72.

12. Murray, J.: Time varying systems and crossed products, (preprint).

13. Sz-Nagy, B., and Foias, C.: Harmonic analysis of operators on Hilbert space, North-Holland Publ. Co., Amsterdam (1970).

Gareth J. Knowles*
Department of Mathematics
Texas A & M University
College Station, Texas 77843
U.S.A.

Richard Saeks*
Department of Electrical Engineering
Texas Tech University
Lubbock, Texas 79409
U.S.A.

*This research supported in part by the Joint Services Electronics Program at Texas Tech University under ONR Contract 76-C-1136.

Operator Theory:
Advances and Applications, Vol. 12

ON COMMUTING INTEGRAL OPERATORS

Naftali Kravitsky

Given an irreducible algebraic curve $D(x_1,x_2) = 0$ in $\mathbb{C}^2$ and a continual set of points $S = \{(\lambda_1(x),\lambda_2(x)) | 0 \leqslant x \leqslant \ell\}$ on it, we construct a pair of commuting integral operators A_1,A_2 which have S as their joint spectrum and satisfy the polynomial equation $D(A_1,A_2) = 0$.

0. INTRODUCTION

The connection between pairs of commuting linear operators and algebraic curves was discovered and further discussed in [1-4]. In [2], [4] a finite system $S = \{(\lambda_j^{(1)},\lambda_j^{(2)}) | 1 \leqslant j \leqslant N\}$ of points on a complex algebraic curve is used to construct a pair of commuting triangular N×N matrices, which have the set S as their joint spectrum and satisfy the equation of the curve.

In the present work we use a *continual set* $S = \{(\lambda_1(x),\lambda_2(x)) | 0 \leqslant x \leqslant \ell\}$ of points on an algebraic curve to construct a pair of *commuting* triangular *integral operators* with joint spectrum S, which satisfy the equation of the curve. The results obtained here are closely related to the problem of constructing a triangular model for two commuting linear operators with continuous joint spectrum.

1. COMMUTATIVITY CONDITIONS

Let A_1 and A_2 be two integral operators in $L^2(0,\ell)$ of the form

$$(A_1 f)(x) = \lambda_1(x)f(x) + \psi(x)\sigma_1 \int_0^x \varphi(t)f(t)dt ,$$

$$(A_2 f)(x) = \lambda_2(x)f(x) + \psi(x)\sigma_2 \int_0^x \varphi(t)f(t)dt . \tag{1}$$

Here σ_1, σ_2 are constant $n\times n$ matrices,

$$\psi(x) = (\psi_1(x)\ \psi_2(x)\ \dots\ \psi_n(x)), \qquad \varphi(x) = \begin{pmatrix}\varphi_1(x)\\ \varphi_2(x)\\ \dots\\ \varphi_n(x)\end{pmatrix}$$

are continuous vector functions and $\lambda_1(x)$, $\lambda_2(x)$ are continuous scalar function on $[0,\ell]$.

Assume that the operators A_1 and A_2 commute. We have

$$(A_1A_2f)(x) = \lambda_1(x)\lambda_2(x)f(x) + \lambda_1(x)\psi(x)\sigma_2\int_0^x \varphi(t)f(t)dt +$$

$$+ \psi(x)\sigma_1\int_0^x \varphi(t)\lambda_2(t)f(t)dt + \psi(x)\sigma_1\int_0^x \varphi(t)\psi(t)\sigma_2\int_0^t \varphi(s)f(s)dsdt =$$

$$= \lambda_1(x)\lambda_2(x)f(x) +$$

$$+ \psi(x)\int_0^x [\lambda_1(x)\sigma_2+\lambda_2(t)\sigma_1+\sigma_1\int_t^x \varphi(s)\psi(s)ds\sigma_2]\varphi(t)f(t)dt .$$

$$(A_2A_1f)(x) = \lambda_2(x)\lambda_1(x)f(x) +$$

$$+ \psi(x)\int_0^x [\lambda_2(x)\sigma_1+\lambda_1(t)\sigma_2+\sigma_2\int_t^x \varphi(s)\psi(s)ds\ \sigma_1]\varphi(t)f(t)dt .$$

$$((A_1A_2-A_2A_1)f)(x) = \int_0^x K(x,t)f(t)dt \tag{2}$$

where

$$K(x,t) = \psi(x)(Q(t)-Q(x))\varphi(t) ,$$

$$Q(x) = \lambda_2(x)\sigma_1-\lambda_1(x)\sigma_2-\sigma_1\int_0^x \varphi(s)\psi(s)ds\sigma_2+\sigma_2\int_0^x \varphi(s)\psi(s)ds\sigma_1 .$$

Therefore, if A_1 and A_2 commute, then $K(x,t) = 0$ for $0 \leqslant t \leqslant x \leqslant \ell$, or

$$\psi(x)Q(t)\varphi(t) = \psi(x)Q(x)\varphi(t), \qquad 0 \leqslant t \leqslant x \leqslant \ell \tag{3}$$

DEFINITION. We say that the system of functions $\varphi(x)$, $\psi(x)$ *has property* (*) if the functions $\varphi(x)$ are linearly independent on each interval of the form $[0,r]$ and the functions

$\psi(x)$ are linearly independent on each interval of the form $[r,\ell]$, $0 < r < \ell$.

LEMMA 1. *Let* $\varphi(x)$, $\psi(x)$ *have property* (*), *let*

$$a(x) = (a_1(x)\ a_2(x)\ \dots\ a_n(x)), \qquad b(x) = \begin{pmatrix} b_1(x) \\ b_2(x) \\ \dots \\ b_n(x) \end{pmatrix}$$

be continuous vector functions and let be

$$\psi(x)b(t) = a(x)\varphi(t), \qquad 0 \leqslant t \leqslant x \leqslant \ell. \tag{4}$$

Then there exists a constant $n\times n$ *matrix* γ_0 *such that*

$$a(x) = \psi(x)\gamma_0, \quad b(x) = \gamma_0\varphi(x), \quad 0 \leqslant x \leqslant \ell .$$

PROOF. From property (*) it follows that for any $0 < r < \ell$ there exist a row-function $\tilde{\varphi}(x;r)$, $0 \leqslant x \leqslant r$ and a column-function $\tilde{\psi}(x;r)$, $r \leqslant x \leqslant \ell$ such that

$$\int_0^r \varphi(t)\tilde{\varphi}(t;r)dt = I_n, \qquad \int_r^\ell \tilde{\psi}(x;r)\psi(x)dx = I_n ,$$

where I_n is the $n\times n$ unit matrix. Denote

$$\gamma_1(y) = \int_0^y b(t)\tilde{\varphi}(t;y)dt, \qquad 0 < y \leqslant \ell ,$$

$$\gamma_2(z) = \int_z^\ell \tilde{\psi}(x;z)a(x)dx , \qquad 0 \leqslant z < \ell .$$

Then it follows from (4) that

$$a(x) = \psi(x)\gamma_1(y) , \qquad 0 < y \leqslant x \leqslant \ell , \tag{5}$$

$$b(t) = \gamma_2(z)\varphi(t) , \qquad 0 \leqslant t \leqslant z < \ell . \tag{6}$$

Choose $0 < y \leqslant z < \ell$, multiply (5) from the left by $\tilde{\psi}(x,z)$ and integrate by x from z to ℓ to obtain

$$\gamma_1(y) = \gamma_2(z), \qquad 0 < y \leqslant z < \ell .$$

This implies that both functions $\gamma_1(y)$ and $\gamma_2(z)$ are constant on $(0,\ell)$ and they possess there a common value which we denote by γ_0. The conclusion of the lemma follows from (5), (6).

THEOREM 1. 1) *Let the operators* A_1 *and* A_2 *commute. Then, if the functions* $\varphi(x)$, $\psi(x)$ *have property* (*), *there exists a constant* $n\times n$ *matrix* γ_0 *such that*

$$\psi(x)(\lambda_1(x)\sigma_2-\lambda_2(x)\sigma_1+\gamma(x)) = 0 \ , \tag{7}$$

$$(\lambda_1(x)\sigma_2-\lambda_2(x)\sigma_1+\gamma(x))\varphi(x) = 0 \ , \tag{8}$$

where

$$\gamma(x) = \gamma_0 + \sigma_1 \int_0^x \varphi(s)\psi(s)ds\ \sigma_2 - \sigma_2 \int_0^x \varphi(s)\psi(s)ds\ \sigma_1 \ . \tag{9}$$

2) *If there exists a matrix* γ_0 *such that* (7)-(9) *hold, then the operators* A_1 *and* A_2 *commute.*

PROOF. 1) Since A_1 and A_2 commute, formula (3) holds. If $\varphi(x)$, $\psi(x)$ have property (*) we may apply Lemma 1 with $a(x) = \psi(x)Q(x)$, $b(t) = Q(t)\varphi(t)$ to obtain

$$\psi(x)(\gamma_0-Q(x)) = 0, \quad (\gamma_0-Q(x))\varphi(x) = 0, \quad 0 \leqslant x \leqslant \ell$$

which is the same as (7)-(9).

2) If there exists γ_0 such that (7)-(9) hold, then

$$Q(t)\varphi(t) = \gamma_0\varphi(t), \quad \psi(x)Q(x) = \psi(x)\gamma_0 \ ,$$

$$K(x,t) = \psi(x)Q(t)\varphi(t) - \psi(x)Q(x)\varphi(t) = 0, \quad 0 \leqslant t, x \leqslant \ell$$

and the operators A_1, A_2 commute by formula (2). The theorem is proved.

2. REGULAR PAIRS OF COMMUTING OPERATORS

2.1. DEFINITION. A pair of commuting operators A_1, A_2 of the form (1) is called *regular* if there exists a constant matrix γ_0 such that equalities (7)-(9) hold.

Let the pair A_1, A_2 be regular. Consider the expression

$$D(x_1,x_2) = \det(x_1\sigma_2-x_2\sigma_1+\gamma_0) \ .$$

This is a polynomial in x_1, x_2 of degree $\leqslant n$. It is called the *discriminant polynomial* [1] of the pair. The set of points $(x_1,x_2) \in \mathbb{C}^2$ which satisfy the equation

$$D(x_1,x_2) = 0 \tag{10}$$

is called the *discriminant manifold* of the pair [3]. This is either an algebraic curve or the whole complex plain $\mathbb{C}^2$.

Consider the subspace

$$\hat{H} = \bigvee_{\substack{k_1,k_2\geqslant 0\\ 1\leqslant j\leqslant n}} A_1^{k_1} A_2^{k_2} \psi_j \tag{11}$$

of $L^2(0,\ell)$, which is obviously invariant under A_1, A_2.

THEOREM 2. *Let* A_1, A_2 *be two commuting integral operators of the form* (1). *If the pair* A_1, A_2 *is regular, then the operators satisfy the algebraic equation*

$$D(A_1,A_2) = 0 \tag{12}$$

on the invariant subspace $\hat{H}$.

REMARK. The equation (12) is related to as the *generalized Cayley-Hamilton equation* [1,4].

PROOF. Denote by $A_k\psi$ the row-vector

$$A_k\psi = (A_k\psi_1 \; A_k\psi_2 \; \dots \; A_k\psi_n), \qquad k = 1,2 .$$

We have

$$(A_k\psi)(x) = \lambda_k(x)\psi(x) + \psi(x)\sigma_k \int_0^x \varphi(t)\psi(t)dt, \qquad k = 1,2 .$$

The equation (7) may be interpreted as

$$0 = \psi(x)(\lambda_1(x)\sigma_2-\lambda_2(x)\sigma_1+\gamma(x)) =$$

$$= \psi(x)[\lambda_1(x)\sigma_2+\sigma_1 \int_0^x \varphi(t)\psi(t)dt\, \sigma_2 -$$

$$- \lambda_2(x)\sigma_1-\sigma_2 \int_0^x \varphi(t)\psi(t)dt\, \sigma_1+\gamma_0] =$$

$$= (A_1\psi)(x)\sigma_2-(A_2\psi)(x)\sigma_1+\psi(x)\gamma_0, \qquad 0 \leqslant x \leqslant \ell . \tag{13}$$

Denote by $\sigma_{ij}^{(k)}$, $\gamma_{ij}^{(0)}$, $i,j = \overline{1,n}$ the elements of the matrices σ_k, $k = 1,2$; γ_0. From (13) it follows

$$0 = \sum_{i=1}^{n} (A_1\psi_i\sigma_{ij}^{(2)}-A_2\psi_i\sigma_{ij}^{(1)}+\psi_i\gamma_{ij}^{(0)}) =$$

$$= \sum_{i=1}^{n} (\sigma_{ij}^{(2)}A_1-\sigma_{ij}^{(1)}A_2+\gamma_{ij}^{(0)})\psi_i , \qquad j = \overline{1,n} . \tag{14}$$

Introduce the $n\times n$ matrix $Q = ||q_{ij}||_{i,j=1}^{n}$ with operator valued entries

$$q_{ij} = \sigma_{ji}^{(2)} A_1 - \sigma_{ji}^{(1)} A_2 + \gamma_{ji}^{(0)} \; ; \qquad i,j = \overline{1,n} \; .$$

The formula (14) is just the same as

$$\begin{pmatrix} q_{11} & q_{12} & \cdots & q_{1n} \\ q_{21} & q_{22} & \cdots & q_{2n} \\ \cdots & \cdots & \cdots & \cdots \\ q_{n1} & q_{n2} & \cdots & q_{nn} \end{pmatrix} \begin{pmatrix} \psi_1 \\ \psi_2 \\ \cdot\cdot \\ \psi_n \end{pmatrix} = \begin{pmatrix} 0 \\ 0 \\ \cdot\cdot \\ 0 \end{pmatrix} .$$

Multiply this matrix equality from the left by the formal adjugate Q^{adj} to obtain

$$(\det Q)\psi_j = 0 \; , \qquad j = \overline{1,n} \; ,$$

where $\det Q$ is the operator

$$\det Q = \det Q^t = \det ||\sigma_{ij}^{(2)} A_1 - \sigma_{ij}^{(1)} A_2 + \gamma_{ij}^{(0)}|| =$$

$$= \det(x_1\sigma_2 - x_2\sigma_1 + \gamma_0)\Big|_{x_1=A_1, x_2=A_2} = D(A_1, A_2)$$

(Q^t is the transposed matrix Q). Since $\det Q$ commutes with A_1, A_2 we have

$$(\det Q)(A_1^{k_1} A_2^{k_2} \psi_j) = A_1^{k_1} A_2^{k_2}((\det Q)\psi_j) = 0, \quad k_1, k_2 \geqslant 0; \quad j = \overline{1,n}$$

i.e. $D(A_1, A_2)h = 0$ for every $h \in \hat{H}$. The proof is complete.

2.2. Denote

$$\Delta(x_1, x_2; x) = x_1\sigma_2 - x_2\sigma_1 + \gamma(x) \; .$$

$$B(x_1, x_2; x) = \det \Delta(x_1, x_2; x) = \det(x_1\sigma_2 - x_2\sigma_1 + \gamma(x)) \; .$$

Observe that if $\varphi(x)$ or $\psi(x)$ is non-trivial, then (7) and (8) imply

$$B(\lambda_1(x), \lambda_2(x); x) = 0 \; . \tag{15}$$

For fixed x, the expression $B(x_1, x_2; x)$ is a polynomial in x_1, x_2,

$$B(x_1,x_2;x) = 0 \tag{16}$$

is an equation of an algebraic curve in the (x_1,x_2)-plane if this polynomial is non-trivial, and equality (15) means that the point $(\lambda_1(x),\lambda_2(x))$ lies on the manifold (16). For $x = 0$ we have $\gamma(0) = \gamma_0$ and $B(x_1,x_2;0) = \det(x_1\sigma_2 - x_2\sigma_1 + \gamma_0) = D(x_1,x_2)$.

THEOREM 3. *If system* (7)-(9) *is satisfied, then the polynomial* $B(x_1,x_2;x)$ *does not actually depend upon* x:

$$B(x_1,x_2;x) = D(x_1,x_2) \tag{17}$$

holds for every $x \in [0,\ell]$, $(x_1,x_2) \in \mathbb{C}^2$.

PROOF. It suffices to show that

$$\frac{\partial B(x_1,x_2;x)}{\partial x} = 0 \ , \qquad 0 < x < \ell \ ,$$

because then $B(x_1,x_2;x)$ is constant in x and equals to its value at $x = 0$ which is $D(x_1,x_2)$.

Let $C(x)$ be a square matrix which is differentiable in x. Then also $\det C(x)$ is differentiable and we have for its derivative the formula

$$\frac{d}{dx} \det C(x) = \operatorname{Tr}(C'(x)C^{adj}(x)) \ . \tag{18}$$

Here C^{adj} is the adjugate matrix of C such that

$$CC^{adj} = C^{adj}C = (\det C)I \ .$$

Formula (18) is another form of the standard differentiation rule for determinants.

In our case $\Delta(x_1,x_2;x)$ is differentiable in x, and by (9)

$$\frac{\partial \Delta}{\partial x} = \frac{d\gamma}{dx} = \sigma_1\varphi\psi\sigma_2 - \sigma_2\varphi\psi\sigma_1$$

(we omit the notation of the variables involved). Therefore

$$\frac{\partial B}{\partial x} = \frac{\partial \det \Delta}{\partial x} = \operatorname{Tr}((\sigma_1\varphi\psi\sigma_2 - \sigma_2\varphi\psi\sigma_1)\Delta^{adj}) \ . \tag{19}$$

From the definition of Δ and formulas (7), (8),

$$\psi\Delta = (\lambda_2 - x_2)\psi\sigma_1 - (\lambda_1 - x_1)\psi\sigma_2 \ ,$$

$$\Delta\varphi = (\lambda_2 - x_2)\sigma_1\varphi - (\lambda_1 - x_1)\sigma_2\varphi .$$

Multiply the first of these equations from the left by $\sigma_1\varphi$, the second from the right by $\psi\sigma_1$ and subtract the first from the second to obtain

$$(\lambda_1 - x_1)(\sigma_1\varphi\psi\sigma_2 - \sigma_2\varphi\psi\sigma_1) = \Delta\varphi\psi\sigma_1 - \sigma_1\varphi\psi\Delta .$$

Then by (19)

$$(\lambda_1 - x_1)\frac{\partial B}{\partial x} = \mathrm{Tr}((\Delta\varphi\psi\sigma_1 - \sigma_1\varphi\psi\Delta)\Delta^{\mathrm{adj}}) =$$

$$= \mathrm{Tr}(\Delta^{\mathrm{adj}}\Delta\varphi\psi\sigma_1 - \sigma_1\varphi\psi\Delta\Delta^{\mathrm{adj}}) = (\det\Delta)\mathrm{Tr}(\varphi\psi\sigma_1 - \sigma_1\varphi\psi) = 0 .$$

Hence $\frac{\partial}{\partial x} B(x_1, x_2; x) = 0$ for $x_1 \neq \lambda_1(x)$, $x \in (0,\ell)$. By continuity this holds for all $(x_1, x_2) \in \mathbb{C}^2$, $x \in (0,\ell)$. The theorem is proved.

Note that a particular case of system (7)-(9) was considered by L.L. Waksman [5,3].

COROLLARY. *If system* (7)-(9) *is satisfied, then*

$$D(\lambda_1(x), \lambda_2(x)) = 0, \qquad 0 \leqslant x \leqslant \ell . \tag{20}$$

The proof is immediate from (15), (17).

Formula (20) means that *for any* $0 \leqslant x \leqslant \ell$ the point $((\lambda_1(x), \lambda_2(x)) \in \mathbb{C}^2$ lies on the same fixed manifold (10), which is the discriminant manifold of the commuting pair A_1, A_2 under consideration.

Note that the set $\{(\lambda_1(x), \lambda_2(x)) \mid 0 \leqslant x \leqslant \ell\} \subset \mathbb{C}^2$ forms the *joint spectrum* [3] of the commuting pair A_1, A_2.

3. THE INVERSE PROBLEM

3.1. In this Section we discuss the problem of recovering a regular pair of commuting operators (1) from the known data σ_1, σ_2, γ_0. A close problem is to find a solution (or all the solutions) of type (1) to the algebraic operator equation $D(A_1, A_2) = 0$, where $D(x_1, x_2)$ is some given polynomial. In view of Theorem 2, the second problem reduces to the first one if we are given a determinantal representation

$$D(x_1, x_2) = \det(x_1\sigma_2 - x_2\sigma_1 + \gamma_0)$$

of the polynomial $D(x_1, x_2)$. Note that for any polynomial

$D(x_1,x_2)$ there exists at least one such representation [6].

By Theorem 1, finding a regular pair of commuting operators (1) with prescribed σ_1, σ_2, γ_0 is equivalent to solving the system of equations (7)-(9) in which σ_1, σ_2, γ_0 have these prescribed values.

3.2. Let the system (7)-(9) be satisfied, and assume that the polynomial $D(x_1,x_2) = \det(x_1\sigma_2 - x_2\sigma_1 + \gamma_0)$ is irreducible. Then

$$D(x_1,x_2) = 0 \tag{21}$$

is an irreducible algebraic curve and it has at most a finite number of singular points. To recall, a point (λ_1,λ_2) is a *singular point* of the curve (21), if $D(\lambda_1,\lambda_2) = \frac{\partial D}{\partial x_1}(\lambda_1,\lambda_2) = \frac{\partial D}{\partial x_2}(\lambda_1,\lambda_2) = 0$. If (λ_1,λ_2) is a non-singular point of the curve (21), then rank $(\lambda_1\sigma_2 - \lambda_2\sigma_1 + \gamma_0) = n-1$ [4,3]. In the same way, in view of Theorem 3, if $(\lambda_1(x),\lambda_2(x))$ is a non-singular point of the curve (21), then $\operatorname{rank}(\lambda_1\sigma_2 - \lambda_2\sigma_1 + \gamma(x)) = n-1$. In this case the adjugate matrix is of rank 1 and it may be written in the form

$$(\lambda_1(x)\sigma_2 - \lambda_2(x)\sigma_1 + \gamma(x))^{\mathrm{adj}} = \tilde{c}(x)\varphi(x)\psi(x),$$

where $\tilde{c}(x)$ is a non-zero scalar. If for all $x \in [0,\ell]$ the points $(\lambda_1(x),\lambda_2(x))$ are non-singular, then equation (9) may be rewritten in the form

$$\frac{d\gamma(x)}{dx} = \sigma_1\varphi(x)\psi(x)\sigma_2 - \sigma_2\varphi(x)\psi(x)\sigma_1 =$$

$$= c(x)(\sigma_1 F\sigma_2 - \sigma_2 F\sigma_1),$$

$$\gamma(0) = \gamma_0,$$

where $F = (\lambda_1(x)\sigma_2 - \lambda_2(x)\sigma_1 + \gamma(x))^{\mathrm{adj}}$ and $c(x) = 1/\tilde{c}(x)$.

3.3. Now assume that σ_1, σ_2, γ_0 are given constant $n\times n$ matrices, $\lambda_1(x)$, $\lambda_2(x)$, $c(x) \neq 0$, $0 \leqslant x \leqslant L$ are given continuous functions, and consider the matrix differential equation

$$\left\{\begin{array}{l} \frac{d\gamma}{dx} = c(x)(\sigma_1 F(x,\gamma)\sigma_2 - \sigma_2 F(x,\gamma)\sigma_1), \quad 0 \leq x \leq L , \\ \gamma(0) = \gamma_0 , \end{array}\right. \tag{22}$$

where

$$F(x,\gamma) = (\lambda_1(x)\sigma_2 - \lambda_2(x)\sigma_1 + \gamma)^{adj} .$$

The right-hand side of the equation is a given continuous function of x and γ, therefore it has a solution in some interval $[0,\ell)$, $0 < \ell \leq L$.

LEMMA 2. *The determinant* $\det(x_1\sigma_2 - x_2\sigma_1 + \gamma)$ *is an integral of the differential equation* (22): *if* $\gamma = \gamma(x)$ *is a solution of the equation, then* $\det(x_1\sigma_2 - x_2\sigma_1 + \gamma(x))$ *does not actually depend upon* x.

PROOF. It suffices to show that

$\frac{\partial}{\partial x} \det(x_1\sigma_2 - x_2\sigma_1 + \gamma(x)) = 0$. Denote

$$\Delta(x_1, x_2; x) = x_1\sigma_2 - x_2\sigma_1 + \gamma(x) , \tag{23}$$

$$\tilde{\Delta}(x) = \lambda_1(x)\sigma_2 - \lambda_2(x)\sigma_1 + \gamma(x) . \tag{24}$$

Then $\tilde{\Delta}(x)F(x,\gamma(x)) = F(x,\gamma(x))\tilde{\Delta}(x) = (\det \tilde{\Delta}(x))I$. Multiply (23) and (24) from the right by $F\sigma_1$:

$$x_1\sigma_2 F\sigma_1 - x_2\sigma_1 F\sigma_1 + \gamma F\sigma_1 = \Delta F\sigma_1 ,$$

$$\lambda_1\sigma_2 F\sigma_1 - \lambda_2\sigma_1 F\sigma_1 + \gamma F\sigma_1 = (\det \tilde{\Delta})\sigma_1 ,$$

hence

$$(\lambda_1 - x_1)\sigma_2 F\sigma_1 - (\lambda_2 - x_2)\sigma_1 F\sigma_1 = (\det \tilde{\Delta})\sigma_1 - \Delta F\sigma_1 \tag{25}$$

In a similar way, if you multiply (23) and (24) from the left by $\sigma_1 F$:

$$(\lambda_1 - x_1)\sigma_1 F\sigma_2 - (\lambda_2 - x_2)\sigma_1 F\sigma_1 = (\det \tilde{\Delta})\sigma_1 - \sigma_1 F\Delta . \tag{26}$$

Subtract (25) from (26) to obtain

$$(\lambda_1 - x_1)(\sigma_1 F\sigma_2 - \sigma_2 F\sigma_1) = \Delta F\sigma_1 - \sigma_1 F\Delta .$$

Therefore

$$(\lambda_1 - x_1)\ \frac{\partial\ \det\ \Delta}{\partial x} = (\lambda_1 - x_1)\mathrm{Tr}(\frac{\partial\Delta}{\partial x}\ \Delta^{adj}) =$$

$$= c(x)\mathrm{Tr}((\lambda_1 - x_1)(\sigma_1 F\sigma_2 - \sigma_2 F\sigma_1)\Delta^{adj}) =$$

$$= c(x)\mathrm{Tr}((\Delta F\sigma_1 - \sigma_1 F\Delta)\Delta^{adj}) = c(x)(\det\ \Delta)\mathrm{Tr}(F\sigma_1 - \sigma_1 F) = 0\ .$$

The proof of the lemma is accomplished by a concluding remark similar to that of Theorem 3.

3.4. Now assume that $\lambda_1(x)$ and $\lambda_2(x)$ in (22) satisfy the algebraic equation

$$\det(\lambda_1(x)\sigma_2 - \lambda_2(x)\sigma_1 + \gamma_0) = 0\ , \qquad 0 \leqslant x \leqslant \ell\ , \tag{27}$$

and for each $x \in [0,\ell]$ $(\lambda_1(x),\lambda_2(x))$ is a non-singular point of the curve $\det(x_1\sigma_2 - x_2\sigma_1 + \gamma_0) = 0$. Then, by Lemma 2.

$$\det(\lambda_1(x)\sigma_2 - \lambda_2(x)\sigma_1 + \gamma(x)) = 0\ , \qquad 0 \leqslant x \leqslant \ell\ .$$

The matrix $F(x,\gamma(x)) = (\lambda_1(x)\sigma_2 - \lambda_2(x)\sigma_1 + \gamma(x))^{adj}$ is of rank 1 and it may be represented in the form

$$F(x,\gamma(x)) = \tilde{\varphi}(x)\tilde{\psi}(x)\ ,$$

where $\tilde{\varphi}(x)$ is a continuous n-column and $\tilde{\psi}(x)$ is a continuous n-row. Choose any two continuous scalar functions $a(x)$, $b(x)$ such that $a(x)b(x) = c(x)$, and define

$$\psi(x) = a(x)\tilde{\psi}(x)\ , \qquad \varphi(x) = b(x)\tilde{\varphi}(x)\ .$$

Then the following system of equalities is satisfied:

$$(\lambda_1(x)\sigma_2 - \lambda_2(x)\sigma_1 + \gamma(x))\varphi(x) = 0\ ,$$

$$\psi(x)(\lambda_1(x)\sigma_2 - \lambda_2(x)\sigma_1 + \gamma(x)) = 0\ ,$$

$$\frac{d\gamma(x)}{dx} = \sigma_1\varphi(x)\psi(x)\sigma_2 - \sigma_2\varphi(x)\psi(x)\sigma_1\ ,$$

$$\gamma(0) = \gamma_0\ ,$$

which is the same as system (7)-(9).

THEOREM 4. *Let* σ_1, σ_2, γ_0 *be constant* $n\times n$ *matrices such that the polynomial*

$$D(x_1,x_2) = \det(x_1\sigma_2 - x_2\sigma_1 + \gamma_0) \tag{28}$$

is irreducible. Let the two continuous functions $\lambda_1(x)$, $\lambda_2(x)$, $0 \leq x \leq L$ *satisfy the algebraic equation* $D(\lambda_1(x),\lambda_2(x)) = 0$ *provided that for any* $x \in [0,L]$ $(\lambda_1(x),\lambda_2(x))$ *is a non-singular point of the curve* $D(x_1,x_2) = 0$. *Then there exist continuous vector functions* $\varphi(x)$, $\psi(x)$, $0 \leq x < \ell \leq L$, *such that* A_1 *and* A_2 *in formulas* (1) *form a regular pair of commuting integral operators with discriminant polynomial* (28) *and joint spectrum* $\{(\lambda_1(x),\lambda_2(x)) | 0 \leq x < \ell\}$.

The PROOF follows from the preceding discussion and Theorem 1.

REMARK. Let $D(x_1,x_2)$ be some irreducible polynomial. Given any determinantal representation (28) of $D(x_1,x_2)$, we are able, in view of Theorem 4 and Theorem 2, to construct pairs of commuting operators of the form (1) which solve the algebraic operator equation $D(A_1,A_2) = 0$, at least on some invariant subspace.

3.5. EXAMPLE. Let be

$$D(x_1,x_2) = 1 - x_1^2 - x_2^2 ,$$

the corresponding algebraic curve $D(x_1,x_2) = 0$ being the unit circle in the (x_1,x_2)-plane. This polynomial has a determinantal representation

$$1 - x_1^2 - x_2^2 = \begin{vmatrix} 1+x_1 & -x_2 \\ -x_2 & 1-x_1 \end{vmatrix} = \det(x_1\sigma_2 - x_2\sigma_1 + \gamma_0) ,$$

where

$$\sigma_1 = \begin{pmatrix} 0 & 1 \\ 1 & 0 \end{pmatrix}, \quad \sigma_2 = \begin{pmatrix} 1 & 0 \\ 0 & -1 \end{pmatrix}, \quad \gamma_0 = \begin{pmatrix} 1 & 0 \\ 0 & 1 \end{pmatrix} .$$

Choose two continuous functions $c(x)$, $\mu(x)$ on $[0,\ell]$ and set

$$\lambda_1(x) = \cos 2\mu(x), \quad \lambda_2(x) = \sin 2\mu(x), \qquad 0 \leq x \leq \ell .$$

Obviously, $D(\lambda_1(x),\lambda_2(x)) = 1 - \lambda_1^2(x) - \lambda_2^2(x) = 0$. Denote

$$\gamma = \begin{pmatrix} \gamma_{11} & \gamma_{12} \\ \gamma_{21} & \gamma_{22} \end{pmatrix} .$$

Then

$$F(x,\gamma) = (\lambda_1(x)\sigma_2 - \lambda_2(x)\sigma_1 + \gamma\ )^{adj} = \begin{pmatrix} \gamma_{22} - \lambda_1(x) & \lambda_2(x) - \gamma_{12} \\ \lambda_2(x) - \gamma_{21} & \gamma_{11} + \lambda_1(x) \end{pmatrix} ,$$

$$\sigma_1 F\sigma_2 - \sigma_2 F\sigma_1 = \begin{pmatrix} \gamma_{12} - \gamma_{21} & -(\gamma_{11} + \gamma_{22}) \\ \gamma_{11} + \gamma_{22} & \gamma_{12} - \gamma_{21} \end{pmatrix} ,$$

and the equation (22) obtains the form

$$\begin{cases} \dfrac{d}{dx}\begin{pmatrix} \gamma_{11} & \gamma_{12} \\ \gamma_{21} & \gamma_{22} \end{pmatrix} = c(x)\begin{pmatrix} \gamma_{12} - \gamma_{21} & -(\gamma_{11} + \gamma_{22}) \\ \gamma_{11} + \gamma_{22} & \gamma_{12} - \gamma_{21} \end{pmatrix} , \\ \gamma(0) = \begin{pmatrix} 1 & 0 \\ 0 & 1 \end{pmatrix} \end{cases}$$

This implies

$$\begin{cases} \gamma_{11} = \gamma_{22}, \quad \gamma_{21} = -\gamma_{12}; \quad \gamma_{11}(0) = 1, \quad \gamma_{12}(0) = 0 , \\ \dfrac{d\gamma_{11}}{dx} = 2c(x)\gamma_{12} ; \quad \dfrac{d\gamma_{12}}{dx} = -\ 2c(x)\gamma_{11} . \end{cases}$$

The solution of this system is

$$\gamma(x) = \begin{pmatrix} \cos\ 2\nu(x) & -\sin\ 2\nu(x) \\ \sin\ 2\nu(x) & \cos\ 2\nu(x) \end{pmatrix} ,$$

where

$$\nu(x) = \int_0^x c(t)dt .$$

Then

$$\lambda_1\sigma_2 - \lambda_2\sigma_1 + \gamma = \begin{pmatrix} \cos 2\nu + \cos 2\mu & -\sin 2\nu - \sin 2\mu \\ \sin 2 - \sin 2\mu & \cos 2\nu - \cos 2\mu \end{pmatrix} =$$

$$= 2 \begin{pmatrix} \cos(\nu+\mu)\cos(\nu-\mu) & -\sin(\nu+\mu)\cos(\nu-\mu) \\ \cos(\nu+\mu)\sin(\nu-\mu) & -\sin(\nu+\mu)\sin(\nu-\mu) \end{pmatrix} .$$

$$F = 2\begin{pmatrix} \sin(\nu+\mu) \\ \cos(\nu+\mu) \end{pmatrix} (-\sin(\nu-\mu) \quad \cos(\nu-\mu)) .$$

Denote for simplicity

$$\nu(x) + \mu(x) = \alpha(x); \qquad \nu(x) - \mu(x) = \beta(x) ,$$

then

$$\varphi(x) = b(x) \begin{pmatrix} \sin \alpha(x) \\ \cos \alpha(x) \end{pmatrix} \qquad \psi(x) = a(x)(-\sin \beta(x) \quad \cos \beta(x))$$

$$\psi(x)\sigma_1\varphi(t) = a(x)b(t)\sin(\alpha(t)-\beta(x)) ,$$

$$\psi(x)\sigma_2\varphi(t) = -a(x)b(t)\cos(\alpha(t)-\beta(x)) ,$$

where

$$a(x)b(x) = 2c(x) = 2\nu'(x) = (\alpha(x)+\beta(x))' .$$

Thus, the final expression is

$$(A_1 f)(x) = \cos(\alpha(x)-\beta(x))f(x) + a(x) \int_0^x \sin(\alpha(t)-\beta(x))b(t)f(t)dt ,$$

$$(A_2 f)(x) = \sin(\alpha(x)-\beta(x))f(x) - a(x) \int_0^x \cos(\alpha(t)-\beta(x))b(t)f(t)dt.$$

Here $\alpha(x)$, $\beta(x)$, $a(x)$, $b(x)$ are continuous functions on $[0,\ell]$ for which

$$(\alpha(x)+\beta(x))' = a(x)b(x) ,$$

$$\alpha(0) + \beta(0) = 0 .$$

A straightforward calculation shows that A_1 and A_2 (commute and) satisfy the operator equation $A_1^2 + A_2^2 = I$ *in the whole space* $L^2(0,\ell)$.

The author is grateful to Professor M.S. Livšic for fruitful discussion.

REFERENCES

1. Livšic, M.S.: Operator waves in Hilbert space and related partial differential equations, Integral Equations and Operator Theory 2/1 (1979), 25-47.

2. Livšic, M.S.: A method for constructing triangular canonical models of commuting operators based on connections with algebraic curves, Integral Equations and Operator Theory 3/4 (1980), 489-507.

3. Livšic, M.S.: Cayley-Hamilton theorem, vector bundles and divisors of commuting operators, Integral Equations and Operator Theory, vol. 6 (1983), 250-273.

4. Kravitsky, N.: Regular colligations for several commuting operators in Banach space, Integral Equations and Operator Theory, vol. 6 (1983), 224-249.

5. Waksman, L.L.: Harmonic analysis of multivariable semi-groups of contractions, Monograph, Kharkov State Univ., USSR, 1979.

6. Dixon, A.C.: Note of the reduction of a ternary quantic to a symmetrical determinant, Proc. Camb. Phil. Soc. 2 (1900-1902), 350-351.

N. Kravitsky,
Department of Mathematics,
Ben-Gurion University of the Negev,
Beer Sheva, Israel

Operator Theory:
Advances and Applications, Vol. 12

INFINITE DIMENSIONAL STOCHASTIC REALIZATIONS OF CONTINUOUS-TIME STATIONARY VECTOR PROCESSES

Anders Lindquist[1] and Giorgio Picci

In this paper we consider the problem of representing a given stationary Gaussian process with nonrational spectral density and continuous time as the output of a stochastic dynamical system. Since the spectral density is not rational, the dynamical system must be infinite-dimensional, and therefore the continuous-time assumption leads to certain mathematical difficulties which require the use of Hilbert spaces of distributions. (This is not the case in discrete time.) We show that, under certain conditions, there correspond to each proper Markovian splitting subspace, two standard realizations, one evolving forward and one evolving backward in time.

1. INTRODUCTION

Let $\{y(t);\ t \in \mathbb{R}\}$ be an m-dimensional stationary Gaussian vector process which is mean-square continuous and purely nondeterministic and which has zero mean. Such a process has a representation

$$y(t) = \int_{-\infty}^{\infty} e^{i\omega t}\, d\hat{y}(i\omega) \tag{1.1a}$$

[5,11]. Here $d\hat{y}$ is an orthogonal stochastic measure with incremental covariance

$$E\{d\hat{y}(i\omega)d\hat{y}(i\omega)^{\dagger}\} = \frac{1}{2\pi}\Phi(i\omega)d\omega \tag{1.1b}$$

where Φ is the $m\times m$-matrix spectral density and $\dagger$ denotes transposition *and* complex conjugation. Let $p \leq m$ be the rank of Φ.

It is well-known that, if Φ is a rational function, y has a (non-unique) representation

[1]This research was supported partially by the National Science Foundation under grant ECS-8215660 and partially by the Air Force Office of Scientific Research under grant AFOSR-78-3519.

$$\begin{cases} dx = Ax\,dt + B\,du & (1.2a) \\ y = Cx & (1.2b) \end{cases}$$

where $\{x(t);\ t \in \mathbb{R}\}$ is a stationary vector process of (say) dimension n, $\{u(t);\ t \in \mathbb{R}\}$ is a vector Wiener process defined on all of $\mathbb{R}$ with incremental covariance

$$E\{du(t)du(t)'\} = I\,dt \tag{1.3}$$

(' denotes transposition), and A, B and C are constant matrices of appropriate dimensions. Such a representation is called a *stochastic realization* of y, and n is its *dimension*. Clearly $\{x(t),\ t \in \mathbb{R}\}$ is a Gaussian Markov process.

It is the purpose of this paper to construct stochastic realizations (with the appropriate systems-theoretical properties) of processes y with nonrational spectral densities. Such realizations are necessarily infinite-dimensional, and the analysis will require methods akin to those used in infinite-dimensional deterministic realization theory [2,3,6]. Our approach, which is based on the geometric theory of stochastic realization theory [8,9,12] and applies Hilbert space constructions common in infinite-dimensional control theory [10], is coordinate-free. These results replace those in Section 9 of [8], which contains an error.

The results of this paper were first presented at the NATO Advanced Study Institute Workshop on Nonlinear Systems in Algarve, Portugal, May 17-28, 1982, and at the Second Bad Honnef Conference on Stochastic Differential Systems in Bonn, West Germany, June 28 - July 3, 1982.

2. PRELIMINARIES AND NOTATIONS

Let H be the Gaussian space of y, i.e. the Hilbert space generated by the random variables $\{y_k(t);\ t \in \mathbb{R},\ k=1,2,\ldots,m\}$ with the inner product $\langle\xi,\eta\rangle = E\{\xi\eta\}$. Let H^- and H^+ be the subspaces generated by $\{y_k(t);\ t \le 0,\ k=1,2,\ldots,m\}$ and $\{y_k(t);\ t \ge 0,\ l=1,2,\ldots,m\}$, respectively. If $\{\eta_\theta;\ \theta \in \Theta\}$ is a set of elements in H, $\mathrm{sp}\{\eta_\theta;\ \theta \in \Theta\}$ denotes the linear span, i.e. the set of all linear combinations of these elements. The word *subspace* will carry with it the assumption that it is closed.

For any two subspaces of H, A and B, $A \vee B$ denotes the closed linear hull and $A \oplus B$ the orthogonal direct sum of A and B. Moreover, E^A denotes orthogonal projection onto A, $\bar{E}^A B$ the closure of $E^A B$, and $A^\perp$ the orthogonal complement of A in H. We write $A \perp B$ to mean that A and B are orthogonal and $A \perp B | X$ to mean that they are *conditionally orthogonal* given the subspace X, i.e.

$$\langle E^X \alpha, E^X \beta \rangle = \langle \alpha, \beta \rangle \qquad \text{for all } \alpha \in A,\ \beta \in B .$$

Since y is stationary there is a strongly continuous group $\{U_t;\ t \in \mathbb{R}\}$ of unitary operators $H \to H$ such that $U_t y_k(0) = y_k(t)$ for $k=1,2,\ldots,m$ [11]. The subspace A is said to be *full range* if $\bigvee_{t \in \mathbb{R}} (U_t A) = H$, i.e. the closed linear hull of the subspace $\{U_t A;\ t \in \mathbb{R}\}$ equals H.

3. BACKGROUND

To provide a setting for our construction we shall briefly review some basic results from our geometric theory of stochastic realization. The reader is referred to [8,9] for full details.

A subspace $X \subset H$ such that

(i) $X^- \perp X^+ | X$

where $X^- := \bigvee_{t \le 0} (U_t X)$ and $X^+ := \bigvee_{t \ge 0} (U_t X)$

(ii) $y_k(0) \in X$ for $k=1,2,\ldots,m$

(iii) $(X^-)^\perp$ and $(X^+)^\perp$ are full range

is called a *proper Markovian splitting subspace*, proper because of condition (iii). If, for the moment, we assume that y is the output process of the finite-dimensional system (1.2), (i)-(iii) is a coordinate-free characterization of (1.2). In fact, if $E\{x(0)x(0)'\} > 0$,

$$X = \{a'x(0) \mid a \in \mathbb{R}^n\} \qquad (3.1)$$

is a proper Markovian splitting subspace. Condition (i) is equivalent to $\{x(t);\ t \in \mathbb{R}\}$ being a Markov process, i.e. that it has a representation (1.2a), (ii) is equivalent to

$$y_k(t) \in \{a'x(t) \mid a \in \mathbb{R}^n\} \qquad \text{for } k=1,2,\ldots,m \,, \tag{3.2}$$

i.e. to the existence of a matrix C such that (1.2b) holds, and (iii) rules out the possibility that x has a deterministic component. The coordinate-free formulation (i)-(iii) enables us to handle also the fact that X is infinite-dimensional and the concept of minimality becomes especially simple: A Markovian splitting subspace is said to be *minimal* if it has no proper subspace which is also a Markovian splitting subspace. We refer the reader to [8,9] for a discussion of what conditions (such as strict noncyclicity) are needed for (iii) to hold.

It is not hard to show that

$$X = \bar{E}^X H^+ \oplus [X \cap (H^+)^\perp] \,. \tag{3.3}$$

An element in $X \cap (H^+)^\perp$ cannot be distinguished from zero by observing the future $\{y(t);\ t \geq 0\}$ and is therefore called *unobservable*. The splitting subspace X is said to be *observable* if the unobservable subspace is trivial, i.e. $X \cap (H^+)^\perp = 0$. Likewise

$$X = \bar{E}^X H^- \oplus [X \cap (H^-)^\perp] \,, \tag{3.4}$$

and we call X *constructible* if the unconstructable subspace $X \cap (H^-)^\perp = 0$. It can be shown that X is minimal if and only if it is both observable and constructible [12].

We shall provide a complete characterization of the class of proper Markovian splitting subspaces. To this end, first consider the class of p×m matrix functions $i\omega \to W(i\omega)$ satisfying

$$W(-i\omega)'W(i\omega) = \Phi(i\omega) \,. \tag{3.5}$$

Such functions exist [11; p.114], and they are called *full-rank spectral factors*. We shall say that W is *stable (completely unstable)* if it can be extended to the right (left) complex half plane and is analytic there. (In the rational case, this means that W has all its poles in the left (right) half plane.)

To each full-rank spectral factor W we may associate a Wiener process

$$u(t) = \int_{-\infty}^{\infty} \frac{e^{i\omega t}-1}{i\omega} W^{-R}(i\omega)'d\hat{y}(i\omega) \tag{3.6}$$

defined on all of $\mathbb{R}$, where W^{-R} is any right inverse of W. (It can be shown that (3.6) is independent of the choice of right inverse.) Let $\mathcal{U}$ be the class of all Wiener processes u generated in this way, and let $\mathcal{U}^-$ ($\mathcal{U}^+$) be the subclass corresponding to a stable (strictly unstable) W. For each $u \in \mathcal{U}$ let $H(du)$, $H^-(du)$ and $H^+(du)$ denote the subspaces generated by $\{u_k(t);\ t \in \mathbb{R},\ k=1,2,\dots,m\}$, $\{u_k(t);\ t \le 0,\ k=1,2,\dots,m\}$ and $\{u_k(t);\ t \ge 0,\ l=1,2,\dots,m\}$, respectively. It is not hard to see that $H(du)=H$ for all $u \in \mathcal{U}$.

THEOREM 3.1 [8]. *The subspace* X *is a proper Markovian splitting subspace if and only if there are* $u \in \mathcal{U}^-$ *and* $\bar{u} \in \mathcal{U}^+$ *such that*

$$H = H^-(d\bar{u}) \oplus X \oplus H^+(du) \quad . \tag{3.7}$$

In that case, $X = H^-(du) \cap H^+(d\bar{u})$, $H^- \vee X = H^-(du)$, *and* $H^+ \vee X = H^+(d\bar{u})$. *The splitting subspace is observable if and only if*

$$H^+(d\bar{u}) = H^+ \vee H^+(du) \tag{3.8}$$

and constructible if and only if

$$H^-(du) = H^- \vee H^-(d\bar{u}) \quad . \tag{3.9}$$

Consequently each X is completely characterized by a pair $(W,\bar{W})$ of full rank-spectral factors, W being stable and $\bar{W}$ completely unstable. The ratio $K := \bar{W}W^{-R}$ is called the *structural function* of X. It plays an important role in the systems-theoretical characterization of X.

Decomposition (3.7) should be compared with the decomposition in terms of ingoing and outgoing subspaces in Lax-Phillips scattering theory [7] and K to the scattering operator.

Next we introduce a semigroup on each X. For $t \geqq 0$ define the operator $U_t(X)\colon X \to X$ by the relation $U_t(X)\xi = E^X U_t \xi$. A proof of the first part of the following theorem can be found in [7, p.62] and a proof of the second part in [8]. (Asterisk denotes adjoint.)

THEOREM 3.2. *The family of operators* $\{U_t(X);\ t \geqq 0\}$ *is a strongly continuous semigroup of contraction operators on* X *which tend strongly to zero as* $t \to \infty$. *Moreover, for all* $\xi \in X$

and $t \geq 0$,

$$E^{H^-(du)} U_t \xi = U_t(X)\xi \tag{3.10}$$

and

$$E^{H^+(d\bar{u})} U_{-t} \xi = U_t(X)^* \xi. \tag{3.11}$$

4. STOCHASTIC REALIZATIONS

The given process $\{y(t);\ t \in \mathbb{R}\}$ will not have a finite-dimensional representation (1.2) unless it has a rational spectral density. Therefore we need to decide how to interpret (1.2) in the infinite-dimensional case. It is natural to require that A is the infinitesimal generator of a strongly continuous semigroup $\{e^{At}; t \geq 0\}$ defined on some Hilbert space X and that $B: \mathbb{R}^p \to X$ and $C: X \to \mathbb{R}^m$ are bounded operators. In the finite-dimensional case, the system equations (1.2) have the unique strong solution

$$\begin{cases} x(t) = \int_{-\infty}^{t} e^{A(t-\sigma)} B\, du(\sigma) & (4.1a) \\ y(t) = Cx(t) & (4.1b) \end{cases}$$

where the integral is defined in quadratic mean (Wiener integral). In the infinite-dimensional case, however, things are more complicated.

To begin with, the integral in (4.1a) may not be well-defined. (This requires that the X-valued function $t \to e^{At}B$ be square-integrable on $[0,\infty)$; cf [13,14].) If not, we shall have to define the *state process* x in some weak sense (to be specified below). Even if the integral is well-defined so that x exists as a strong Hilbert space valued random process, (4.1) may not be a strong solution of (1.2) [14]. In this case, (1.2) will simply be defined as (4.1); this is known as "mild solution".

If (1.2) does have a strong solution, y must satisfy the integral equation

$$y(t) = y(0) + \int_0^t CAx(\sigma)\,d\sigma + \int_0^t CB\,du(\sigma) \tag{4.2}$$

[13,14] and therefore

$$E^{H^-(du)}[y_k(h) - y_k(0)] = O(h) \tag{4.3}$$

for all $k = 1,2,\ldots,m$. In view of (3.10), this is equivalent to

$$y_k(0) \in \mathcal{D}(\Gamma) \qquad \text{for } m=1,2,\ldots,m \tag{4.4}$$

where $\mathcal{D}(\Gamma)$ is the domain of the infinitesimal generator Γ of the semigroup $\{U_t(X);\ t \geq 0\}$. From now on, we shall assume that X is a proper Markovian splitting subspace such that (4.4) holds. This turns out to be a natural assumption even in those cases that our construction does not produce a system (1.2) with a strong solution. Condition (4.4) is nontrivial only if X is infinite-dimensional, for, if $\dim X < \infty$, $\mathcal{D}(\Gamma) = X$.

Then, following a standard construction [1,4], we define the space Z to be $\mathcal{D}(\Gamma)$ equipped with the graph topology

$$\langle\xi,\eta\rangle_Z = \langle\xi,\eta\rangle_X + \langle\Gamma\xi,\Gamma\eta\rangle_X. \tag{4.5}$$

Since Γ is a closed operator whose domain is dense in $\mathcal{X}$ (see eg [7, p.247]), Z is a Hilbert space, which is continuously imbedded in X. The topology of Z is stronger than that of X, and therefore all continuous linear functionals on X are continuous on Z as well. Consequently we can think of the dual space X* as imbedded in the dual space Z*. Then, identifying X* with X, we have

$$Z \subset X \subset Z^* \tag{4.6}$$

where Z is dense in X, which in turn is dense in Z*. We shall write (z,z^*) to denote the value of the functional $z^* \in Z^*$ evaluated at $z \in Z$ (or, by reflexivity, the value at z^* of z regarded as a functional on Z*). Clearly the bilinear form (z,z^*) coincides with the inner product $\langle z,z^*\rangle_X$ whenever $z^* \in X$.

Since $X \subset H^-(du)$, any $\xi \in X$ has a unique representation

$$\xi = \int_{-\infty}^{0} f(-\sigma)'du(\sigma) \tag{4.7}$$

where $f \in L_2(\mathbb{R}_+, \mathbb{R}^p)$ and the integral is defined in quadratic mean. Define the (real) Hilbert space

$$\mathcal{X} = \left\{f \,\middle|\, \int_{-\infty}^{0} f(-\sigma)'du(\sigma) \in X\right\} \tag{4.8}$$

with inner product $\langle f,g\rangle_{\mathcal{X}} = \int_0^\infty f(t)'g(t)dt$. It is a well-known property of Wiener integrals that the mapping $I_u\colon \mathcal{X} \to X$ defined by (4.7) $[\xi = I_u f]$ is an isometry, and therefore we have established an isometric isomorphism between X and $\mathcal{X}$.

Clearly $\{I_u^{-1}U_t(X)^*I_u; t \geq 0\}$ is a strongly continuous semigroup on X. Let A be its infinitesemal generator, i.e.

$$e^{At} = I_u^{-1}U_t(X)^*I_u. \tag{4.9}$$

The operator A is (in general) unbounded and densely defined in X. The adjoint A* is the infinitesimal generator of the adjoint semigroup $\{I_u^{-1}U_t(X)I_u ; t \geq 0\}$. Since

$$U_t\xi = \int_{-\infty}^{t} f(t-\sigma)'du(\sigma) , \tag{4.10}$$

(3.10) yields

$$U_t(X)\xi = \int_{-\infty}^{0} f(t-\sigma)'du(\sigma) . \tag{4.11}$$

Therefore,

$$\frac{1}{h}[U_h(X)\xi - \xi] = \int_{-\infty}^{0} \frac{1}{h}[f(h-\sigma) - f(-\sigma)]'du(\sigma) \tag{4.12}$$

and consequently A*f is the L_2 derivative of f (i.e. the limit in L_2 topology of the difference quotient).

Hence we have a functional representation of (4.6), namely

$$Z \subset X \subset Z^* \tag{4.13}$$

where $Z := I_u^{-1}Z$ is $\mathcal{D}(A^*)$ equipped with the inner product

$$\langle f,g\rangle_Z = \langle f,g\rangle_X + \langle A^*f,A^*g\rangle_X \tag{4.14}$$

i.e. a Hilbert space continuously imbedded in X, and Z^* is its dual, constructed as above. Here Z is a subspace of the Sobolev space $H^1(0,\infty)$, and Z^* is a space of distributions [1]. As before we write (f,f^*) to denote the scalar product between Z and Z^* extending $\langle f,f^*\rangle_X$ from $Z \times X$ to $Z \times Z^*$.

Next, define $D: Z \to X$ to be the differentiation operator. Then $Df = A^*f$ for all $f \in Z$, but, since $\|Df\|_X \leq \|f\|_Z$, D is a bounded operator (in Z-topology). Its adjoint $D^*: X \to Z^*$ is the extension of A to X, because $(f,D^*g) = \langle A^*f,g\rangle_X$. Since $\{e^{A^*t}; t \geq 0\}$ is a completely continuous contraction semigroup (Theorem 3.2), D is dissipative, i.e. $\langle Df,f\rangle_X \leq 0$ for all $f \in Z$, and I-D maps Z onto X, i.e.

$$(I-D)Z = X \tag{4.15}$$

[17; p.250]. Moreover, I - D is injective. In fact, in view of the dissipative property,

$$\| (I - D) f \|_X^2 \geq \| f \|_X^2 + \| Df \|_X^2 . \tag{4.16}$$

Consequently, $(I - D)^{-1}: X \to Z$ is defined on all of X, and, as can be seen from (4.16), it is a bounded operator. Likewise, the adjoint $(I - D^*)^{-1}$ is a bounded operator mapping Z^* onto X.

Now, assume that $f \in Z$, and let ξ be defined by (4.7), i.e. $\xi = I_u f$. Then it follows from (4.11) that $f(t+\sigma) = [I_u^{-1} U_t(X)\xi](\sigma) = (e^{A^*t}f)(\sigma)$ for $\sigma \geq 0$, i.e.

$$f(t) = (e^{A^*t}f)(0) . \tag{4.17}$$

Since Z is a *bona fide* function space and e^{A^*t} maps Z into Z, (4.17) is well-defined. In fact, as Z is a subspace of the Sobolev space $H^1(0,\infty)$, the evaluation functionals $\delta_k \in Z^*$ defined by $(f,\delta_k) = f_k(0)$, $k = 1,2,\dots,m$, are continuous, because the evaluation operator in $H^1(0,\infty)$ is [1,4]. (Note that, since δ_k is restricted to Z, it is not the Dirac function.)

Consequently, we have

$$f_k(t) = (e^{A^*t}f, \delta_k) \tag{4.18}$$

We wish to express this in terms of the inner product in X, which from now on we shall denote $\langle \cdot , \cdot \rangle$, dropping the subscript X, whenever there is no risk for misunderstanding. To this end, note that (4.18) can be written

$$f_k(t) = \langle (I-D)e^{A^*t}f,\ (I-D^*)^{-1}\delta_k \rangle \tag{4.19}$$

Since $Df = A^*f$ and A^* and e^{A^*t} commute, this yields

$$f_k(t) = \langle (I-D)f,\ e^{At}Be_k \rangle , \tag{4.20}$$

where $B : \mathbb{R}^p \to X$ is the bounded operator

$$Ba = \sum_{k=1}^{p} (I - D^*)^{-1}\delta_k a_k \tag{4.21}$$

and e_k is the k:th axis unit vector in $\mathbb{R}^p$. Therefore, in view of (4.10), we have

$$U_t\xi = \int_{-\infty}^{t} \sum_{k=1}^{p} \langle (I-D)f, e^{A(t-\sigma)}Be_k \rangle du_k(\sigma) . \tag{4.22}$$

If the integral (4.1a) is well-defined, i.e. $t \to e^{At}B$ belongs to $L_2(\mathbb{R}_+, X)$, the usual limit argument yields

$$U_t\xi = \langle g, x(t) \rangle \tag{4.23}$$

where $g := (I-D)f = (I-D)I_u^{-1}\xi$. As can be seen from the following theorem, $x(t)$ being a strong X-valued random variable is a property of the structural function K of the splitting subspace X.

THEOREM 4.1. *Let* A *and* B *be given by (4.9) and (4.21) respectively. A necessary condition for the integral (4.1a) to define a (strong)* X*-valued random variable* $x(t)$ *is that the structural function* K *be meromorphic in the whole complex plane and analytic along the imaginary axis. A sufficient condition is that* K *be analytic in some strip* $-\alpha < \text{Re}(s) \leq 0$ *(where* s *is the complex variable).*

PROOF. To establish the necessary condition, assume that (4.1a) is well-defined. It is no restriction to set $t = 0$. Let $g_1, g_2 \in X$ be arbitrary, and define, for $i = 1,2$, $f_i := (I-D)^{-1}g_i$ and $\xi_i := I_u f_i$. Then, since $E\{\xi_1\xi_2\} = \langle f_1,f_2\rangle$, (4.23) yields

$$E\{\langle g_1,x(0)\rangle\langle g_2,x(0)\rangle\} = \langle g_1,\Lambda g_2\rangle \tag{4.24}$$

where $\Lambda := [(I-D)(I-D^*)]^{-1}$ is the correlation operator. This operator must be nuclear [15; p.9], and therefore $(I-D)^{-1}$ is compact [4; p.34], and so is $e^{Dt}(I-D)^{-1}$ for all $t > 0$, since e^{Dt} is bounded. Then the rest follows from [7; p.83]. In the sufficiency part, the assumption on K implies that the spectrum of A^* lies in the region $\text{Re}(s) \leq -\alpha < 0$ [7; p. 70]. Therefore there are positive numbers k and $\beta < \alpha$ such that $\|e^{A^*t}\| \leq k e^{-\beta t}$ [17]. But $\|e^{At}\| = \|e^{A^*t}\|$, and hence $\|e^{At}\| \in L_2(0,\infty)$. □

If the stochastic integral (4.22) is not well-defined, we can nevertheless think of x as a generalized (weakly defined) random process [4; p.242]. In fact, in view of (4.15), (4.22) assigns to each pair $(t,g) \in \mathbb{R} \times X$ a unique random variable $U_t\xi$. Thought of in this way, (4.23) makes perfect sense, and we shall take this as our definition of $\{x(t);\ t \in \mathbb{R}\}$ whenever this process is not strongly defined.

Since there is a one-one correspondence in (4.23) between $\xi \in Z$ and $g \in X$, it follows from (4.15) that $\{\langle g,x(0)\rangle \mid g \in X\} = Z$. But Z is dense in X, and therefore

$$X = \text{cl}\{\langle g,x(0)\rangle \mid g \in X\} \tag{4.25}$$

where cl stands for closure (in the topology of H). This is the infinite-dimensional counterpart of (3.1). In fact, in the special case that $\dim X < \infty$, $Z = X$, and then no closure is needed.

It remains to construct a counterpart of (1.2b). In view of (4.4), $y_k(0) \in Z$, and therefore $w_k := I_u^{-1} y_k(0) \in Z$ for $k = 1,2,\ldots,m$. Therefore, defining $C : X \to \mathbb{R}^m$ by

$$(Cg)_k = \langle (I-D)w_k, g\rangle \tag{4.26}$$

for $k = 1,2,\ldots,m$, (4.22) yields

$$y(t) = \int_{-\infty}^{t} Ce^{A(t-\sigma)}Bdu(\sigma). \tag{4.27}$$

We may write this as

$$\begin{cases} dx = Axdt + Bdu & (4.28a) \\ y = Cx & (4.28b) \end{cases}$$

if we interpret this system as described above.

The above construction is in several respects similar to those found in (infinite-dimensional) deterministic realization theory [2, 3, 6]. Note, however, that in comparison with the shift realizations in, for example, [3], our set-up has been transposed. This is necessary in order to obtain results such as those in Theorems 5.1 and 5.2 and is quite natural if we think of stochastic realization theory as representations of *functionals* of data.

5. OBSERVABILITY AND REACHABILITY

The system (4.28) is said to be *observable* if $\cap_{t\geq 0} \ker C e^{At} = 0$ and *reachable* if $\cap_{t\geq 0} \ker B^* e^{A^*t} = 0$ [3].

THEOREM 5.1. *The system (4.28) is reachable.*

PROOF. Since $\langle (I-D)f, Ba\rangle = f(0)'a$ for all $f \in Z$, the adjoint operator $B^* : X \to \mathbb{R}^p$ is given by

$$B^*g = [(I-D)^{-1}g](0). \tag{5.1}$$

Therefore, since e^{A^*t} and $(I-D)^{-1}$ commute,

$$B^*e^{A^*t}g = [e^{A^*t}(I-D)^{-1}g](0), \tag{5.2}$$

which, in view of (4.17), can be written

$$B^*e^{A^*t}g = f(t), \tag{5.3}$$

where $f := (I-D)^{-1}g$. Hence $g \in \cap_{t\geq 0}\ker B^*e^{A^*t}$ if and only if $f(t) = 0$ for all $t \geq 0$, i.e. $f = 0$, or, equivalently, $g = 0$. This establishes reachability. □

THEOREM 5.2. *The system (4.28) is observable if and only if the splitting subspace (4.25) is observable (in the sense of Section 3).*

For the proof we need a few concepts. Define M to be the vector space

$$M = \mathrm{sp}\{E^X y_k(t)\ ;\ t \geq 0,\ k = 1,2,\dots,m\}. \tag{5.4}$$

Since $E^X y_k(t) = U_t(X) y_k(0)$, M is invariant under the action of $U_t(X)$, i.e. $U_t(X) M \subset M$ for all $t \geq 0$. Moreover, $\mathcal{D}(\Gamma)$ is invariant under $U_t(X)$; this is a well-known property of a semigroup. Hence, it follows from (4.4) that $M \subset Z$. Now, if X is observable, M is dense in X, but this does not automatically imply that M is dense in Z (in graph topology). In the present case, however, this is true, as can be seen from the following lemma. In the terminology of [1; p.101], this means that the Hilbert space Z containing the vector space M and continuously embedded in the Hilbert space X is *normal*.

LEMMA 5.1. *Let* X *be observable. Then* M *is dense in* Z.

A proof of this lemma provided by A. Gombani will be given below. Setting $\mathcal{M} := I_u^{-1} M$, we have

$$\mathcal{M} = \mathrm{sp}\{e^{A^*t} w_k\ ;\ t \geq 0,\ k = 1,2,\dots,m\} \tag{5.5}$$

and therefore we may, equivalently, state Lemma 5.1 in the following way: *If* X *is observable, then* $\mathcal{M}$ *is dense in* $\mathcal{Z}$.

LEMMA 5.2. *The vector space* $\mathcal{M}$ *is dense in* $\mathcal{Z}$ *if and only if* $(I-D)\mathcal{M}$ *is dense in* $\mathcal{X}$.

PROOF. (if): Assume that $(I-D)\mathcal{M}$ is dense in $\mathcal{X}$. Then (4.15), i.e. $(I-D)\mathcal{Z} = \mathcal{X}$, and (4.16), i.e. $\|f\|_{\mathcal{Z}} \leq \|(I-D)f\|_{\mathcal{X}}$, imply that $\mathcal{M}$ is dense in $\mathcal{Z}$.
(only if): This part follows from (4.15) and the trivial relation $\|(I-D)f\|_{\mathcal{X}}^2 \leq 2\|f\|_{\mathcal{Z}}^2$. □

PROOF OF THEOREM 5.2. First note that, since e^{A^*t} and (I-D) commute,

$$(Ce^{At}g)_k = \langle (I-D)e^{A^*t} w_k, g\rangle. \tag{5.6}$$

Hence $g \in \bigcap_{t\geq 0} \ker Ce^{At}$ if and only if

$$\langle h,g\rangle = 0 \quad \text{for all } h \in (I-D)\mathcal{M}. \tag{5.7}$$

Now, if (4.28) is observable, only $g = 0$ satisfies (5.7). Hence $(I-D)M$ is dense in X. Therefore, M is dense in Z (Lemma 5.2) and hence in X (weaker topology), or, equivalently M is dense in X, i.e. X is observable. Conversely, assume that X is observable. Then M is dense in Z (Lemma 5.1), and consequently $(I-D)M$ is dense in X (Lemma 5.2). But then only $g = 0$ can satisfy (4.34) and therefore (4.28) is observable. □

PROOF OF LEMMA 5.2. Assume that M is dense in X, and let $\bar{M}$ be the closure of M in graph topology. We know that $\bar{M} \subset Z$, and we want to show that $\bar{M} = Z$. To this end, define $\bar{D}$ to be the restriction of D to $\bar{M}$. Then $\bar{D}$ is an unbounded operator defined on a dense subset of X, and, like D, it is closed and dissipative. Hence the range of $(I-\bar{D})$ is closed [3; Thm 3.4, p.79]. Therefore, if we can show that the range of $(I-\bar{D})$ is dense in X, we know that it is all of X. This would mean that $\bar{D}$ is maximal dissipative [3; Thm 3.6, p.81]. However, D is a dissipative extension of $\bar{D}$ and hence $\bar{D} = D$. Then $\mathcal{D}(\bar{D}) = \mathcal{D}(D)$, i.e. $\bar{M} = Z$ as required.

Consequently it remains to prove that $(I-\bar{D})\bar{M}$ is dense in X. Since $\bar{M}$ is dense in X, we only need to show that the equation $(I-\bar{D})f = g$, i.e.

$$\dot{f} - f = -g \tag{5.8}$$

has a solution $f \in \bar{M}$ for each $g \in \bar{M}$. But, for such a g, (5.8) has the L_2 solution

$$f(t) = \int_0^\infty e^{-\sigma} g(t+\sigma)\,d\sigma = \int_0^\infty (e^{A^*\sigma} g)(t)\,dm(\sigma) \tag{5.9}$$

where $dm = e^{-\sigma} d\sigma$, so it remains to show that this f belongs to $\bar{M}$. It follows from (5.5), that $e^{A^*\sigma} M \subset M$, and therefore, by continuity, $e^{A^*\sigma} g \in \bar{M}$ for each $\sigma \geq 0$. The function $\sigma \to e^{A^*\sigma} g$ is therefore mapping $\mathbb{R}_+$ into $\bar{M}$. It is clearly strongly measurable, and, since $e^{A^*\sigma}$ is a contraction, $\| e^{A^*\sigma} g \|_{\bar{M}} \leq \| g \|_{\bar{M}}$. Hence

$$\int_0^\infty \| e^{A^*\sigma} g \|_{\bar{M}}^2 \,dm(\sigma) < \infty \tag{5.10}$$

and consequently (5.9) is a Bochner integral [17, p.133]. Hence, by definition, $f \in \bar{M}$ as required. □

6. BACKWARD REALIZATIONS

As can be seen from (3.7), the splitting subspace X is orthogonal to $H^+(du)$. Therefore, in the model (4.28), the future increments of the generating process u are independent of present state. A system with this property is said to evolve *forward* in time.

Replacing condition (4.4) by

$$y_k(0) \in \mathcal{D}(\Gamma^*) \qquad \text{for } m = 1,2,\ldots,m \tag{6.1}$$

we can proceed as in Section 4 with obvious modifications. such as replacing $U_t(X)$, H^+ and $H^-(du)$ by $U_t(X)^*$, H^- and $H^+(d\bar{u})$ respectively, to construct a system

$$\begin{cases} d\bar{x} = \bar{A}\bar{x}dt + \bar{B}d\bar{u} & (6.2a) \\ y = \bar{C}\bar{x} & (6.2b) \end{cases}$$

having the same properties as (4.28), except that it evolves *backwards* in time. By this we mean that $X \perp H^-(d\bar{u})$, i.e. the past increments of the generating process $\bar{u}$ are independent of present state. That this is so again follows from (3.7). A backward realization (6.2) is said to be *constructible* if $\cap_{t\geq 0}\bar{C}e^{\bar{A}t} = 0$ and *controllable* if $\cap_{t\geq 0}\bar{B}^*e^{\bar{A}^*t} = 0$. Then, the backward counterparts of Theorems 5.1 and 5.2 say that (6.2) is always controllable and constructible if and only if X is constructible in the sense defined in Section 3.

Consequently, if X is such that

$$y_k(0) \in \mathcal{D}(\Gamma) \cap \mathcal{D}(\Gamma^*), \qquad k = 1,2,\ldots,m, \tag{6.3}$$

X has both a forward and a backward realization. Such an X will be called *regular*. A process y may have both regular and nonregular splitting subspaces, regularity depending on the position of X in the natural partial ordering of minimal Markovian splitting subspaces [8,9]. An investigation of these questions can be found in [16].

It sufficies to mention here that, if $(W,\bar{W})$ is the pair of spectral factors of X, (4.4) holds if and only if there is a constant matrix N such that $sW(s) - N$ is square integrable on the imaginary axis. Noting that W is the Laplace transform of

$(w_1, w_2, \ldots, w_m)$ and that (4.4) is equivalent to $w_k \in \mathcal{D}(A^*)$, this follows from Lemma 3.1 in [7]. Likewise (6.1) holds if and only if there is a constant matrix $\bar{N}$ such that $s\bar{W}(s) - \bar{N}$ is square integrable on the imaginary axis.

REFERENCES

1. J.-P. Aubin, *Applied Functional Analysis*, Wiley-Interscience, 1979.
2. J.S. Baras, R.W. Brockett, and P.A. Fuhrmann, State space models for infinite dimensional systems, *IEEE Trans. Automatic Control* AC-19 (1974), 693-700.
3. P.A. Fuhrmann, *Linear Systems and Operators in Hilbert Space*, McGraw Hill, 1981.
4. I.M.Gelfand and N. Ya. Vilenkin, *Generalized Functions*, Vol. 4, Academic Press, 1964.
5. I.I.Gikhman and A.V. Skorokod, *Introduction to the Theory of Random Processes*, Saunders, 1965.
6. J.W. Helton, Systems with infinite-dimensional state spaces: The Hilbert space approach, *Proc. IEEE* 64 (1979), 145-160.
7. F.D. Lax and R.S. Phillips, *Scattering Theory*, Academic Press, 1967.
8. A. Lindquist and G. Picci, State space models for gaussian stochastic processes, *Stochastic Systems: The Mathematics of Filtering and Identification and Applications*, M. Hazewinkel and J.C. Willems (eds.), Reidel Publ. Co., 1981, 169-204.
9. A. Lindquist, M. Pavon, and G. Picci, Recent trends in stochastic realization theory, *Prediction Theory and Harmonic Analysis*, The Pesi Masani Volume, V. Mandrekar and H. Salehi (eds.), North Holland, 1983.
10. J.L. Lions, *Optimal Control of System Governed by Partial Differential Equations*, Springer Verlag, 1971.
11. Yu.A. Rozanov, *Stationary Random Processes*, Holden-Day, 1967.
12. G. Ruckebusch, Théorie géométrique de la représentation markovienne, *Ann. Inst. Henri Poincaré* XVI3 (1981), 225-297.
13. A. Bensoussan, *Filtrage Optimal des Systèmes Linéaires*, Dunod, 1971.
14. R.F. Curtain and A.J. Pritchard, *Infinite Dimensional Linear Systems*, Springer Verlag, 1978.
15. I.A. Ibraganov and Y.A. Rozanov, *Gaussian Random Processes*, Springer Verlag, 1978.
16. A. Lindquist and G. Picci, Forward and backward semimartingale representations for stationary increment processes, *Proc.*

1983 Intern. Symp. Math. Theory Networks & Systems, Beer Sheva, Israel, Springer Verlag, (to be published).

17. K. Yosida, *Functional Analysis*, Springer Verlag, 1965.

Anders Lindquist
Department of Mathematics
Royal Institute of Technology
S-100 44 Stockholm
Sweden

Giorgio Picci
LANDSEB-CNR
Corso Stati Uniti
35100 Padova
Italy

Operator Theory:
Advances and Applications, Vol. 12

THE ALGEBRAIC MATRIX RICCATI EQUATION

A.C.M. Ran and L. Rodman *)

We review some recent results concerning the symmetric algebraic matrix Riccati equation and especially its hermitian solutions. The main idea is description of such solutions in terms of invariant subspaces of a certain matrix, which is self-adjoint in an indefinite scalar product. Some new results on this subject are presented as well.

1. INTRODUCTION

In this paper we shall consider the algebraic matrix Riccati equation

$$XDX + XA + A^{*}X = C, \tag{1}$$

where A, C, D are $n \times n$ (complex) matrices with D and C hermitian, and X is an $n \times n$ matrix to be found.

Such Riccati equations play an important role in many branches of mathematics and engineering. We shall describe two problems where the Riccati equation appears naturally. The first is disconjugacy of Hamiltonian equations (see [6,23,24]). Consider the Hamiltonian differential equation

$$\begin{cases} y' = By + Cz, \\ z' = -Ay - B^{*}z, \end{cases} \tag{2}$$

where A, B, C are $n \times n$ (complex) matrices, and $y = y(t)$ and $z = z(t)$ are n-dimensional vector functions in the real variable t $(-\infty < t < \infty)$. The system (2) is called *disconjugate* if one of the

*) This paper was written while the second author was a senior visiting fellow at the Vrije Universiteit at Amsterdam.

two following equivalent conditions holds:

(i) for every nontrivial solution $y(t)$, $z(t)$, of (2) the n-vector $y(t)$ vanishes at most once;

(ii) for every pair of real numbers $t_1 \neq t_2$ and every pair of n-dimensional vectors y_1, y_2 there is a solution $y(t)$, $z(t)$ of (2) for which $y(t_i) = y_i$, $i = 1,2$.

It turns out that, assuming $A = A^*$, C is nonnegative definite and

$$\operatorname{rank}[C, BC, \ldots, B^{n-1}C] = n,$$

the Hamiltonian equation (2) is disconjugate if and only if the equation

$$WCW + WB + B^*W + A = 0$$

has a hermitian solution W.

The second problem is the quadratic cost optimal control in linear systems, which is important in engineering and to which a vast literature is dedicated (see, e.g., books [4,15]). Consider the linear system

$$(3) \qquad x' = Ax + Bu, \qquad x(0) = x_0,$$

where $x = x(t)$, $u = u(t)$ are n-dimensional and p-dimensional vector functions, respectively, and A (resp. B) is an $n \times n$ (resp. $n \times p$) constant matrix. Let

$$J(u) = \int_0^\infty (x^*C^*Cx + u^*Ru)\, dt,$$

where C is a $q \times n$ constant matrix and R is a $p \times p$ positive definite matrix. The problem is to find a $u = u(t)$ such that $J(u)$ is minimal (here, of course, $x = x(t)$ is assumed to satisfy (3)). If the system is controllable and observable (i.e.

$$\operatorname{rank}[B, AB, \ldots, A^{n-1}B] = \operatorname{rank}\begin{bmatrix} C \\ CA \\ \vdots \\ CA^{n-1} \end{bmatrix} = n\ ),$$

then there exists a unique positive definite solution $\hat{X}$ of the equation

$$(4) \qquad \hat{X}BR^{-1}B^*\hat{X} - \hat{X}A - A^*\hat{X} = C^*C,$$

and for $u(t) = -R^{-1}B^*\hat{X}x(t)$ the value $J(u)$ is finite and minimal.

For a more general version of this result (with weakened controllability and observability assumptions) see, e.g., [13,15, 35].

Among other problems where the algebraic matrix Riccati equations play an important role, are optimal filtering (see [1]), stochastic control [35] and differential games [31].

The applications to the disconjugacy of Hamiltionian systems and to the optimal control problem make it natural to study the Riccati equation (1) with the extra assumptions that D is nonnegative definite and (A,D) is controllable, i.e. rank $[D, AD, \dots, A^{n-1}D] = n$. Of special interest are hermitian solutions of (1). Note that equation (4) under the controllability and observability assumptions is a particular case of (1). The main purpose of this paper is to present a concise account of a theory of this equation, especially about its hermitian solutions. An extensive literature on this subject exists (see expositions in [7,12,14,30]), so here we will focus mainly on recent results, including some new ones, which have not been dealt with in these expositions.

We remark that the interest in equation (1) and its hermitian solutions is by no means restricted to the case when D is nonnegative definite and (A,D) is controllable. In differential games a Riccati equation (1) appears with indefinite D (see, e.g., [31]). An important case is when D and C are nonnegative definite (without the controllability assumption); see, e.g., [19,20]. Recently some results concerning hermitian solutions were obtained under a relaxed controllability assumption [34]. However, complete results on equation (1) and especially its hermitian solutions are available only assuming that D is nonnegative definite and (A,D) is controllable. These assumptions will be maintained throughout the paper.

The paper consists of 9 sections. In the second section we state the basic results on existence and description of hermitian solutions of equation (1). The existence is given in terms of the Jordan structure of the matrix $M = i\begin{bmatrix} A & D \\ C & -A^* \end{bmatrix}$, as well as in terms of nonnegativeness of certain rational matrix functions.

The description is given by establishing a one-to-one correspondence between hermitian solutions of (1) and certain sets of M-invariant subspaces in $\mathbb{C}^{2n}$. Also, extremal and spectral hermitian solutions are described here. Another description of hermitian solutions of (1) in terms of minimal factorizations of a rational matrix function is given in Section 3. In Section 4 we show that the one-to-one correspondence mentioned above preserves the naturally defined partial order on the set of all hermitian solutions. Further in the fifth section we show that this correspondence is continuous (with respect to the natural topologies) and analytic in a certain sense. As an application, in the next section the isolated hermitian solutions are characterized. The algebraic Riccati equation with real coefficients is discussed in Section 7. In Section 8 we expose some results concerning behavior of the hermitian solutions of (1) under perturbations of the coefficients A,D,C. Finally, in the last section we describe certain non-hermitian solutions of (1) in terms of invariant subspaces of M.

Many results and facts exposed in this paper are not new; as a rule, their proofs are omitted here, and instead a proper reference is given. For the results which to our knowledge are new, full proofs are usually provided.

The following notations and conventions are used throughout the paper. The $m \times p$ complex matrices Z, as well as the real ones (if not stated otherwise), are considered, when convenient, as linear transformations $\mathbb{C}^p \to \mathbb{C}^m$ in a natural way. For such a matrix $\operatorname{Im} Z = \{Zx \mid x \in \mathbb{C}^p\} \subset \mathbb{C}^m$. The $m \times m$ unit matrix is denoted I_m or I. We denote by $\sigma(Z)$ the set $\{\lambda \in \mathbb{C} \mid \det(\lambda I - Z) = 0\}$ of all eigenvalues of the $m \times m$ matrix Z. $R_{\lambda_0}(Z)$ stands for the root subspace of Z corresponding to its eigenvalue λ_0: $R_{\lambda_0}(Z) = \operatorname{Ker}(\lambda_0 I - Z)^m$. The sizes of the Jordan blocks with eigenvalue λ_0 in the Jordan normal form of Z are called multiplicities of Z corresponding to λ_0. The spectral subspace of Z corresponding to a set C in the complex plane is, by definition, $\Sigma R_{\lambda_0}(Z)$, where the sum is taken over all eigenvalues λ_0 of Z which are in C. Finally, $Z|_N$ stands for the restriction of Z to a Z-invariant subspace $N \subset \mathbb{C}^m$.

2. EXISTENCE AND DESCRIPTION OF HERMITIAN SOLUTIONS

Consider the Riccati equation

$$XDX + XA + A^*X = C, \tag{5}$$

where A, C, D are $n \times n$ matrices with the following properties:

$$C = C^*;\ D \text{ is nonnegative definite}; \tag{6}$$
$$\text{rank}\,[D, AD, \ldots, A^{n-1}D\,] = n.$$

Define the matrices

$$M = i\begin{bmatrix} A & D \\ C & -A^* \end{bmatrix}, \qquad H = \begin{bmatrix} -C & A^* \\ A & D \end{bmatrix}.$$

Throughout the rest of this paper (with exception of Theorems 15, 16, 17) we shall assume that H is invertible and the signature (i.e. the difference between the number of positive eigenvalues and the number of negative eigenvalues, in both cases counted with multiplicities) of H is zero. There is no loss of generality in this assumption, because one can replace A by $A + \alpha iI$ with real α without altering the solutions of (5).

The following theorem gives necessary and sufficient conditions for existence of hermitian solutions of (5) in terms of the spectral structure of M, as well as in terms of certain rational matrix functions. A subspace $N \subset \mathbb{C}^{2n}$ is called H-*neutral* if $\langle Hx, y\rangle = 0$ for all $x, y \in N$. Here $\langle \cdot, \cdot \rangle$ is the standard scalar product in $\mathbb{C}^{2n}$.

THEOREM 1. *The following statements are equivalent:*

(i) there exists a hermitian solution X *of* (5);

(ii) the multiplicities of M *corresponding to its real eigenvalues (if any) are all even;*

(iii) there exists an n*-dimensional* M*-invariant* H*-neutral subspace;*

(iv) the $n \times n$ *rational matrix function*

$$Z(\lambda) = I + D_0^*(\lambda I + iA^*)^{-1}C(\lambda I - iA)^{-1}D_0,$$

where D_0 *is such that* $D = D_0D_0^*$*, is nonnegative, i.e.*

$$\langle Z(\lambda)x, x\rangle \geq 0$$

for all $x \in \mathbb{C}^n$ *and all real* λ *which are not poles of* $Z(\lambda)$*;*

(v) *the rational matrix function*

$$W(\lambda) = I + (\lambda I - iA)^{-1} D_0 Z(\lambda) D_0^* (\lambda I + iA^*)^{-1}$$

is nonnegative.

The equivalence of (i), (ii) and (iii) was proved in [8,16]; the equivalence of (i) and (iv) is due to J. Willems [33]. The equivalence of (i) and (v) was shown in Chapter II.4 in [12], where also a proof of the whole Theorem 1 can be found.

The crucial fact for the proof of Theorem 1 and many other results presented in this paper is that the matrix M is selfadjoint in the indefinite scalar product determined by the matrix H, in short, M is H-*selfadjoint*, i.e.

$$[Mx,y] = [x,My] \quad \text{for all } x,y \in \mathbb{C}^{2n},$$

where $[x,y] = \langle Hx,y\rangle$. We refer to [12] for the basic facts concerning selfadjoint matrices in indefinite scalar products.

In connection with Theorem 1 let us mention that an M-invariant subspace is H-neutral if and only if it is J-neutral, where $J = i\begin{bmatrix} 0 & I \\ -I & 0 \end{bmatrix}$. Also, the function $W(\lambda)$ from (v) has the following minimal realization (see, e.g., [3] for the definition of this notion):

$$W(\lambda) = I + [0 \ I]\, J\, (\lambda I - M)^{-1} \begin{bmatrix} 0 \\ I \end{bmatrix} .$$

Existence of hermitian solutions of (5) implies actually more properties of invariant subspaces of M than stated in Theorem 1:

THEOREM 2. *Let* (5) *have hermitian solutions, and let* M_0 *be the spectral subspace of* M *corresponding to its real eigenvalues* ($M_0 = (0)$ *if* $\sigma(M) \cap \mathbb{R} = \phi$)*. Then there exists in* M_0 *a unique* $\frac{1}{2}(\dim M_0)$*-dimensional* M*-invariant* H*-neutral subspace* S_0*. Moreover,* $\sigma(M|_{S_0}) = \sigma(M) \cap \mathbb{R}$ *, and the multiplicities of the restriction* $M|_{S_0}$ *at its eigenvalue* λ_0 *are half the multiplicities of* M *at* λ_0*.*

Theorem 2 follows from [12], Theorem I.3.22 taking into account part (ii) of Theorem 1 and Corollary II.4.7 in [12].

The next theorem describes the hermitian solutions of (5) in terms of the invariant subspaces of M. Here and in the sequel M_+ stands for the spectral subspace of M corresponding to the open upper half-plane.

THEOREM 3. *Assume there is a hermitian solution of* (5). *Then:*

(i) *for every solution* $X = X^*$ *of* (5) *the subspace*

(7) $$L = \operatorname{Im} \begin{bmatrix} I \\ X \end{bmatrix}$$

is n-*dimensional,* M-*invariant and* H-*neutral; conversely, every* n-*dimensional* M-*invariant* H-*neutral subspace has the form* (7) *for some solution* $X = X^*$ *of* (5).

(ii) *for every hermitian solution* X *of* (5) *there is a unique* M-*invariant subspace* N *with* $\sigma(M|_N)$ *lying in the open upper half-plane and with*

(8) $$\operatorname{Im} \begin{bmatrix} I \\ X \end{bmatrix} \cap M_+ = N ;$$

conversely, if N *is an* M-*invariant subspace such that* $\sigma(M|_N)$ *lies in the open upper half-plane, then there exists a unique solution* $X = X^*$ *of* (5) *for which* (8) *holds.*

Part (i) of Theorem 3 was established in [16], and part (ii) in [28](see also [27]). The idea of describing hermitian solutions of (5) in terms of M-invariant subspaces goes back to [17,19].

It is worth mentioning that $M|_L$ is similar to $i(A + DX)$, where the hermitian solution X of (5) and the n-dimensional M-invariant H-neutral subspace L are in correspondence as in (i).

Using, for instance, the canonical form for the H-self-adjoint matrix M (see, e.g., Chapter I.3 in [12], Chapter S.5 in [11],[32]), we obtain from Theorem 3(i) that for any hermitian solution X of (5) the equality

(9) $$\operatorname{Im} \begin{bmatrix} I \\ X \end{bmatrix} \cap M_0 = S_0$$

holds, where M_0 and S_0 are defined as in Theorem 2. In connection with this property note that

(10) $$\dim S_0 + \dim M_+ = \dim S_0 + \dim M_- = n ,$$

where M_- is the spectral subspace of M corresponding to the open lower half-plane.

As a consequence of Theorem 3(ii) one can count the hermitian solutions of (5) as follows (assuming they exist). If $\dim \operatorname{Ker}(\lambda I - M) = 1$ for every eigenvalue λ of M with $\operatorname{Im} \lambda > 0$, then the number of hermitian solutions of (5) is finite and is equal to $\Pi_{i=1}^{\alpha} (1 + \dim R_{\lambda_i}(M))$, where $\lambda_1, \ldots, \lambda_\alpha$ are all the different eigenvalues of M in the open upper half-plane; otherwise, there is a continuum of hermitian solutions of (5).

Of particular interest are extremal hermitian solutions of equation (5). A hermitian solution X of (5) is called *maximal* (resp. *minimal*) if for any hermitian solution Y of (5) the difference $X - Y$ is nonnegative definite (resp. nonpositive definite). One can prove (see [7,23], also Chapter II.4 in [12]) that the maximal and the minimal hermitian solutions exist always, provided the set of hermitian solutions is not empty. Extremal solutions can be characterized as follows.

THEOREM 4. *The following statements are equivalent for a hermitian solution* X *of* (5)*:*

(i) X *is the maximal (resp. the minimal) hermitian solution;*

(ii) $\sigma(A + DX)$ *lies in the closed right (resp. left) half-plane;*

(iii) $\operatorname{Im} \begin{bmatrix} I \\ X \end{bmatrix} \supset M_+$ *(resp.* $\operatorname{Im} \begin{bmatrix} I \\ X \end{bmatrix} \cap M_+ = (0)$ *).*

For the proof of Theorem 3 see Chapter II.4 in [12]. The equivalence of (i) and (ii) is proved also in [7].

Using Theorems 3, 4 and equalities (9) and (10) it is easily seen that for the maximal and minimal solutions X_+ and X_-, respectively, the equalities

$$(11) \qquad \operatorname{Im} \begin{bmatrix} I \\ X_+ \end{bmatrix} = S_0 \dotplus M_+ , \quad \operatorname{Im} \begin{bmatrix} I \\ X_- \end{bmatrix} = S_0 \dotplus M_-$$

hold, where M_- and S_0 are defined as in (10) and Theorem 2.

In particular, if M does not have real eigenvalues, then a hermitian solution X of (5) is maximal (resp. minimal) if and only if the spectrum of $A + DX$ lies in the open right (resp. open left) half-plane.

An important case of equation (5) is when, in addition to (6), C is nonnegative definite. In this case, assuming that M has no real eigenvalues, there exists a unique nonnegative definite solution (see, e.g., [19,30]; this conclusion holds true even if one drops the assumption $\operatorname{rank}[D, AD, \ldots, A^{n-1}D] = n$); in particular, the maximal hermitian solution is nonnegative definite.

Another interesting class of hermitian solutions of (5) is obtained in the following way. Let C be a set of non-real eigenvalues of M with the property that $\lambda_0 \in C$ implies $\overline{\lambda_0} \notin C$ and which is maximal with respect to this property. Then, assuming (5) has a hermitian solution, there exists a unique hermitian solution X of (5) with the property that $\sigma(i(A + DX)) \setminus \mathbb{R} = C$. Such hermitian solutions will be termed *spectral*. In particular, the maximal and the minimal hermitian solution are spectral, they correspond to the cases when $C = \{\lambda \in \sigma(M) \mid \operatorname{Im} \lambda > 0\}$ and $C = \{\lambda \in \sigma(M) \mid \operatorname{Im} \lambda < 0\}$, respectively. The spectral solutions can be characterized as follows:

PROPOSITION 5. *The following statements are equivalent for a solution* $X = X^*$ *of* (5):

(i) X *is spectral;*

(ii) $i(A + DX)$ *and its adjoint have no common eigenvalues;*

(iii) for every non-real eigenvalue λ *of* M *either*

$$R_\lambda(M) \subset \operatorname{Im} \begin{bmatrix} I \\ X \end{bmatrix} \qquad R_\lambda(M) \cap \operatorname{Im} \begin{bmatrix} I \\ X \end{bmatrix} = (0)$$

holds.

Proposition 5 can be easily deduced from Theorem 3 and remarks thereafter.

3. HERMITIAN SOLUTIONS AND MINIMAL FACTORIZATIONS OF RATIONAL MATRIX FUNCTIONS

Consider the Riccati equation (5), where A, D and C satisfy the same conditions as in Section 2. Put

$$M = i\begin{bmatrix} A & D \\ C & -A^* \end{bmatrix}, \quad J = i\begin{bmatrix} 0 & I \\ -I & 0 \end{bmatrix}, \quad R = \begin{bmatrix} 0 \\ I \end{bmatrix},$$

and

$$W(\lambda) = I_n + R^* J(\lambda I_{2n} - M)^{-1} R . \tag{12}$$

The realization (12) is minimal, and a minimal realization of the inverse function $W(\lambda)^{-1}$ is given by

(13) $$W(\lambda)^{-1} = I_n - R^*J(\lambda I_{2n} - M^\times)^{-1}R ,$$

where

$$M^\times = M - RR^*J = i\begin{bmatrix} A & D \\ C+I & -A^* \end{bmatrix} .$$

Assume now (5) has hermitian solutions. Then by Theorem 1 the function $W(\lambda)$ is nonnegative; consequently, $W(\lambda)^{-1}$ is also nonnegative. Since (13) is a minimal realization, the partial multiplicities of $M^\times$ corresponding to its real eigenvalues (if any) are all even (see [20]). In view of Theorem 1 this leads us to the following conclusion: if the Riccati equation (5) has a hermitian solution, then the Riccati equation

(14) $$XDX + A^*X + XA - (C+I) = 0$$

has a hermitian solution as well.

Note that if $X = X^*$ is a solution of (5) and $\tilde{X} = \tilde{X}^*$ is a solution of (14), then the difference $X - \tilde{X}$ is invertible. Indeed, let $x \in \mathbb{C}^n$ be such that $Xx = \tilde{X}x$. Then

$$x^*XDXx + x^*XAx + x^*A^*Xx = x^*\tilde{X}D\tilde{X}x + x^*\tilde{X}Ax + x^*A^*\tilde{X}x .$$

But the left-hand side of this equality is x^*Cx, while the right-hand side is $x^*(C+I)x$. So $x = 0$.

There is a one-to-one correspondence between minimal factorizations of $W(\lambda)$ of the type $W(\lambda) = (L(\bar{\lambda}))^*L(\lambda)$ where $L(\lambda)$ is an $n \times n$ rational matrix function with $L(\infty) = I_n$ and pairs of matrices $(X,\tilde{X})$, where X is a hermitian solution of (5) and $\tilde{X}$ is a hermitian solution of (14) (see, e.g., [3] for the definition and a comprehensive treatment of minimal factorizations of rational matrix functions). This correspondence is decribed in the following theorem.

THEOREM 6. *Assume that the rational matrix function* $W(\lambda)$ *given by* (12) *is nonnegative. Let* X *and* $\tilde{X}$ *be hermitian solutions of* (5) *and* (14), *respectively. Then* $W(\lambda)$ *admits a minimal factorization*

(15) $$W(\lambda) = (L(\bar{\lambda}))^*L(\lambda)$$

with

$$(16) \qquad L(\lambda) = I_n + R^* J \begin{bmatrix} ZX & -Z \\ \tilde{X}ZX & -\tilde{X}Z \end{bmatrix} (\lambda I_{2n} - M)^{-1} R ,$$

where $Z = (X - \tilde{X})^{-1}$. *Conversely, any rational matrix function* $L(\lambda)$ *with* $L(\infty) = I$ *and for which* (15) *is a minimal factorization of* $W(\lambda)$, *has the form* (16) *for some pair* $X, \tilde{X}$ *of hermitian solutions of* (5) *and* (14), *respectively. Moreover, the zeros of* $L(\lambda)$ *are the eigenvalues of* $i(A + D\tilde{X})$, *and the poles of* $L(\lambda)$ *are the eigenvalues of* $-i(A + DX)^*$, *in both cases multiplicities counted. More exactly: the partial zero (resp. pole) multiplicites at its zero (resp. pole)* λ_0 *are precisely the multiplicities of* $i(A + D\tilde{X})$ *(resp.* $-i(A + DX)^*$*) corresponding to* λ_0.

For the definitions and basic properties of partial zero multiplicities and partial pole multiplicities, see, e.g., [3].

PROOF. As shown in [20], all minimal factorizations of $W(\lambda)$ of the type $W(\lambda) = (L(\bar{\lambda}))^* L(\lambda)$ with a rational matrix function $L(\lambda)$ such that $L(\infty) = I$ are given by the formula

$$W(\lambda) = [\, I + R^* J(\lambda I - M)^{-1}(I - \pi)R \,] \cdot$$
$$\cdot [\, I + R^* J\pi(\lambda I - M)^{-1} R \,] ,$$

where $\pi : \mathbb{C}^{2n} \to \mathbb{C}^{2n}$ is any projection with the properties that $\operatorname{Ker} \pi$ is M-invariant, $\operatorname{Im} \pi$ is $M^\times$-invariant and both subspaces $\operatorname{Ker} \pi$ and $\operatorname{Im} \pi$ are n-dimensional and J-neutral.

Now let (15) be a minimal factorization, where $L(\infty) = I$. Then the subspace $\operatorname{Ker} \pi$ is M-invariant J-neutral of dimension n. According to Theorem 3 and a remark preceding this theorem there is a hermitian solution X of (5) such that $\operatorname{Ker} \pi = \operatorname{Im} \begin{bmatrix} I \\ X \end{bmatrix}$. Similarly one shows that $\operatorname{Im} \pi = \operatorname{Im} \begin{bmatrix} I \\ \tilde{X} \end{bmatrix}$ for a hermitian solution $\tilde{X}$ of (14). Note that the matrix

$$(18) \qquad \begin{bmatrix} ZX & -Z \\ \tilde{X}ZX & -\tilde{X}Z \end{bmatrix}, \qquad Z = (X - \tilde{X})^{-1}$$

is a projection with image equal to $\operatorname{Im} \begin{bmatrix} I \\ \tilde{X} \end{bmatrix}$ and kernel equal to $\operatorname{Im} \begin{bmatrix} I \\ X \end{bmatrix}$, and therefore this projection is π. Now compare (16) with (17) to obtain the converse statement of the theorem.

Assume now that X and $\tilde{X}$ are hermitian solutions of (5) and (14), respectively. Then the subspace $\operatorname{Im}\begin{bmatrix} I \\ X \end{bmatrix}$ is M-invariant J-neutral of dimension n, and the subspace $\operatorname{Im}\begin{bmatrix} I \\ \tilde{X} \end{bmatrix}$ is $M^{\times}$-invariant J-neutral of dimension n. We claim that

$$\mathbb{C}^{2n} = \operatorname{Im}\begin{bmatrix} I \\ X \end{bmatrix} \dotplus \operatorname{Im}\begin{bmatrix} I \\ \tilde{X} \end{bmatrix}$$

(direct sum). Indeed, assume

$$\begin{bmatrix} x \\ Xx \end{bmatrix} = \begin{bmatrix} y \\ \tilde{X}y \end{bmatrix}$$

for some $x,y \in \mathbb{C}^n$. Then $x = y$ and $Xx = \tilde{X}x$; as $X - \tilde{X}$ is invertible, $x = y = 0$. Let π be the projection onto $\operatorname{Im}\begin{bmatrix} I \\ \tilde{X} \end{bmatrix}$ along $\operatorname{Im}\begin{bmatrix} I \\ X \end{bmatrix}$, then π is given by (18). According to the remark stated in the beginning of this proof, the factorization (17) is of the desired form.

It remains to prove the statements about the poles and zeros of $L(\lambda)$. It is well known (see [3]) that the zeros of $L(\lambda)$ are the eigenvalues of $M^{\times}|_{\operatorname{Im}\pi}$, multiplicities counted. Analogously, the poles of $L(\lambda)$ coincide (including multiplicities) with the complex conjugates of the eigenvalues of $M|_{\operatorname{Ker}\pi}$. Applying the remark after Theorem 3 gives the desired result. □

In the case when M does not have real eigenvalues, a correspondence between hermitian solutions of (5) and factorizations of a nonnegative rational matrix function was treated in [10]. Connections between hermitian solutions of the algebraic Riccati equation and nonnegative rational matrix functions and their factorizations have been observed before in [18,33].

Choosing a fixed solution $\tilde{X}$ of (14) one obtains from Theorem 6 a one-to-one correspondence between the set of hermitian solutions of (5) and a set of certain minimal factorizations of type (15). For instance:

COROLLARY 7. *There is a one-one correspondence between hermitian solutions* X *of* (5) *and minimal factorizations of* $W(\lambda)$ *of the form* $W(\lambda) = (L(\bar{\lambda}))^* L(\lambda)$ *such that* $L(\infty) = I_n$ *and* $L(\lambda)$ *has all its zeros in the closed upper half plane. This correspondence is given by formula* (16), *where* $\tilde{X}$ *is the maximal hermitian solution of* (14).

Assume now D is positive definite (and not only nonnegative definite). Then one can describe hermitian solutions of (5) in terms of factorizations of a nonnegative matrix polynomial, as the next theorem shows. Consider together with (5) the matrix polynomial

$$K(\lambda) = \lambda^2 I + \lambda K_1 + K_0,$$

where $K_1 = i(\hat{A} - \hat{A}^*)$, $K_0 = \hat{C} + \hat{A}^*\hat{A}$ and $\hat{A} = D^{-\frac{1}{2}}AD^{\frac{1}{2}}$, $\hat{C} = D^{\frac{1}{2}}CD^{\frac{1}{2}}$. Note that $K_i = K_i^*$, $i = 0,1$.

THEOREM 8. *Assume* D *is positive definite. If* X *is a hermitian solution of* (5), *then* $K(\lambda)$ *admits a factorization*

$$K(\lambda) = (\lambda I - T^*)(\lambda I - T) , \tag{19}$$

where

$$T = -iD^{-\frac{1}{2}}(A + DX)D^{\frac{1}{2}} . \tag{20}$$

Conversely, if $K(\lambda)$ *admits a factorization of type* (19), *then* T *is given by* (20) *for a unique hermitian solution* X *of* (5).

The proof of Theorem 8, and some related results, can be found in [16].

4. PARTIAL ORDER STRUCTURE OF THE SET OF HERMITIAN SOLUTIONS

We continue to consider the Riccati equation (5) with the conditions stipulated in (6). Assume there exists a hermitian solution of (5). Let Her(A,D,C) be the set of hermitian solutions of (5), and let $I_+(M)$ be the set of M-invariant subspaces N such that $N \subset M_+$. By Theorem 3, there is a one-to-one correspondence $\tau : \mathrm{Her}(A,D,C) \to I_+(M)$, given by the formula

$$\tau(X) = \mathrm{Im}\begin{bmatrix} I \\ X \end{bmatrix} \cap M_+ .$$

The set Her(A,D,C) has a natural partial order: $X \leq Y$, where $X,Y \in \mathrm{Her}(A,D,C)$, means that $Y - X$ is nonnegative definite. Also the set $I_+(M)$ has a natural partial order determined by the inclusion relation. The next theorem shows that τ and its inverse preserve these partial orders.

THEOREM 9. *Let* $X, Y \in \mathrm{Her}(A,D,C)$. *Then* $X \leq Y$ *if and only if* $\tau(X) \subset \tau(Y)$

PROOF. First we show that we can reduce the theorem to the special case $C = 0$ and $\sigma(iA)$ lies in the closed upper half plane. Let X_0 be the maximal solution of (5); then $\sigma(i(A + DX_0))$ lies in the closed upper half plane. Consider the Riccati equation

$$YDY + (A + DX_0^*)Y + Y(A + DX_0) = 0 \tag{21}$$

A straightforward calculation shows that X is a solution of (5) if and only if $Y = X - X_0$ is a solution of (21). Put

$$\hat{M} = i\begin{bmatrix} A+DX_0 & D \\ 0 & -(A+DX_0)^* \end{bmatrix} .$$

Note that $\hat{M} = EME^{-1}$, where $E = \begin{bmatrix} I & 0 \\ -X_0 & I \end{bmatrix}$. By Theorem 3 we have the one-to-one and onto map $\hat{\tau} : \mathrm{Her}(A + DX_0, D, 0) \to I_+(\hat{M})$ determined by the equality

$$\hat{\tau}(Y) = \mathrm{Im}\begin{bmatrix} I \\ Y \end{bmatrix} \cap \hat{M}_+ ,$$

where $\hat{M}_+$ is the spectral subspace of $\hat{M}$ corresponding to its eigenvalues in the upper half-plane. Also define the maps $q : \mathrm{Her}(A,D,C) \to \mathrm{Her}(A+DX_0,D,0)$, where $q(X) = X - X_0$, and $r : I_+(M) \to I_+(\hat{M})$, where $r(L) = EL$. Clearly q and r are one-to-one and onto maps, and q and r and their inverses preserve the partial orders. One checks easily that $\hat{\tau}q = r\tau$. So τ and τ^{-1} preserve the partial orders if and only if $\hat{\tau}$ and $\hat{\tau}^{-1}$ do, and the proof of Theorem 9 is reduced to the case when $C = 0$ and $\mathrm{Im}\,\sigma(iA) \geq 0$.

Assume $C = 0$ and $\sigma(iA)$ lies in the closed upper half-plane. Then

$$M_+ = \mathrm{Im}\begin{bmatrix} I \\ 0 \end{bmatrix} \cap \begin{bmatrix} A_+ \\ 0 \end{bmatrix} ,$$

where A_+ is the spectral subspace of iA corresponding to its eigenvalues in the open upper half-plane. By Theorem 4, the zero matrix is the maximal hermitian solution of the equation

$$XDX + A^*X + XA = 0 . \tag{22}$$

Now suppose X_1 and X_2 are two hermitian solutions with $X_1 \le X_2 \le 0$. Then for i = 1,2:

$$N_i \stackrel{\text{def}}{=} \operatorname{Im}\begin{bmatrix} I \\ X_i \end{bmatrix} \cap M_+ = \left\{\begin{bmatrix} x \\ 0 \end{bmatrix} \mid x \in \operatorname{Ker} X_i\right\} \cap \begin{bmatrix} A_+ \\ 0 \end{bmatrix}$$

So in order to prove that $N_1 \subset N_2$ it suffices to show that $\operatorname{Ker} X_1 \subset \operatorname{Ker} X_2$. But this follows easily from the inequalities $X_1 \le X_2 \le 0$.

Conversely, assume $N_1 \subset N_2$, i.e., $\operatorname{Ker} X_1 \subset \operatorname{Ker} X_2$. With respect to the orthogonal decomposition $\mathbb{C}^n = \operatorname{Ker} X_1 \dot{+} \operatorname{Im} X_1$ write

$$X_1 = \begin{bmatrix} 0 & 0 \\ 0 & \tilde{X}_1 \end{bmatrix}, \quad X_2 = \begin{bmatrix} 0 & 0 \\ 0 & \tilde{X}_2 \end{bmatrix}, \quad D = \begin{bmatrix} D_1 & D_2 \\ D_2^* & D_3 \end{bmatrix}.$$

As the subspace $\left\{\begin{bmatrix} x \\ 0 \end{bmatrix} \mid x \in \operatorname{Ker} X_1\right\}$ is M-invariant, the subspace $\operatorname{Ker} X_1$ is A-invariant, so with respect to the same decomposition we have

$$A = \begin{bmatrix} A_{11} & A_{12} \\ 0 & \tilde{A} \end{bmatrix}.$$

Now the equalities

$$X_i D X_i + A^* X_i + X_i A = 0, \qquad i = 1,2$$

imply

$$\tilde{X}_i D_3 \tilde{X}_i + \tilde{A}^* \tilde{X}_i + \tilde{X}_i \tilde{A} = 0, \qquad i = 1,2. \tag{23}$$

First we shall show that $\tilde{A}$ has no pure imaginary eigenvalues. Indeed, we have

$$M = i \begin{bmatrix} A_{11} & A_{12} & D_1 & D_2 \\ 0 & \tilde{A} & D_2^* & D_3 \\ 0 & 0 & -A_{11}^* & 0 \\ 0 & 0 & -A_{12}^* & -\tilde{A}^* \end{bmatrix}.$$

Let X_+ (resp. X_-) be the maximal (resp. minimal) solution of (5), and let S_0 be the subspace described in Theorem 2. We have

$$S_0 = \left\{\begin{bmatrix} x \\ 0 \end{bmatrix} \mid x \in \operatorname{Ker} X_-\right\} = \operatorname{Im}\begin{bmatrix} I \\ X_+ \end{bmatrix} \cap \operatorname{Im}\begin{bmatrix} I \\ X_- \end{bmatrix}$$

(the last equality follows from (11)).

Since $S_0 \subset \text{Im}\begin{bmatrix} I \\ X \end{bmatrix}$ for every hermitian solution X of (22), it follows that $\text{Ker } X_- \subset \text{Ker } X_1$. So $M|_{S_0}$ is similar to $iA_{11}|_{\text{Ker } X_-}$. Hence iA_{11} has the same real eigenvalues as M, with multiplicities precisely half of the multiplicites of M corresponding to real eigenvalues (see Theorem 2). As

$$\det(\lambda - M) = \det(\lambda - iA_{11})\det(\lambda - i\tilde{A})\det(\lambda + iA_{11}^*)\det(\lambda + i\tilde{A}^*),$$

by our preceding remark we obtain that $\det(\lambda - i\tilde{A})\det(\lambda + i\tilde{A}^*)$ has no real zeros. Hence $\tilde{A}$ has no pure imaginary eigenvalues.

Let us check that the pair $(\tilde{A}, D_3)$ is controllable. Indeed, applying a congruence transformation to D_3 (and the corresponding similarity to $\tilde{A}$), and taking into account that D, and therefore D_3, are nonnegative definite, we can assume that $D_3 = \begin{bmatrix} I & 0 \\ 0 & 0 \end{bmatrix}$ with respect to some orthogonal decomposition of $\text{Im } X_1$. But then D_2^* must be of the form $\begin{bmatrix} Z \\ 0 \end{bmatrix}$ for some matrix Z, so $D_2^* = D_3 D_2^*$. Now

$$[D, AD, \ldots, A^{n-1}D] =$$

$$= \begin{bmatrix} & & & * & & & \\ 0 & D_3 & 0 & \tilde{A}D_3 & \ldots 0 & \tilde{A}^{n-1}D_3 \end{bmatrix} \left(\begin{bmatrix} I & 0 \\ D_2^* & I \end{bmatrix} \oplus \begin{bmatrix} I & 0 \\ D_2^* & I \end{bmatrix} \oplus \ldots \oplus \begin{bmatrix} I & 0 \\ D_2^* & I \end{bmatrix} \right),$$

where $*$ denotes a part of the matrix which is not of interest to us. As (A,D) is controllable the rows of $[D_3, \tilde{A}D_3, \ldots, \tilde{A}^{n-1}D_3]$ are linearly independent, and the controllability of $(\tilde{A}, D_3)$ follows.

Since $\tilde{X}_1$ is invertible, we can rewrite (23) in the form $\tilde{X}_1^{-1}\tilde{A}^* + \tilde{A}\tilde{X}_1^{-1} = -D_3$. Since $\tilde{A}$ has no pure imaginary eigenvalues the equation $Y\tilde{A}^* + \tilde{A}Y = -D_3$ has a unique solution. It follows that the equation

$$(24) \qquad XD_3X + \tilde{A}^*X + X\tilde{A} = 0$$

has only one invertible solution, namely $\tilde{X}_1$. This implies that $\tilde{X}_1$ is the minimal solution of the Riccati equation (24). As $\tilde{X}_2$ is also a hermitian solution of (24), we get $\tilde{X}_1 \le \tilde{X}_2$ and consequently $X_1 \le X_2$. □

In the case of real A,D,C and real symmetric solutions Theorem 9 was proved in [7].

As a consequence of Theorem 9 we obtain that in Her(A,D,C) there exist the least upper bound X_u and the greatest lower bound X_ℓ for any set Q of hermitian solutions of (5). Here X_u and X_ℓ are defined by the properties that $X_u \geq X$ (resp. $X_\ell \leq X$) for all $X \in Q$ and if for some $Y \in Her(A,D,C)$ the inequalities $Y \geq X$ (resp. $Y \leq X$) hold for all $X \in Q$, then $Y \geq X_u$ (resp. $Y \leq X_\ell$).

The following corollary is worth mentioning (cf. Theorem 5 in [33]):

COROLLARY 10. *Let* X_+ *and* X_- *be the maximal and the minimal solutions of* (5), *respectively. Then* $X_+ - X_-$ *is invertible if and only if* M *has no real eigenvalues.*

PROOF. As in the proof of Theorem 9 we can assume that $C = 0$ and $\operatorname{Im} \sigma(iA) \geq 0$. Then $X_+ = 0$. Note that X_- is invertible if and only if the subspace S_0 introduced in the proof of Theorem 9 is the zero subspace. By definition, $S_0 = (0)$ if and only if M has no real eigenvalues. □

5. CONTINUITY AND ANALYTICITY

We maintain the notation introduced in the preceding section. Denoting by In(M,H) the set of all n-dimensional M-invariant H-neutral subspaces, Theorem 3 establishes the map $\phi : Her(A,D,C) \to In(M,H)$, $\phi(X) = \operatorname{Im} \begin{bmatrix} I \\ X \end{bmatrix}$, which is one-to-one and onto. The set Her(A,D,C) has a natural topology induced by the topology in the set of all $n \times n$ matrices. Also, the sets In(M,H) and $I_+(M)$ have a natural topology induced by the gap metric in the set of all subspaces in $\mathbb{C}^{2n}$ (see, e.g., Chapter S.4 in [11]). Namely, the gap between two subspaces M_1 and M_2 in $\mathbb{C}^{2n}$ is defined by

$$\operatorname{gap}(M_1,M_2) = \max_{\|x\|=1} \|(P_{M_1} - P_{M_2})x\| ,$$

where P_{M_i} is the orthogonal projection on M_i, $i = 1,2$, and the norms of vectors in $\mathbb{C}^{2n}$ is euclidean. It is easy to see that ϕ is a homeomorphism (in these topologies). It turns out that the one-to-one and onto map $\tau : Her(A,D,C) \to I_+(M)$ (introduced in the preceding section) is a homeomorphism as well (see [29], Theorem 4.2 in [22]; also Theorem 2.7 in [21]).

As a consequence of the continuity of τ and τ^{-1} it follows that the set of hermitian solutions of (5) is compact.

Indeed, it is well known that the set of invariant subspaces of a given matrix is compact in the gap topology. Another way to see the compactness of Her(A,D,C) is by using the extremal hermitian solutions.

The question about analyticity of ϕ and τ is more involved. Let Ω be an open connected set in $\mathbb{R}^k$. A $p \times q$ matrix function $X(t)$, $t \in \Omega$ is said to be *analytic* in Ω if in a neighborhood of each point $t_0 = (t_{01}, t_{02}, \ldots, t_{0k}) \in \Omega$ $X(t)$ is represented by a power series

$$X(t) = \sum \gamma(\alpha_1, \alpha_2, \ldots, \alpha_k)(t_1 - t_{01})^{\alpha_1}(t_2 - t_{02})^{\alpha_2} \ldots (t_k - t_{0k})^{\alpha_k},$$
$$t = (t_1, t_2, \ldots, t_k),$$

where the summation is taken over all k-tuples of nonnegative integers $(\alpha_1, \ldots, \alpha_k)$, and $\gamma(\alpha_1, \alpha_2, \ldots, \alpha_k)$ is a $p \times q$ matrix depending on α_i, $i = 1, \ldots, k$. A subspace valued function $L(t)$, $t \in \Omega$, so that $L(t)$ is a subspace in $\mathbb{C}^m$ for each $t \in \Omega$, is *analytic* in Ω if the orthogonal projection on $L(t)$ is an analytic matrix function in Ω.

THEOREM 11. *The maps ϕ and τ are analytic in the following sense. Given an analytic hermitian matrix function $X(t) \in$ Her(A,D,C), $t \in \Omega$, the subspace valued functions $\phi(X(t))$ and $\tau(X(t))$ are analytic as well. Conversely, assume that a subspace valued function $L(t) \in$ In(M,H) (resp. $N(t) \in I_+(M)$) is analytic in $t \in \Omega$. Then the hermitian matrix function $X(t) \in$ Her(A,D,C) determined by*

$$\phi(X(t)) = L(t)$$

(resp. $\tau(X(t)) = N(t)$) is also analytic.

PROOF. The analyticity of ϕ and its inverse is not difficult to prove. Indeed, the orthogonal projection on the subspace $\mathrm{Im}\begin{bmatrix} I \\ X \end{bmatrix}$ is

$$\begin{bmatrix} Z & ZX^* \\ XZ & XZX^* \end{bmatrix} \qquad \text{with } Z = (I + X^*X)^{-\frac{1}{2}} .$$

Hence if $X = X(t)$ is an analytic matrix function, then so is the orthogonal projection on $\mathrm{Im}\begin{bmatrix} I \\ X \end{bmatrix}$, and vice versa.

Let us prove the analyticity of τ. Assume $X(t) \in \text{Her}(A,D,C)$ is an analytic hermitian matrix function in $t \in \Omega$. The subspace $\tau(X(t))$ which is by definition $\text{Im}\begin{bmatrix} I \\ X(t) \end{bmatrix} \cap M_+$, is equal to $\text{Im}\left(P_+\begin{bmatrix} I \\ X(t) \end{bmatrix}\right)$, where P_+ is the projection on M_+ along the spectral subspace of M corresponding to its eigenvalues in the set $\{\lambda \mid \text{Im}\,\lambda \le 0\}$. Indeed, the inclusion

$$\text{Im}\begin{bmatrix} I \\ X(t) \end{bmatrix} \cap M_+ \subset \text{Im}\left(P_+\begin{bmatrix} I \\ X(t) \end{bmatrix}\right)$$

is immediate; the opposite inclusion follows easily taking into account that the subspace $\text{Im}\begin{bmatrix} I \\ X(t) \end{bmatrix}$ is M-invariant, and therefore also P_+-invariant. From the continuity of τ and from the fact that $\dim V_1 = \dim V_2$ as long as $\text{gap}(V_1,V_2) < 1$ for subspaces V_1, V_2 in $\mathbb{C}^{2n}$, it follows that the rank of the $2n \times n$ matrix $P_+\begin{bmatrix} I \\ X(t) \end{bmatrix}$ does not depend on t (here also the connectedness of Ω was used). Therefore, for every $t \in \Omega$, there exists a basis $x_1(t),\dots,x_k(t)$ in $\text{Im}\left(P_+\begin{bmatrix} I \\ X(t) \end{bmatrix}\right)$ such that the vectors $x_i(t)$ are analytic in $t \in \Omega$ (see, e.g., Chapter S.6 in [11]). Performing the Gram-Schmidt orthogonalization (which does not effect the analyticity of $x_i(t)$ because t represents *real* variables), we can assume that $x_i(t)$ are orthogonal. By an analogous argument, choose for each $t \in \Omega$ an orthogonal basis $x_{k+1}(t),\dots,x_{2n}(t)$ in $\left\{\text{Im}\left(P_+\begin{bmatrix} I \\ X(t) \end{bmatrix}\right)\right\}^{\perp} = \text{Ker}([I\ X(t)]P_+^*)$ in such a way that $x_j(t)$ is analytic in $t \in \Omega$, $j = k+1,\dots,2n$. Now the matrix

$$[x_1(t),x_2(t),\dots,x_{2n}(t)]\begin{bmatrix} I_k & 0 \\ 0 & 0 \end{bmatrix}[x_1(t),x_2(t),\dots,x_{2n}(t)]^{-1}$$

is the orthogonal projection on $\tau(X(t))$ for each $t \in \Omega$. This matrix is clearly analytic in Ω.

The converse statement of the theorem concerning τ follows from the already proved converse statement about ϕ taking into account the following lemma. □

It will be convenient to state the lemma in general terms of selfadjoint matrices in an indefinite scalar product.

LEMMA 12. *Let* $H = H^*$ *be an invertible* $2n \times 2n$ *matrix, and let* B *be a* $2n \times 2n$ *matrix which is selfadjoint in the indefinite scalar product determined by* H, *i.e.* $HB = B^*H$. *Assume that for every* B*-invariant subspace* N *such that* $\text{Im}\,\sigma(B|_N) > 0$ *there* *exists a unique* n*-dimensional* B*-invariant* H*-neutral subspace* $L \subset \mathbb{C}^{2n}$ *such that* $L \cap B_+ = N$, *where* B_+ *is the spectral subspace of* B *corresponding to its eigenvalues in the open upper half-plane. If* $N(t)$, $t \in \Omega$, *is an analytic subspace valued function such that for each* $t \in \Omega$ *the subspace* $N(t)$ *is* B*-invariant and* $N(t) \subset B_+$, *then the subspace valued function* $L(t)$ *defined by the properties that for each* $t \in \Omega$ $L(t)$ *is* n*-dimensional* B*-invariant* H*-neutral and* $L(t) \cap B_+ = N(t)$, *is analytic in* $t \in \Omega$ *as well.*

PROOF. Using the canonical form of the pair (H,B) (see, e.g., [32] or Chapter S.5 in [11]; also the proof of Theorem I.3.22 in [12]) we can assume that H and B are partitioned as follows:

$$B = \begin{bmatrix} B_1 & 0 & 0 \\ 0 & B_0 & 0 \\ 0 & 0 & B_1^* \end{bmatrix}, \qquad H = \begin{bmatrix} 0 & 0 & I_k \\ 0 & H_0 & 0 \\ I_k & 0 & 0 \end{bmatrix},$$

where B_1 is a $k \times k$ matrix such that $\operatorname{Im} \sigma(B_1) > 0$, B_0 is a $2(n-k) \times 2(n-k)$ matrix with real eigenvalues $(0 \le k \le n)$. There is a unique $(n-k)$-dimensional B_0-invariant H_0-neutral subspace N_0. Indeed, if there were two such subspaces $N_0^{(1)}$ and $N_0^{(2)}$, then the subspaces

$$L^{(i)} = \left\{ \begin{bmatrix} x \\ y \\ 0 \end{bmatrix} \;\middle|\; x \in \mathbb{C}^k,\ y \in N_0^{(i)} \right\}, \qquad i=1,2$$

are n-dimensional B-invariant H-neutral such that $L^{(1)} \cap B_+ = L^{(2)} \cap B_+$, which contradicts the assumption of the theorem.

Given $N(t)$ as in the statement of the theorem, it is easily seen that

$$L(t) = \left\{ \begin{bmatrix} x \\ y \\ z \end{bmatrix} \;\middle|\; x \in N(t),\ y \in N_0,\ z \in [N(t)]^\perp \right\}.$$

Here we identify B_+ with $\mathbb{C}^k$, and $[N(t)]^\perp$ stands for the orthogonal complement to $N(t)$ in $\mathbb{C}^k$. So $L(t)$ is the image of some $2n \times 2n$ matrix function $Q(t)$ which is analytic in Ω. As $\dim L(t) = n$ is independent of t, it follows that $L(t)$ is analytic (cf. the proof of analyticity of τ). □

We can state informally the description of hermitian solutions of (5) in terms of invariant subspaces as follows. The set of all hermitian solutions of (5) (assuming it is not empty) has the structure of all invariant subspaces of the restriction of the matrix M to its spectral subspace corresponding to the eigenvalues in the open upper half-plane, with respect to the partial order, topology and analyticity. This description allows one to reduce many questions about hermitian solutions of (5) to the corresponding questions about invariant subspaces. In the Sections 6 and 8 we shall describe some results on the hermitian solutions which are obtained in this way.

6. ISOLATED HERMITIAN SOLUTIONS

A hermitian solution X of (5) is called *isolated* if there is no other hermitian solution of (5) in a neigborhood of X.

The following characterization of isolated hermitian solutions was obtained in [22], Theorem 4.4. It follows easily from the description of isolated invariant subspaces of a matrix (see [2,5]), taking into account that τ is a homeomorphism.

THEOREM 13. *Let* X *be a hermitian solution of* (5). *Then the following statements are equivalent:*

(i) X *is isolated;*

(ii) each common non-real eigenvalue λ *of* $i(A+DX)$ *and its adjoint* $-i(A^*+XD)$ *is an eigenvalue of* M *with* $\dim \operatorname{Ker}(\lambda I - M) = 1$;

(iii) for all non-real eigenvalues λ *of* M *such that* $\dim \operatorname{Ker}(\lambda I - M) > 1$, *we have either* $R_\lambda(M) \subset \operatorname{Im}\begin{bmatrix} I \\ X \end{bmatrix}$ or $R_\lambda(M) \cap \operatorname{Im}\begin{bmatrix} I \\ X \end{bmatrix} = (0)$.

In particular, the maximal and minimal, and more generally, the spectral solutions of (5) are isolated.

The property of being an isolated hermitian solution of (5) is a stable property (under some restrictions) in the following sense.

THEOREM 14. *Assume* X *is an isolated solution of* (5). *Then there exists* $\varepsilon > 0$ *such that any hermitian solution* Y *of the equation*

$$YD'Y + YA' + A'^*Y = C'$$

is isolated in Her(A',D',C') *as long as* D' *is nonnegative definite,* $C' = C'^*$, *the number of real eigenvalues of* $i\begin{bmatrix} A' & D' \\ C' & -A'^* \end{bmatrix}$ *(counting multiplicities) is equal to that of* M, *and*

$$||D' - D|| + ||A' - A|| + ||C' - C|| + ||Y - X|| < \varepsilon.$$

The proof of Theorem 14 is reduced, via Theorem 13, to the following statement: Given an $m \times m$ (complex) matrix S and an S-invariant subspace N such that for every eigenvalue λ of S with $\dim \operatorname{Ker}(\lambda I - S) > 1$ either $R_\lambda(S) \subset N$ or $R_\lambda(S) \cap N = (0)$ holds, there exists $\varepsilon > 0$ such that for any $m \times m$ matrix T and any T-invariant subspace L with

$$||T - S|| + \operatorname{gap}(L, N) < \varepsilon$$

either $R_\mu(t) \subset L$ or $R_\mu(T) \cap L = 0$ holds for each eigenvalue μ of T with $\dim \operatorname{Ker}(\mu I - T) > 1$. This statement is actually proved in the proof of Theorem 8.14 in [3] (see also Theorem 8.1 there).

An important particular case of Theorem 14 appears when M does not have real eigenvalues. In such case the condition on the number of real eigenvalues of $i\begin{bmatrix} A' & D' \\ C' & -A'^* \end{bmatrix}$ is fulfilled automatically provided $\varepsilon > 0$ is small enough.

7. REAL CASE

Consider now the Riccati equation (5) with the matrices A,D,C satisfying (6) and in addition assume that they are real. The results stated in Section 2 have analogues in this case. Now it is more convenient to work with matrices $\tilde{M} = \begin{bmatrix} A & D \\ C & -A^T \end{bmatrix}$ and $\tilde{J} = \begin{bmatrix} 0 & -I \\ I & 0 \end{bmatrix}$ instead of M and H, respectively.

THEOREM 15. *Assume* A,D,C *satisfy* (6) *and are, in addition, real matrices. Then the following statements are equivalent:*

(i) there exists a real symmetric solution of (5);

(ii) there exists a hermitian solution of (5);

(iii) the sizes of Jordan blocks with pure imaginary (including zero) eigenvalues in the Jordan form of $\tilde{M}$ *are all even;*

(iv) there exists an n*-dimensional* $\tilde{M}$*-invariant* $\tilde{J}$*-neutral subspace.*

Actually, if A,D,C are real and if there is a hermitian solution of (5), then the maximal and the minimal hermitian solutions are real (see, e.g., Chapter II.4 in [12]). So (i) and (ii) are equivalent. The rest of the theorem is easily seen to be a reformulation of part of Theorem 1.

Assume there exists a real symmetric solution of (5) (with real A,D,C). Then there is a one-to-one correspondence between the set of all real symmetric solutions X of (5) and the set of all n-dimensional $\widetilde{M}$-invariant $\widetilde{J}$-neutral subspaces L in $\mathbb{R}^{2n}$ given by the formula

$$L = \left\{ \begin{bmatrix} x \\ Xx \end{bmatrix} \middle| x \in \mathbb{R}^n \right\} .$$

Also, there exists a one-to-one and onto map $\rho : Her_r\,(A,D,C) \to$ $\to In_r(\widetilde{M})$, where $Her_r(A,D,C)$ is the set of all real symmetric solutions of (5) and $In_r(\widetilde{M})$ is the set of all $\widetilde{M}$-invariant subspaces N (in $\mathbb{C}^{2n}$) with the property that $\sigma(M|_N)$ lies in the quadrant $\{\lambda \in \mathbb{C} | \operatorname{Re} \lambda > 0,\ \operatorname{Im} \lambda \leq 0\}$. Here

$$\rho(X) = \operatorname{Im} \begin{bmatrix} I \\ X \end{bmatrix} \cap \widetilde{M}_+ ,$$

where $\widetilde{M}_+$ is the spectral subspace of $\widetilde{M}$ corresponding to the above mentioned quadrant. See [12] for the proof and further details; in another form this result appears in [7].

As in the complex case the existence of ρ allows us to count real symmetric solutions of (5) (assuming A,D,C are real and (5) has real symmetric solutions): if $\dim \operatorname{Ker}(\lambda I - \widetilde{M}) = 1$ for every eigenvalue λ of $\widetilde{M}$ with $\operatorname{Re} \lambda > 0$ and $\operatorname{Im} \lambda \leq 0$, then the number of real symmetric solutions of (5) is equal to $\Pi_{i=1}^{\alpha} (1 + \dim R_{\lambda_i}(\widetilde{M}))$, where $\lambda_1, \ldots, \lambda_\alpha$ are all such different eigenvalues; otherwise, there is a continuum of real symmetric solutions of (5).

The map ρ and its inverse are partial order preserving, continuous and analytic (the analyticity is understood in the same way as in Theorem 11). These facts are easily obtained from the corresponding properties of τ (Theorems 9 and 11) and the following relationship between τ and ρ. We denote by $\widetilde{M}_+$ (resp. M_{++}) the spectral subspace of $\widetilde{M}$ corresponding to its eigenvalues in the set $\{\lambda \in \mathbb{C} | \operatorname{Re} \lambda > 0,\ \operatorname{Im} \lambda \leq 0\}$ (resp. in the set $\{\lambda \in \mathbb{C} | \operatorname{Re} \lambda > 0,$ $\operatorname{Im} \lambda < 0\}$). If X is a real symmetric solution of (5), then

(25) $\rho(X) = \tau(X) \cap \tilde{M}_+$

and

(26) $\tau(X) = \rho(X) \dot{+} \overline{(\rho(X) \cap M_{++})}$.

Here $\overline{(\rho(X) \cap M_{++})}$ stands for the subspace (in $\mathbb{C}^{2n}$) consisting of the vectors obtained by coordinatewise complex conjugation from the vectors in $\rho(X) \cap M_{++}$. See Section II.4.7 in [12] for the proof of (25) and (26).

In [7] the partial order preservation of ρ was proved using the representation of any real symmetric solution as a combination of the extremal ones, which is described in the next theorem.

THEOREM 16. *Let* X_+ *(resp.* X_-*) be the maximal (resp. minimal) real symmetric solution of* (5), *where* A,D,C *are real and satisfy* (6). *If* N *is any* $(A + DX_+)$*-invariant subspace in* $\mathbb{R}^n$ *such that* $\sigma(A + DX_+)|_N$ *lies in the open right halfplane, then*

$$\tilde{N} \stackrel{\text{def}}{=} \{x \in \mathbb{R}^n \mid (X_+ - X_-)x \text{ is orthogonal to } N\}$$

is a direct complement to N *in* $\mathbb{R}^n$, *and the matrix*

(27) $X = X_+P_N + X_-(I - P_N)$,

where P_N *is the projection on* N *along* $\tilde{N}$, *is a real symmetric solution of* (6). *Conversely, for every real symmetric solution* X *of* (6) *there is a unique subspace* N *with the above properties such that* (27) *holds.*

For the proof of Theorem 16 and other properties of representation (27) see [7]. As it is shown in [29], the one-to-one correspondence (27) between real symmetric solutions of (5) and $(A + DX_+)$-invariant subspaces N such that $\text{Re}\, \sigma(A + DX_+)|_N > 0$ can be given a meaning of isomorphism of projective varieties.

As a corollary from the continuity of ρ and ρ^{-1}, let us state the following criterium for isolatedness of real symmetric solutions.

THEOREM 17. *Assume* A,D,C *are real. Then the following statements are equivalent for a real symmetric solution* X *of* (5):

(i) X *is isolated in the set of all real symmetric solutions of* (5);

(ii) for every eigenvalue λ *of* $\tilde{M}$ *such that* $\operatorname{Re}\lambda > 0$, $\operatorname{Im}\lambda \le 0$ *and* $\dim \operatorname{Ker}(\lambda I - \tilde{M}) > 1$, *either* $R_\lambda(\tilde{M}) \subset \operatorname{Im}\begin{bmatrix} I \\ X \end{bmatrix}$ *or* $R_\lambda(\tilde{M}) \cap \operatorname{Im}\begin{bmatrix} I \\ X \end{bmatrix} = (0)$ *holds;*

(iii) each common not purely imaginary eigenvalue λ *of* $A + DX$ *and its transpose* $A^T + XD$ *is an eigenvalue of* $\tilde{M}$ *with* $\dim \operatorname{Ker}(\lambda I - \tilde{M}) = 1$.

8. BEHAVIOR OF HERMITIAN SOLUTIONS UNDER PERTURBATIONS OF A,D AND C

Until now it was assumed (except for Theorem 14) that the coefficients A,D,C of the Riccati equation are fixed. Here we change this point of view and present some results on the behavior of hermitian solutions of (5) when A,D,C are allowed to change.

First, we consider stability of hermitian solutions. A hermitian solution X of (5) will be called *stable* if for every $\varepsilon > 0$ there exists $\delta > 0$ such that every Riccati equation

$$YD'Y + YA' + A'^*Y - C' = 0 \tag{28}$$

with $\|A - A'\| + \|D - D'\| + \|C - C'\| < \delta$ and A', D', C' satisfy (6) has a hermitian solution Y with $\|X - Y\| < \varepsilon$ provided (28) has a hermitian solution. The following theorem has been proved in [22] (Theorem 4.4).

THEOREM 18. *A hermitian solution of* (5) *is stable if and only if it is isolated.*

The property of having a stable hermitian solution is stable in the following sense. Let X be a stable hermitian solution of (5). Then for every $\varepsilon > 0$ there exists $\delta > 0$ such that every equation (28) with (A', D', C') satisfying (6) and $\|A - A'\| + \|D - D'\| + \|C - C'\| < \delta$ which has a hermitian solution also has a stable hermitian solution Y with $\|X - Y\| < \varepsilon$ (see Theorem 4.8 in [22]). Also, there exists $\zeta > 0$ such that if (A', D', C') satisfy (6), the number of real eigenvalues of $i\begin{bmatrix} A' & D' \\ C' & -A'^* \end{bmatrix}$ (counting multiplicities) is the same as that for M, and

$$\|Y - X\| + \|A - A'\| + \|D - D'\| + \|C - C'\| < \zeta$$

for a hermitian solution Y of (28), then Y is stable (cf. Theorem 14).

In particular, the extremal solutions of (5) are stable. Actually, they are continuous in the following sense.

THEOREM 19. *Let* $\mathcal{R}$ *be the set of all triples of* $n \times n$ *matrices* (A,D,C) *satisfying condition* (6) *and for which there exists a hermitian solution of* (5). *Then the maximal and the minimal hermitian solutions of* (5) *are continuous functions on* $(A,D,C) \in \mathcal{R}$, *where* $\mathcal{R}$ *is regarded in the natural topology as a subset of* $\mathbb{C}^{3n^2}$.

For the proof of this theorem see [25], Chapter III.4 in [12].

A stronger type of stability is the Lipschitz stability. A hermitian solution X of (5) is called *Lipschitz stable* if there exist positive constants δ and K such that if A', D', C' satisfy (6), $\|A - A'\| + \|D - D'\| + \|C - C'\| < \delta$ and (28) has a hermitian solution it also has a hermitian solution Y with

$$\|X - Y\| < K(\|A - A'\| + \|D - D'\| + \|C - C'\|) .$$

It turns out that, assuming M has no real eigenvalues, a hermitian solution of (5) is Lipschitz stable if and only if it is spectral (Theorem 4.9 in [22]).

We pass now to analytic perturbations of A,D,C. Here the spectral hermitian solutions are the ones which behave nicely:

THEOREM 20. *Suppose* $A(t)$, $D(t)$ *and* $C(t)$ *satisfy* (6) *for each* $t \in U$ *and depend analytically on a parameter* $t \in U$, *where* U *is an open interval in* $\mathbb{R}$. *Assume that for each* $t \in U$ *the Riccati equation*

$$XD(t)X + A(t)^*X + XA(t) - C(t) = 0 \tag{29}$$

has a hermitian solution. Suppose also the number of real eigenvalues (counting multiplicities) of the matrix

$$M(t) = i\begin{bmatrix} A(t) & D(t) \\ C(t) & -A(t)^* \end{bmatrix} \tag{30}$$

is constant for $t \in U$. *Let* X_0 *be a spectral hermitian solution of* (29) *with* $t = t_0 \in U$. *Then there exists a matrix function* $X(t)$ *which is analytic in some neighborhood* U_0 *of* t_0 *such that* $X(t)$ *is a spectral hermitian solution of* (29) *for each* $t \in U_0$ *and* $X(t_0) = X_0$.

The proof of this theorem follows the line of argument of the proof of Theorem 6 in [26] based on Theorem 3 in [26], where analyticity of the extremal solutions was proved. We omit the details.

For analytic dependence on several real variables an analogue of Theorem 20 is valid under the additional assumption that M has no real eigenvalues.

THEOREM 21. *Suppose* $A(t)$, $D(t)$ *and* $C(t)$ *satisfy* (6) *for each* $t \in \Omega$ *and depend analytically on* $t \in \Omega$ *where* Ω *is an open set in* $\mathbb{R}^k$. *Assume that* $M(t)$ *given by* (30) *has no real eigenvalues for* $t \in \Omega$. *Then for any spectral hermitian solution* X_0 *of* (29) *with* $t = t_0 \in \Omega$ *there esists a matrix function* $X(t)$ *which is defined and analytic in a neighborhood* Ω_0 *of* t_0 *and such that* $X(t)$ *is a spectral hermitian solution of* (29) *for* $t \in \Omega_0$ *and* $X(t_0) = X_0$.

PROOF. Let Γ be a contour (possibly consisting of two closed rectifiable Jordan curves) in the complex plane such that $\Gamma \cap \{\overline{\lambda} \mid \lambda \in \Gamma\} = \emptyset$, the eigenvalues of $M(t_0)\big|_{\operatorname{Im}\begin{bmatrix} I \\ X_0 \end{bmatrix}}$ are inside Γ, and $M(t_0)$ has no other eigenvalues inside or on Γ. Choose a neighborhood Ω_0 of t_0 such that $M(t)$ has no eigenvalues on Γ for $t \in \Omega_0$. The spectral subspace $L_\Gamma(t)$, $t \in \Omega_0$, of $M(t)$ corresponding to the eigenvalues of $M(t)$ inside Γ is J-neutral and of dimension n. Indeed, the J-neutrality of $L_\Gamma(t)$ follows from the fact that any spectral subspace of the J-sefadjoint matrix $M(t)$ corresponding to a set of eigenvalues of $M(t)$ which does not contain any pair of complex conjugate numbers, is J-neutral (see the canonical form of a selfadjoint matrix in an indefinite scalar product; e.g., Chapter I.3 in [12]). Moreover, the subspace valued function $L_\Gamma(t)$ is analytic in $t \in \Omega_0$, because

$$L_\Gamma(t) = \operatorname{Im}\left[\frac{1}{2\pi i}\int_\Gamma (\lambda I - M(t))^{-1}\,d\lambda\right], \qquad t \in \Omega_0$$

and the matrix in the square brackets is analytic in $t \in \Omega_0$. By Theorem 3, $L_\Gamma(t) = \operatorname{Im}\begin{bmatrix} I \\ X(t) \end{bmatrix}$, $t \in \Omega_0$, for some spectral hermitian solution $X(t)$ of (29). As in the proof of Theorem 11, one shows $X(t)$ is analytic, and, of course, $X(t_0) = X_0$. □

In particular, if the coefficients of (5) depend analytically on a real parameter t, then the extremal solutions of (5) depend analytically on t as long as they exist and the number of real eigenvalues of M stays fixed. The conclusion is true also for analytic dependence on several real variables provided M does not have real eigenvalues (see [25,9]). This result was used in [25] to prove that the solution of the optimal control problem stated in Section 1 depends analytically on the initial data.

9. NON-HERMITIAN SOLUTIONS

We have seen that hermitian solutions X of (5) correspond in a one-to-one way to M-invariant H-neutral subspaces L of dimension n by means of the formula

$$L = \mathrm{Im}\begin{bmatrix} I \\ X \end{bmatrix} .$$

Such solutions do not exist always. However, M is H-selfadjoint and the signature of H is zero, hence there always exists an n-dimensional M-invariant H-nonpositive subspace N (H-nonpositivity means that $\langle Hx,x \rangle \leq 0$ for all $x \in N$); see, e.g., Section I.3.12 in [12]. It turns out that to such subspaces correspond certain, in general non-hermitian, solutions of (5):

THEOREM 22. *There is a one-to-one correspondence between the set of all solutions* X *of* (5) *such that the matrix* $(X^* - X)(A + DX)$ *is nonpositive definite and the set of all* n-*dimensional* M-*invariant* H-*nonpositive subspaces* L, *given by the formula*

$$L = \mathrm{Im}\begin{bmatrix} I \\ X \end{bmatrix} .$$

We refer to Chapter II.4 in [12] (see also [16]) for the proof of Theorem 22. In particular, there always exists a solution X of (5) for which $(X^* - X)(A + DX)$ is nonpositive definite. A result analogous to Theorem 22 holds for solutions X with $(X^* - X)(A + DX)$ nonnegative definite (then the corresponding subspaces L are H-nonnegative). Certain results on stability of such solutions are found in [22].

REFERENCES

1. Anderson, B.D.O., Moore, J.B.: Optimal filtering, Prentice-Hall, Englewood Cliffs, N.J., 1979.

2. Bart, H., Gohberg, I., Kaashoek, M.A.: Stable factorizations of monic matrix polynomials and stable invariant subspaces, Int. Eq. Op. Th. 1 (1978), 496-517.

3. Bart, H., Gohberg, I., Kaashoek, M.A.: Minimal factorizations of matrix and operator functions. Birkhäuser Verlag, Basel, 1979.

4. Brockett, R.: Finite dimensional linear systems. John Wiley, New York, 1970.

5. Campbell, S., Daughtry, J.: The stable solutions of quadratic matrix equations, Proc. Amer. Math. Soc. 74 (1979), 19-23.

6. Coppel, W.A.: Disconjugacy. Lecture Notes in Math. 220, Springer-Verlag, Berlin-Heidelberg-New York, 1971.

7. Coppel, W.A.: Matrix quadratic equations, Bull. Austral. Math. Soc. 10 (1974), 377-401.

8. Čurilov, A.N.: On the solutions of quadratic matrix equations, Nonlinear vibrations and control theory 2 (1978), Udmurt State University, Izhevsk. (Russian).

9. Delshamps, D.F.: A note on the analiticity of the Riccati metric, Amer. Math. Soc. Lectures in Appl. Math. 18 (1980), 37-42.

10. Finesso, L., Picci, G.: A characterization of minimal square spectral factors, IEEE Trans. on Autom. Control 27 (1982), 122-127.

11. Gohberg, I., Lancaster, P., Rodman, L.: Matrix polynomials. Academic Press, New York etc., 1982.

12. Gohberg, I., Lancaster, P., Rodman, L.: Matrices and indefinite scalar products. Birkhäuser-Verlag, Basel, 1983.

13. Kučera, V.: A contribution to matrix quadratic equations, IEEE Trans. on Automat. Control 17 (1972), 344-347.

14. Kučera, V.: A review of the matrix Riccati equation, Kybernetika 9 (1973), 42-61.

15. Kwakernaak, H., Sivan, R.:Linear Optimal Control Systems. Wiley, New York, 1972.

16. Lancaster, P., Rodman, L.: Existence and uniqueness theorems for algebraic Riccati equations, Int. J. Control 32 (1980), 285-309.

17. Martensson, K.: On the matrix Riccati equation, Inf. Sciences 3 (1971), 17-49.

18. Molinari, B.P.: Equivalence relations for the algebraic Riccati equation, Siam J. of Control and Opt. 11 (1973), 272-285.

19. Potter, J.E.: Matrix quadratic solutions, Siam J. Appl. Math. 14 (1966), 496-501.

20. Ran, A.C.M.: Minimal factorizations of selfadjoint rational matrix functions, Int. Eq. Op. Th. 5 (1982) 850-869.

21. Ran, A.C.M., Rodman, L.: Stability of invariant maximal semidefinite subspaces I (submitted to Lin. Alg. Appl.).

22. Ran, A.C,M., Rodman, L.: Stability of invariant maximal semidefinite subspaces II. Applications: selfadjoint rational matrix functions, algebraic Riccati equations (submitted to Lin. Alg. Appl.).

23. Reid, W.T.: Riccati matrix differential equations and non-oscillation criteria for associated linear differential systems, Pacific J. Math. 13 (1963), 665-685.

24. Reid, W.T.: Riccati differential equations. Academic Press, New York, 1972.

25. Rodman, L.: On extremal solutions of the algebraic Riccati equations, A.M.S. Lectures on Applied Math. 18 (1980), 311-327.

26. Rodman, L.: On nonnegative invariant subspaces in indefinite scalar product spaces, Lin. and Multilin. Alg. 10 (1981), 1-14.

27. Rodman, L.: Maximal invariant neutral subspaces and an application to the algebraic Riccati equation, Manuscripta Math. 43 (1983), 1-12.

28. Shayman, M.A.: Geometry of the algebraic Riccati equation I, Siam J. Contr. Opt. 21 (1983), 375-394.

29. Shayman, M.A.: Geometry of the algebraic Riccati equation II, Siam J. Contr. Opt. 21 (1983), 395-409.

30. Singer, M.A., Hammarling, S.J.: The algebraic Riccati equation: a summary review of some available results. Nat. Phys. Lab. Report DITC 23/83, 1983.

31. Swieten, A.C.M. van: Qualitative behaviour of dynamical games with feedback strategies, Ph.D. Thesis, University of Groningen, The Netherlands, 1977.

32. Thompson, R.C.: The characteristic polynomial of a principal subpencil of a Hermitian matrix pencil, Lin. Alg. Appl. 14 (1976), 135-177.

33. Willems, J.C.: Least squares stationary optimal control and the algebraic Riccati equation, IEEE Trans. on Autom. Contr. 16 (1971), 621-634.

34. Wimmer, H.K., The algebraic Riccati equation without complete controllability, Siam J. Alg. Discr. Meth. 3 (1982), 1-12.

35. Wonham, W.M.: On a matrix Riccati equation of stochastic control, Siam J. Contr. 6 (1968), 681-697. Erratum, ibid. 7 (1969), 365.

A.C.M. Ran,
Vrije Universiteit
Wiskundig Seminarium
Postbus 7161
1007 MC Amsterdam, Holland

L. Rodman,
School of Mathematical Sciences
Tel-Aviv, Ramat Aviv,
Israel.

WORKSHOP PROGRAM

MONDAY, JUNE 13, 1983

9.00 M. Kaashoek, *Similarity of matrix blocks and canonical forms.*

10.00 L. de Branges, *The expansion theorem for Hilbert spaces of analytic functions.*

11.00 Coffee.

11.20 H. Gauchmann, *On the coupling of linear representations of a Lie algebra.*

12.30 Lunch, Weizmann Institute Faculty Club.

14.00 P. Dewilde, *Lossless inverse scattering and digital filters.*

15.00 H. Bart, *The coupling method for solving integral equations.*

16.00 Coffee.

19.00 Dinner, Weizmann Institute Faculty Club.

TUESDAY, JUNE 14, 1983

8.30 G. Knowles, *Invertibility for nest subalgebras of Von Neumann algebras.*

9.30 E. Jonckheere, *The linear quadratic optimal control problem - the operator theoretic viewpoint.*

10.30 Coffee.

10.50 I. Gohberg, *Maximal entropy principles and related interpolation problems.*

12.00 Lunch, Weizmann Institute Faculty Club.

14.00 Visit the Diaspora Museum and Tel-Aviv University.

16.30 Coffee - School of Mathematical Sciences, Tel-Aviv University.

17.30 N. Cohen, *Factorization of matrix polynomials and rational matrix functions.*

18.00 H. Dym, *Covariance extensions and canonical systems of differential equations.*

19.00 Dinner, Tel-Aviv University Faculty Club.

20.30 Evening tour of Jaffa.

WEDNESDAY, JUNE 15, 1983

8.30 J. Ball, *Invariant subspaces, unitary interpolants and factorization indices.*

9.30 A. Lindquist, *Differential equations representations of stationary stochastic processes.*

10.30 Coffee.

10.50 S. Levin, *Multivariable systems theory.*

12.00 Lunch, Weizmann Institute Faculty Club.

13.00 Depart for Jerusalem tour, and dinner in Jerusalem.

THURSDAY, JUNE 16, 1983

9.00 A. Ran, *Stability of invariant maximal semi-definite subspaces.*

10.00 L. Lerer, *Wiener-Hopf factorization of piecewise matrix functions.*

11.00 Coffee.

11.20 N. Kravitsky, *On commuting integral operators.*

12.30 Lunch, Weizmann Institute Faculty Club.

14.00 G. Zames, *The optimally robust servomechanism problem.*

15.00 L. Rodman, *The algebraic matrix Riccati equation.*

16.00 Coffee.

18.45 Closing dinner, and Concert.

PARTICIPANTS

ALPAY, D.,	The Weizmann Institute of Science, Israel.
BALL, J.A.,	The Weizmann Institute of Science, Israel, and Virginia Polytechnic Institute & State University, U.S.A.
BART, H.,	Free University of Amsterdam, The Netherlands.
BEN-ARTSI, A.,	Tel-Aviv University, Israel.
DE BRANGES, L.,	Purdue University, U.S.A.
COHEN, N.,	The Weizmann Institute of Science, Israel.
DEWILDE, P.M.,	Delft University of Technology, The Netherlands.
DYM, H.,	The Weizmann Institute of Science, Israel.
FEINTUCH, A.,	Ben-Gurion University of the Negev, Israel.
FUHRMANN, P.,	Ben-Gurion University of the Negev, Israel.
GAUCHMANN, H.,	Ben Gurion University of the Negev, Israel.
GOHBERG, I.,	The Weizmann Institute of Science, and Tel-Aviv University, Israel.
IACOB, A.,	The Weizmann Institute of Science, Israel.
JONCKHEERE, E.A.,	University of Southern California, U.S.A.
KAASHOEK, M.A.,	Free University of Amsterdam, The Netherlands.
KNOWLES, G.J.,	Texas A & M University, U.S.A.
KRAVITSKY, N.,	Ben-Gurion University of the Negev, Israel.
LEVIN, S.,	Ben-Gurion University of the Negev, Israel.
LERER, L.,	Technion, Israel Institute of Technology, Israel.
LINDQUIST, A.G.,	The Royal Institute of Technology, Sweden.
PERELSON, A.,	Tel-Aviv University, Israel.
PIATETSKI-SHAPIRO, E.,	Tel-Aviv University, Israel.
PICCI, G.,	University of Padova, Italy.
RAN, A.,	Free University of Amsterdam, The Netherlands.
RODMAN, L.,	Tel-Aviv University, Israel.
RUBINSTEIN, S.,	Tel-Aviv University, Israel.
SAEKS, R.,	Texas Tech University, U.S.A.
SONTAG, E.D.,	Rutgers University, U.S.A.
TAMIR, S.,	Tel-Aviv University, Israel.
ZAMES, G.,	McGill University, Canada.